FLOOD CONTROL AND DRAINAGE ENGINEERING

T0239704

Flood Control and Drainage Engineering

Fourth Edition

S.N. Ghosh
Formerly, Professor
Department of Civil Engineering
Indian Institute of Technology Kharagpur
India

CRC Press
Taylor & Francis Group
Boca Raton   London   New York   Leiden

CRC Press is an imprint of the
Taylor & Francis Group, an **informa** business

A BALKEMA BOOK

First issued in paperback 2018

CRC Press/Balkema is an imprint of the Taylor & Francis Group, an inform a business

© 2014 Taylor & Francis Group, London, UK

Published by: CRC Press/Balkema
                  P.O.Box 11320, 2301 EH, Leiden, The Netherlands
                  e-mail: Pub.NL@taylorandfrancis.com
                  www.crcpress.com - www.taylorandfrancis.com

ISBN 13: 978-1-138-07715-7 (pbk)
ISBN 13: 978-1-138-02627-8 (hbk)

Preface to the Fourth Edition

This fourth edition of the book has been thoroughly revised and upgraded to include two new chapters *Flood Damage Management Due to Tsunamis and Storm Surges* and *Flooding Due to Collapse of Dams*. The latter also covers flooding due to breaching of flood levees and flooding in urban and coastal areas. The other notable features include the incorporation of a number of worked out practical design and analysis problems, including several case studies.

In the preparation of this edition I have drawn on publications of the Journal of Hydraulic Engineering, published by the Indian Society of Hydraulics Pune, Internet websites of National Hurricane Center, NOAA, USA, Wikipedia, and other journals which have been duly acknowledged both in the text as well as in the illustrations. To all these organizations and authors, I express my sincere gratitude.

I would like to acknowledge help rendered by my wife Smt. Dyuti Ghosh, son Archishman Ghosh, daughter in law Soumya Ghosh in the preparation of manuscript.

In revising the book my primary objective was to present to the students community and professional various aspects of flooding and the methodology on how to tackle them purely from engineering point of view without going into the extremely complex nature of the physics of the phenomena. I will appreciate receiving constructive suggestions and criticism to help me to improve the quality and usefulness of the book.

As mentioned in the first edition of the book, my work is dedicated to the memory of my mother Sm Golaprani Ghosh who left us more than thirty years ago and I wish to reiterate the dedication.

S.N. Ghosh

Kolkata,
November, 2013

Preface to the First Edition

Floods mean many things to many people depending on the profession they are engaged in. To the poorer village people who are caught in the swirling mudladen waters it represents terror and unlimited suffering. To the urban people it means suffering in indirect ways because of disruptions of communication due to breaching of embankments, roads, railways, culverts and other essential supplies. To the government floods mean additional expenditure for rescue, relief, rehabilitation measures and damages to national properties including losses of crops, domesticated animals and human lives. It also means sanctioning of flood protection measures to reduce the flood damage and promote general welfare of the people. To the planning commission and other planning bodies it calls for a need to provide rules and regulations whereby land prone to floods can be identified for proper utilisation and to insurance people it represents a challenge of how to underwrite flood damages. In general to all the citizens of a flood affected country it represents a natural phenomena of which millions of rupees are spent annually for providing protection against floods.

To the civil engineers who are responsible for designing flood protection measures, it means how to build up a design flood based on the knowledge of hydrology of the catchment area. Thereafter, he is required to plan engineering structures, such as storage reservoirs, its schedule of operations, so that the flood cannot cause serious damage downstream. Further as the flood wave passes through a stream it is necessary to know how the stage varies with respect to time and distance for the design of river engineering works as well as for issuance of flood warning by the civil authorities. Another consequence of floods is the problem of drainage and water logging of low lying lands particularly of deltaic origin, where the drainage is affected due to tidal fluctuations in the river.

In recent years there has been considerable development in the hydraulic design of engineering structures needed for flood control, particularly in the field of hydrology, reservoir planning and operation, river engineering and tidal hydraulics. To the student community such new knowledge should be available in simplified manner in the form of a suitable book. There exists other books particularly under irrigation engineering devoting one or two chapters on flood control. In this book an attempt has been made to provide a broad coverage of the whole subject matter including recent developments as an integrated whole. It is hoped that this will provide a clear overview of the whole engineering aspect of flood control and associated land drainage problem.

The book is designed primarily as a textbook for both undergraduate and postgraduate students and it can as well be adopted as a reference book for practising engineers. The book is organised in ten chapters. First three chapters deal with basic information needed for the design of flood control structures; chapter IV and V deal with spillway design and planning and operation of reservoirs; chapter VI deals with river engineering and design principles of flood protection measures; chapter VII deals with flood forecasting and warming; chapter VIII deals with cost benefit analysis of flood control project and finally chapter IX and X deal with drainage of flood affected areas.

In the preparation of the manuscript the author has referred to lectures and notes delivered by eminent hydraulic engineers in seminars, international courses, various journals published by American Society of Civil Engineers, Institution of Civil Engineers (Lond.), publications of the Indian Standards Institution, Central Board of Irrigation and Power, various National and International seminars and symposiums, publication of various state Irrigation Departments such as West Bengal, Orissa, Karnataka etc. and other source materials including those furnished under references and bibliography. To them the author acknowledges his indebtedness and gratitude.

The author is deeply conscious of his serious limitations in writing a book of this nature and complexity and it is likely that there will occur lapses here and there inspite of best efforts taken. The author will deeply appreciate constructive criticism and suggestions from teachers, students and professional engineers which will help to improve further the subject matter in future. Finally the author expresses his sincere thanks and gratitude to his wife, to the members of the Editorial Board of Oxford-IIT series. This humble work is dedicated to the memory of my beloved mother Sm. Golap Rani Ghosh who had been the greatest source of inspiration of all of my work and who left us suddenly for her heavenly abode in April 26th, 1983.

S.N. GHOSH

IIT, Kharagpur
March, 1986.

Contents

1

Flood Problems

1.1 INTRODUCTION

In Webster's New International Dictionary, a 'flood' is defined as a 'great flow of water... especially, a body of water, rising, swelling, and over-flowing land not usually thus covered; a deluge; a freshet; an inundation'. Commonly, it is considered to be a phenomenon associated with an unusually high stage or flow over land or coastal area, which results in severe detrimental effects. 'Flood control' implies all measures taken to reduce the detrimental effects of flood.

There exist several types of flooding such as:

- **River Flooding:** This is the major cause of flooding extensive areas as a result of heavy rains in the catchment areas as well as local areas thereby increasing the river levels.
- **Flash Floods:** This results due to heavy rains in hilly areas which cause local rivers and small streams to rise to dangerous level within a short period of time say 6 to 12 hours. Heavy and continuous rains in local areas can cause flash floods.
- **Urban Flooding:** Local heavy rains up to 100 mm or more in a day over the city and larger towns can cause damaging and disruptive flooding due to poor or chocked drainage and rapid runoff.
- **Strom Surge or Tidal Flooding:** This results mostly due to tropical disturbances, developing to cyclones and crossing surrounding coastlines. Cyclone induced storm surges have devastating consequences in coastal areas and such surge induced floods may extend many kilometers inland.
- **Floods Arising due to Failure of Dam:** A large number of large and small dams are constructed to store water for various purposes. Due to poor maintenance and due to exceptionally high precipitation a severe flood may result causing failure of the dam. This causes a surging water front travelling with high velocity causing destruction of properties and loss of life.

Floods result from a number of causes as mentioned above. However, one due to heavy and prolonged rainfall is the most frequent one. For systematic studies it is necessary to classify such flood events as flash floods, single event floods, multiple event floods and seasonal floods. Flash floods have sharp peak, the rise and fall are almost equal and rapid. Single event floods have a single main peak and have more duration than flash floods. Multiple event flood are caused by more severe complex weather situations where successive flood peaks follow closely. The floods occurring during rainy season are known as seasonal floods. Excessive snow melt also results in frequent flooding in many countries. Floods not directly connected to rainfall may result due to failure of dam causing thereby sudden escape of huge volume of water stored in the reservoir. When a dam fails it causes severe flooding of the down stream areas resulting in heavy damages and loss of lives. Floods may also result due to landslides, which may temporarily block the water passage and later on gives away as a result of built up pressure etc. The landslides may trigger of due to tectonic movement of earthen surface or instability of soil mass.

In nature, the problems associated with floods are diverse and extremely complicated. Floods inundate built-in property, endanger lives, and prolonged high flood stages delay rail and highway traffic. Further, it interferes with efficient drainage and economic use of lands for agricultural or industrial purposes. Due to high rate of flow or runoff from the catchment areas of the streams, there occurs large-scale erosion of lands and consequently to sediment deposition problems downstream. Floods also cause damages to drainage channels, bridge abutments, sewer outfalls and other structures. It further interferes with navigation as well as hydroelectric power generation. In short, floods cause severe strain and hardship to the civilised life of a community. Apart from loss of human lives, economic losses associated with the above combined effects, run to several thousands billions of dollars on a global basis.

1.2 CAUSES OF FLOODING AND ECONOMIC LOSSES

The causes of flooding in all the major river systems, therefore, are more or less the same. They are: (a) spilling over the banks, resulting in flooding other areas; (b) bank erosion; (c) rising of river beds caused by deposition of silt; and (d) changing of the river course from time to time.

Generally speaking, a quantitative estimate of the losses is rather difficult to provide since the intangible component of the flood losses is a dominating factor. On an average several crore people are affected annually and a few hundred lives are lost. Besides, there occur huge losses from the death of domestic animals. Overall several hundred crores worth of property is lost which does not take into account the losses and privations arising out of break-down of communications, disruption of essential services, environmental deterioration, etc.

The expected annual damages are normally computed in probabilistic terms. The procedure followed includes routing of the reaches from the dam site to the downstream control point where the computed discharge is converted to respective elevations using rating curve made at the control point cross section. The inundated areas corresponding to the computed elevation are marked on the map. Nowadays this can be done with the help of data collected through remote sensing and GIS technology. An exhaustive survey is made in the delineated areas for the census of inhabitants, live stock, valuable properties including valuable agricultural lands and the important structures such as bridges, culverts etc. An assessment on the likely damages is made for several flood events and elevation vs annual flood damage curve is plotted. Using long-term historical annual flood series a frequency analysis is carried out and probabilities are assigned to the above flood events observed at the control point. Then a plot between elevation and probability of exceedence is prepared. It is then converted to the plot between probability of exceedence and annual flood damages and non-dimensionalised using the potential damage. The probabilistic damage curve thus prepared is then used for making financial compensation decision by appropriate authorities.

Despite flood hazards mankind has always shown preference to settle near the reaches due to assured supply of water, facility of navigation and fertility of river valleys. A considerable portion of world population lives in areas adjacent to rivers and often becomes victim of misery due to devastating nature of floods. Floods are always a part of mankind's life throughout the history and extensive literature exists giving accounts of man's struggle to cope with this natural phenomena.

1.3   FLOOD MANAGEMENT MEASURES

Flood management measures can be classified as (i) short-term and (ii) long-term measures. The nature and extent of flood damages as well as local conditions determine the measures to be taken up. However, short-term measures are dependent for their effectiveness on long-term measures.

Short-term measures

These measures are adopted for giving quick results when immediate relief to some pockets or locations is felt necessary and they are respectively:
- Construction of embankments along the low level banks that are subject to frequent flood spells.
- Construction of raised platforms for temporary shelter during flood.
- Dewatering by pumps of flooded pockets, towns when gravity discharge of floodwater is not possible.
- Construction of floodwalls near congested areas of cities, towns and industrial belts.

Long-term measures

- Construction of storage reservoirs to moderate the flow peak thereby ensuring regulation of flood downstream.
- Integrated watershed management in the hilly area catchment, which ensures reduction in surface runoff, erosion and increase of infiltration capacity thereby reducing the impact of flood.
- Flood forecasts and warning based on hydro-geomorphological studies which can be given with some lead period, thereby ensuring minimization of property loss and loss of human life by shifting them to safer places.

1.4 FLOOD CONTROL STRATEGIES

To reduce losses due to flood the strategies to be followed can be stated as follows:

I. *Modify flooding by structural means*:
   Herein the strategies to be followed involves construction of dams, dikes, levees, channel alterations, high flow diversions and land treatment. The main idea is to keep water away from the potential damage areas.

II. *Flood forecasting: This is a non-structural measure*:
   Here forecast of flooding is provided at the potential damage points. The population both human as well as livestock and the movable properties are shifted to a safer place if there is a chance of flood damage. Herein the objective is to keep people away from inundated areas.

III. *Modify susceptibility to flood damage*:
   Regulations are framed to avoid undesirable or unwise rise of flood plains. Necessary steps are taken to modify the impact of flooding through individual or group action designed for assisting people in the preparatory, survival and recovery phase of floods which are namely through education and information on floods, flood insurance, taxation relief, etc.

Structural measures

The structural measures are aimed to mitigate flood damage by regulating the movement of flood water and these include:

- Dams, reservoirs and high flow diversions, their purpose is to store flood water temporarily or to divert it from the area to be protected.
- Channel improvement works to increase the carrying capacity of a river channel and to pass the flood water quickly through the channel reach.
- Embankments, levees and flood walls to stop the flood water from entering the areas to be protected.
- Catchment treatments to induce holding of water in the catchment temporarily.

Non-structural measures

The main idea is to keep the general civil and industrial activities undiminished during flood which can be ensured by flood forecasting warning systems, flood regulation through zoning, emergency plans, modifying building codes, flood proofing, disaster preparedness and assistance.

Automatic rain gauge stations provide reliable picture of the rainfall events in the basin. Radar data provide details with regard to movement and dynamic characteristics of storms on a large scale. A useful input to the forecast is the meteorological satellite data. Real time forecasts are issued nowadays in many countries including India for important rivers. The data requirements for forecast are for flood arising out of rainfall—rainfall details, catchment details, river geometry, discharge, water, level.

1.5 ALLEVIATION OF FLOODING

It would thus be apparent that flood control is and will continue to remain one of the major requirements of a comprehensive water resources development project.

Voluminous and extensive literature is available on different methods of flood control. These can be divided mainly under the following categories:

(a) Detention and storage or use of existing lakes or construction of a number of tanks in the catchments for the purpose of flood moderation.

(b) Or ring bunds around important towns, properties and estates to prevent flooding.

(c) (i) **Enlargement of existing channels or rivers:** Training of rivers to provide local protection at critical points from erosion, scour or flooding by spurs, revetment, dykes, etc. to form artificial cuts to lower the flood level and improve the river regime, widen and deepen the river bed artificially by dredging or by other training measures such as bandalling; groynes, etc.

(ii) **Construction of by-pass channels on rivers:** Natural or artificial flood diversion through subsidiary channels of the parent river or in another river system, to selected depressions, lakes, etc. with the aim of relieving intensity of floods in the main rivers. Flood water can also be diverted by constructing overflow weirs in the flood embankment at a predetermined flow stage.

(d) By flood plain zoning which is purely an administrative measure.

(e) By provision of spreading grounds.

(f) By providing suitable drainage arrangement by installing pumping facilities which comes under drainage engineering.

(g) Soil conservation measures in the catchments by various known methods.

(h) Preparation of detailed action plans for mobilisation of the local resources for supervision of embankments during flood, flood relief works and other emergency measures. Further plan for collection of materials such as earth bags, stones, brushwood mattresses to reinforce or add to the protection works including mobilisation of local people for execution of such works.

(i) Flood forecasting and warning system to keep alert all concerned people in advance and to take timely action for evacuation in case of impending danger.

The above methods can be used either singly or in combination. This, however, requires knowledge of the capabilities, limitations and relative advantage of such measure singly or in combination. Associated with this is an economic study to determine the potential damages averted by the projected works and whether the benefits justify the cost.

Finally this chapter ends by providing the reader the various methods of flood management in a nutshell as proposed by the National Commission on Floods, 1980 in Fig. 1.1.

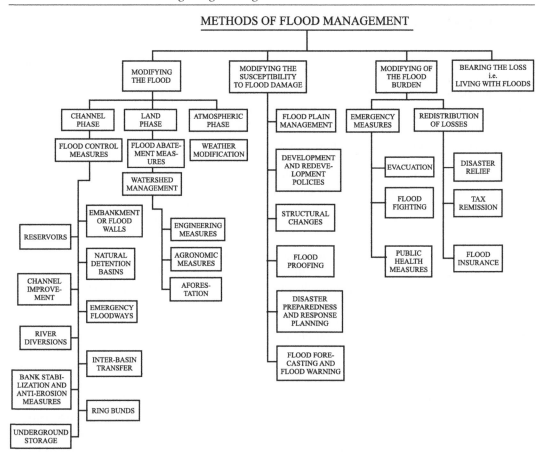

Figure 1.1 Method of flood management (National Commission on Floods, 1980).

2

Estimation of Design Flood

2.1 INTRODUCTION

Estimation of design flood is one of the important components of planning, design and operation of water resources projects. Information on flood magnitudes and their frequencies is needed for design of hydraulic structures such as dams, spillways, road and railway bridges, culverts, urban drainage systems, flood plain zoning, economic evaluation of flood protection projects. According to Pilgrim et al [35], the estimation of peak flows on small and medium sized plains is generally the common application as they are required for the design of conservation works, etc. Although many different methods are available for estimating floods, the three widely used ones are rational method, US Soil Conservation Service Curve Number method and regional flood frequency methods.

Hydrologic design is important for the safety, economy and proper functioning of the hydraulic structure. The purpose of hydrologic design is to estimate the maximum average or minimum flood, which the structure is expected to handle. This estimate has to be made quite accurately in order that the project functions properly.

To quote an example, consider the estimation of the spillway design flood. This is a very important preliminary study required to be carried out before the actual design of a dam or detention reservoir is taken in hand. Its importance is heightened by the fact that in any such project there should be a balance between economy and efficiency. Collectively the reservoir and the spillway should be able to accommodate the critical or the worst possible flood conditions (within a specified period depending on the structure) in the catchment—partly by storage in the reservoir resulting in rise of the reservoir level and partly, by spillage through the spillway. At the same time the flood waters released through the spillway should not create flooding down below in the lower reach.

Thus, on the whole, a balance is to be worked out between the economy, efficiency in regard to flood moderation and safety, the reservoir capacity and the required spillway capacity. Hence, for designing the spillway, a knowledge is necessary about the maximum intensity of the critical flood, or the 'design' flood as it is called, as well as the duration and volume of this flood.

The generally adopted methods of flood estimation are based on two types of approaches viz., deterministic and statistical. The deterministic approach is based on the hydro meteorological techniques which requires design storm and the unit hydrograph for catchment. The statistical method is based on the flood frequency analysis using the observed annual maximum peak flood data. Another alternative of estimating the frequency based floods is to carryout frequency of rainfall data and convolute the excess rainfall of the derived frequency with the unit hydrograph or some rainfall run-off model appropriate for the basin.

2.2 METHODS OF DESIGN FLOOD COMPUTATIONS

There are various methods by which the estimate of design flood can be made. Some of them are purely empirical and some are based on statistical analysis of the previous records, apart from the rational methods based on unit hydrograph principle. These are:

  (i) Observation of the highest flood level or maximum historical flood,
 (ii) Empirical formulae,
(iii) Enveloping curves,
 (iv) Flood frequency studies,
  (v) Derivation of design flood from storm studies, and
     (a) application of unit hydrogaph principle, or
     (b) by application of instantaneous unit hydrograph principle.

2.2.1 Observation of Highest Flood

In the absence of any data, an idea of the magnitude of the maximum flood that might have occurred may be worked out by the highest point that the flood water has reached at the project site or at a nearby site on the same stream.

The high flood having been established, the following methods are possible for estimating the corresponding discharge:

  (i) If a weir or dam exists near the site, the head on its crest will determine the discharge (see Appendix I for details).
 (ii) If afflux at a flumed bridge has been observed, approximate discharge can be calculated from it (see Appendix I for details).
(iii) If a fairly long straight reach of the river is available, velocity can be calculated with the help of Lacey's regime equation. The cross-section can be surveyed and plotted up to high-flood level and from that, the values of hydraulic radius R and flow area A, can be determined. Discharge can then be worked out by multiplying the cross-sectional flow area with mean velocity.

This method is only applicable to channels with mobile beds. For streams with rock beds Manning's formula may be used for determination of velocity but it would be subject to the uncertainty involved in the value of the rugosity coefficient. In channels with mobile beds, there is scour of bed during floods. The calculated discharge will, therefore, be somewhat less than the actual value. If a few soundings can be taken during the flood, the error can be considerably reduced (see Appendix I for details).

2.2.2 Empirical Flood Formulae

Empirical formulae should only be used as a last resort in cases where no runoff or rainfall data are available. Several types of empirical relationships have been established based on the catchment properties (mainly area) and in some cases on the rainfall characteristics. The empirical formulae commonly used in this country are the Dickens', Ryves and Inglis formulae. A summary of various empirical formulae are furnished in the following:

(a) Involving the basin area only—Most of these formulae are of the type

$$Q = CA^n \quad \text{(peak discharge)} \tag{2.1}$$

where

C = empirical constant,
A = drainage area,
n = exponent ($1 > n > 0$)

The above equation defines the upper limit or the enveloping curve of the basin.

There are over 30 such widely known formulae and they are used in many countries. In India, the following formulae are mostly referred:

Inglis $$Q = \frac{7000\, A}{(A + 4)^{1/2}} \quad \text{(fan-shape area)} \tag{2.2}$$

Dickens $$Q = 1795\, A^{0.75} \quad \text{(rainfall approximately 100 inch)} \tag{2.3}$$

Dickens $$Q = 149\, A^{0.75} \quad \text{(rainfall 30 to 40 inch)} \tag{2.4}$$

Ryves $$Q = 675\, A^{0.67} \quad \text{(for areas near hills)} \tag{2.5}$$

Ryves $$Q = 560\, A^{0.67} \quad \text{(for areas from 50 to 100 miles from sea)} \tag{2.6}$$

Ryves $$Q = 450\, A^{0.67} \quad \text{(for areas within 50 miles from sea)} \tag{2.7}$$

Madras

$$Q = 2000\, A^{(0.92\, -\, 1/15\, \log A)} \quad \text{(Tungbhadra river)} \tag{2.8}$$

Hyderabad

$$Q = 1750\, A^{(0.92\, -\, 1/14\, \log A)} \quad \text{(Tungbhadra river)} \tag{2.9}$$

where

A = drainage area in sq. miles,
Q = flow in cfs.

Besides these there is a curve proposed by Whiting for Bombay area.

Baird and McIllwraith suggested a formula for maximum recorded flood throughout the world

$$Q_M = \frac{131000\,A}{(107 + A)^{0.78}}$$ (2.10)

where

A = drainage area in sq. miles,
Q_M = maximum flow in cfs.

(b) Involving the basin area and other basin characteristics

Burge (India) $$Q = \frac{CA}{l^{2/3}}$$ (2.11)

where

C = 1300,
l = average length of area in miles,

Kinnison and Colby (New England, USA)

$$Q = \frac{(0.000036\,h^{2.4} + 124)\,A^{0.95}}{r^{1.7}}$$ (2.12)

where

h = median altitude of basin in ft. above the outlet,
r = percentage of lake, pond and reservoir area.

(c) Involving basin area and rainfall features

Rational formula $$Q = CIA$$ (2.13)

where

I = intensity of rainfall in inch/hr,
A = basin area in acres,
Q = flow in cfs.

Isakowski (Austria)

$$Q_m = Km\overline{H}A$$ (2.14)

where

Q_m = peak flow in m³/sec,
$\overline{H}$ = average annual precipitation in metre,
m = coefficient varying 10 to 1 for basin areas from 1 to 25000 km²,
K = coefficient characterising the morphology of the basin and varying from 0.017 to 0.80,
A = basin area in km².

(d) Involving basin characteristics and rainfall features

Pettis (USA)

$$Q = CPB^{-1.25}$$ (2.15)

where

Q = once in 100 years flood peak,
P = once in 100 hours one day rainfall, inches,
B = average width of basin $x = A/l$ in miles,
C = varying from 310 in humid to 40 in desert regions.

Potter (USA)

$$Q = 0.038\ ALR \tag{2.16}$$

where

A = basin area in acres,
L = length of principal waterway divided by the square root of slope in feet per mile,
R = rainfall factor.

Mac Math (St. Louis, USA)

$$Q = CiA^{4/5}S^{1/5} \tag{2.17}$$

where

C = 0.75,
i = rainfall intensity in cm/hr,
A = basin area, acres,
S = average basin slope.

Possentis (Italian mountainous basins)

$$Q_m = \frac{H_m}{l}\left(A_m + \frac{A_p}{3}\right)K \tag{2.18}$$

where

Q_m = flow in m^3/sec,
A_m = area of mountainous part in km^2,
A_p = area of flat part in km^2,
l = length of main water course in km,
K = coefficient varying from 700 to 800; the greater l, the smaller the value of K.

(e) Involving basin area and flood frequency

Fuller (USA)

$$Q_T = Q_1\ (1 + 0.8\ \log T) \tag{2.19}$$

and

$$Q_1 = CA_1^{0.8} \tag{2.20}$$

where

Q_T = T-yr average 24-hr flood,
Q_1 = average of the annual maximum floods during the period of observation,
T = recurrence interval,
A_1 = basin area in sq. miles,

Q_1 and Q_T are average daily discharge; to arrive at peak discharge Q_T is multiplied by

$$\left(1 + \frac{266}{A^{0.3}}\right),\ \text{where } A = \text{basin area in km}^2$$

Central Water and Power Commission had made extensive studies and proposed the following peak flood formula

$$Q_R = CA^{3/4}S_0^{1/4}\frac{C_p}{C_t}\left(\frac{0.6L}{L_c}\right)^{\frac{1}{4}} \tag{2.21}$$

where

C = $C_t/3.0$
Q_R = peak flow for a particular return of R years, C_p and C_t are coefficients similar to that proposed by Snyder for 1 hr unit hydrograph,

S_0 = mean slope of the basin,
A = area of catchment,
L = main channel length, and
L_c = distance to the centre of area of the catchment.

The values of C are given in the form of generalised atlas contour of the coefficients for the whole country [1]. The above method is suggested for estimating design floods for railways and road bridges, designing small catchments, spillways and small tanks.

The Irrigation Research Institute, Roorkee has also given a design flow equation for estimating design flood Q of any recurrence interval T years.

$$Q = Q_{\text{min}} + 2.303 \left(\frac{\Sigma Q}{N} - Q_{\text{min}}\right) \log_{10} \left(\frac{NT}{t_r}\right) \tag{2.22}$$

where

Q_{min} = minimum recorded flood peak,
ΣQ = summation of recorded yearly flood peaks,
N = number of hazard events on record,
t_r = length of the recorded data in years.

Sometimes maximum flood is obtained from the envelope curve of all the observed maximum floods for a number of catchments in a homogeneous meteorological region plotted against drainage area on a log-log paper. Figure 2.1 shows the world enveloping curve proposed by Creager et al. and envelope curves for Indian rivers proposed by Sawin and Kapoor et al. This method cannot be relied upon for estimating maximum probable floods and serves only as a guard to check the values obtained by other methods. Envelope curves are useful to obtain flood values for comparison with those derived by other methods. Design flood values purporting to be the maximum probable in general should be higher than those obtained from the envelope curves. These curves are useful for quick preliminary estimates of design floods. Figure 2.1 is based on the equation

$$Q = 46 \ C \ A \ (0.894 \ A^{-0.018})$$

where

Q = flood in cusecs for sq. miles,
A = drainage area in sq. miles,
C = coefficient depending on the characteristics of the drainage basin with $C = 100$.

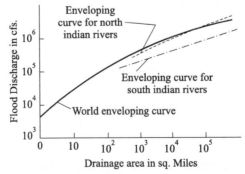

Figure 2.1   Enveloping curves for flood peaks.

COMMENTS

When applying these relationships, it should be kept in mind that they were derived for specific conditions and, therefore, are valid only under similar conditions. Such formulae do not take into account the critical duration and distribution of storm precipitation or its areal distribution pattern and antecedent ground conditions. The data of these formulae could come from records of varying length, floods having entirely different shapes of hydrographs or from meteorologically non-homogeneous areas. In deriving such formulae, streams are assumed to have flood characteristics equal to the worst flood on record; while the possibility of still bigger flood is ignored. The contention that many of these formulae are based on a long period of record is not tenable because a long period of record does not necessarily reflect the critical flood potentialities of a region. As an example it may be cited that three of the greatest floods within approximately the last 300 years in northeastern USA occurred during the years 1936–1938. Another serious criticism of such formulae is that they give the maximum flood without any frequency consideration.

2.2.3   Flood Frequency Study

Flood frequency studies are generally made for the following purpose:

(i) As a guide to judgement in determining the capacity of a structure such as highway bridge, or a coffer dam when it is considered permissible to take a calculated risk.

(ii) As a means of estimating the probable flood damage prevented by a system of flood protection works over a period of years usually equal to the estimated economic life of the work under consideration.

In the first case, the magnitude of the flood discharge that will be equalled or exceeded in a certain period of years is desired. In the second case, it may be necessary to consider in addition to the peak flood discharge, factors such as duration of flood. Depending on the nature of the problem, hydrologists usually employ suitable statistical techniques for estimation of the flood which will equal or exceed once say in a period t-year. For an extremely long periods of record the frequency distribution can be expressed as in Fig. 2.2. In the figure, the ordinates are the probability density and abscissae are the magnitudes of floods, the area under the curve being unity. The area under the curve below any magnitude X_1 is the probability that X_1 will be equalled or less in any year. Usually the steps followed in deriving a frequency relation for peak discharge consist of (a) computation of flood peaks in order of magnitude, (b) computation of recurrence interval, and (c) plotting. The recurrence interval of each flood peak is computed by one of the following formulae, i.e.

$$T = \frac{N}{M} \tag{2.23}$$

$$T = \frac{N}{M - 0.5} \tag{2.24}$$

$$T = \frac{(N - 1)}{M} \tag{2.25}$$

where

N = Number of years of observation,

M = order number of events,

T = recurrence interval.

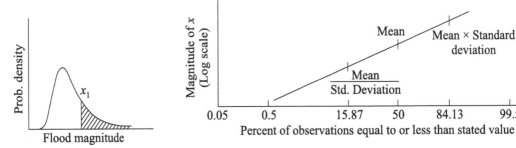

Figure 2.2   Flood frequency distribution.

When streamflow records are available for a number of years, probability studies for the frequency of future floods may be instituted in two ways, viz.

(i) Basic stage method (or partial duration series method) which considers the frequency of exceedance of a particular basic stage during the period of record; and

(ii) Yearly flood method which makes use of only the annual maximum flood during each year of record.

The first method aims at finding the maximum peak flood likely to occur, while the second method is used to ascertain the maximum of the highest annual flood discharge. The two methods would, however, ultimately estimate the same thing, since the highest flood peak in 'N' years is also the highest of the maximum annual floods during these 'N' years.

In the basic stage method the data of all the flood peaks during the 'N' years of record are to be taken and classified into groups according to their intensities, so as to give frequency distribution of all these floods. If a particular flood has been equalled, or exceeded 'M' times during these 'N' years, then its observed return period is N/M years while its percentage of occurrence is $100 \times M/N$.

On the other hand, in the Yearly Flood Method the maximum annual flood discharges of different years are to be arranged in descending order of magnitude and the observed frequencies of the annual floods of different years may be calculated similarly. Thus, the highest of these has an observed return period of N years (equal to the period of record), the second highest which has been equalled or exceeded twice in N years has an observed return period of $N/2$ years and so on. A flood having the M-th position in the descending order of magnitude has been exceeded or equalled M times in N years, and has an observed return period of N/M years.

These observed frequencies; or return periods plotted against the corresponding discharge would indicate a trend of the flood frequency. Extension of the curve indicated by this trend to rarer frequencies, i.e., higher return periods, would estimate the magnitudes of the flood with this return period in the light of these observed data. Such extension to a great length much beyond the range of the observed data is, however, a hazardous task. Individual bias would enter into such extension or extrapolation of the frequency curves. For the purpose of this extension the frequency curve is graduated with a probability equation based on statistical principles, and the extrapolation is done with the help of this equation. Such methods of fitting frequency curves have been given by many investigators. Of these, Gumbel's method based on his theory of extreme values has perhaps the most theoretical support and hence is discussed below.

Statistical study of floods refers to frequency analysis of extreme values. The highest daily mean flow or the maximum flood peak during a year are extreme values, one value being for each year. The annual maximum of a flood series also a homogeneous series of independent variables. Fisher and Trippett [2] showed that if one selected the largest event from each of many large samples, the distribution of these extreme values was independent of the original distribution and conformed to a limiting function. Gumbel [3] described the frequency distribution of a series consisting of the extreme values of a number of homogeneous and independent group of observations. He assumed that the shape of the frequency distribution for floods conforms to the theoretical distribution of extreme values and suggested that this distribution of extreme values was appropriate for flood analysis since the annual flood could be assumed to be the largest of the sample of 3 to 5 possible values for each year.

The general equation for frequency analysis of hydrologic events as given by Chow [36] can be expressed as

$$X_T = \overline{X} + K\sigma \qquad (2.26)$$

where

    X_T = the variate X of a random hydrological series with a return period T,
    $\overline{X}$ = the mean value of the variate,
    σ = the standard deviation of the variate,
    K = a frequency factor.

Which value depends on the return period T and the assumed distribution of frequency.

$$P = 1 - e^{-e^{-b}} \qquad (2.27)$$

where

    e = base of Naperian logarithm

and
$$b = \frac{1}{0.78\sigma}(X - \overline{X} + 0.45\sigma), \text{ i.e. (the reduced variate)} \qquad (2.28)$$

$$\sigma = \Sigma\sqrt{\frac{\Sigma(X - \overline{X^2})}{N - 1}} \qquad (2.29)$$

where

    X = flood magnitude,
    $\overline{X}$ = arithmetic average of all floods in the series,
    N = number of items of the series, i.e. number of years of record.

Taking the natural logarithm of both sides of Eq. (2.27) we obtain,

$$\ln P = e^{-\frac{1}{0.78\sigma}(X - \overline{X} + 0.45\sigma)} \qquad (2.30)$$

Taking the natural Logarithm of the above expression once more we obtain,

$$-\ln[-\ln(1-P)] = \frac{1}{0.78\sigma}(X - \overline{X} + 0.45\sigma) \qquad (2.31)$$

and solving for X, we obtain,

$$X = \overline{X} - \sigma[0.78 \ln\{-\ln(1 - P)\} + 0.45] \qquad (2.32)$$

Further the graphical solution of Eq. (2.27) can be done as the probability P is related to the recurrence interval viz. $T = \frac{1}{P}$ on a special Gumbel's skew-probability paper the relationship (2.27) plots in a straight line i.e., peak flood vs recurrence interval T or b. The plotting paper can be constructed (Fig. 2.3) by laying out on a linear scale of b, the corresponding value of

$$T = \frac{1}{P} = \frac{1}{1 - e^{-e^{-b}}}$$

VALUE OF b

−1	0	1	2	3	4
1.1	1.5	2	3	5	10	20

T = RETURN PERIOD

Figure 2.3   Parameter b vs T.

The computed line will be a straight line and it is sufficient to calculate the return period corresponding to two flood flows and to draw a straight line defined by this point. A third point can be used as a check.

For an infinite series, i.e. $N \rightarrow \alpha$, the reduced variate b is calculated by the formula

$$b = \frac{1.2825(X - \overline{X})}{\sigma} + 0.577 \tag{2.33}$$

The relationship governing reduced variate b to probability P can be expressed as

$$b_p = -\ln[-\ln(1 - P)] \tag{2.34}$$

Alternately the value of b for a given return period T in terms of natural logarithm can be expressed as

$$b_T = -\left[\ln \ln \frac{T}{T-1}\right] \tag{2.35}$$

or,

$$b_T = 0.834 + 2.303 \log \log \frac{T}{T-1} \tag{2.36}$$

The relationship to calculate the value of variate T for the return period T, follows from equation, i.e.

$$X - \overline{X} = (b_T - 0.577) \frac{\sigma}{1.28}$$

or,

$$X = \overline{X} + \left(\frac{b_T - 0.577}{1.28}\right)\sigma \tag{2.37}$$

i.e.

$$X = \overline{X} + K\sigma \tag{2.38}$$

For series of finite length the modified equation can be expressed as

$$X_T = \overline{X} + K\sigma_{n-1}, \text{ where} \tag{2.39}$$

$$\sigma_{n-1} = \sqrt{\frac{(X - \overline{X})^2}{N - 1}} \text{ and} \tag{2.40}$$

$$K = \frac{b_T - b_n}{\sigma_n} \tag{2.41}$$

where

b_T = the reduced variate for a given return period T which is as given by Eq. (2.36)

$\overline{b_n}$ = the reduced mean as a function of sample series N ($\overline{b_n} \to 0.577$ when $N \to \infty$)

σ_n = the reduced standard deviation as a function of sample size N ($\sigma_n \to 1.2825$ when $N \to \infty$)

These values can be obtained from tables provided by Gumbel.

The risk criteria, $P(E)$, i.e. the probability of a higher than X flood occurring within the life period L years of the structure is given by

$$P(E) = 1 - \left[1 - \frac{1}{T(X)}\right]^L \tag{2.42}$$

where $P(E)$ is the probability that at least one flood which equals or exceeds the $T(X)$ year flood will occur in any series of L-years, and $T(X)$ is the true return period of the event X.

COMMENTS

Statistical analysis requires that all data in a series be gathered under similar conditions. The construction of reservoirs, flood embankments, by-passes, or other works which might alter flood flows on a stream, result in a non-homogeneous series. If the change caused by the works is large, the analysis should be limited to the period before or after the change, depending upon the purpose of the study.

Further, the probability method when applied to derive design floods for long recurrence intervals, several times larger than the length of data, has many limitations. In certain cases, however, as in the case of very large catchments when the unit hydrograph method is not applicable and where sufficient long-term discharge-data is available, the frequency method may be the only course available. In such a case, the design to be adopted for major structures should have a frequency of not less than once in 1000 years. When assumed flood values of adequate length are available they are analysed by Gumbel's method and when data is meagre either partial duration method or regional frequency technique is to be adopted as a tentative approach and the results verified and checked by hydrological approaches. Sometimes, when the flood data are inadequate, frequency analysis of recorded storms is made and the storm of a particular frequency is applied to the unit hydrograph to derive the flood. This flood usually has a return period greater than that of the storm.

2.2.4 Derivation from Storm Studies and Application of Unit Hydrograph Principle

The steps involved in these methods are briefly as follows:

(i) Analysis of rainfall versus runoff data for derivation of loss rates under critical conditions.

(ii) Derivation of unit hydrograph by analysis (or by synthesis, in case data are not available).

(iii) Derivation of the design storm, and

(iv) Derivation of design flood flow from the design storm by the application of the rainfall excess increments to the unit hydrograph.

2.2.4.1 Rainfall analysis—Mass rainfall curves

One of the most convenient methods of estimating the intensity and sequence of actual rainfall during past storms at various rain gauge stations in drainage basin is the preparation of 'mass rainfall curves'. A mass rainfall curve is a plot of accumulated total rainfall up to known time intervals against these time intervals (Fig. 2.4). For stations fitted with recording raingauges, which automatically maintain a record of rainfall against time, a mass rainfall curve is directly obtained. At ordinary raingauge stations the raingauge is not observed continuously but only at certain fixed times each day. Hence, only the total rainfall between two successive observations is known and the variation in its intensity during this period cannot be directly ascertained. However, if there are a number of recording raingauges dispersed over the area, the mass rainfall curves for nonrecording gauges can be interpolated on the basis of the mass rainfall curves of the surrounding recording gauges after making allowance, as far as possible, for factors which might cause the rainfall pattern at such nonrecording stations to be different from the neighbouring recording stations.

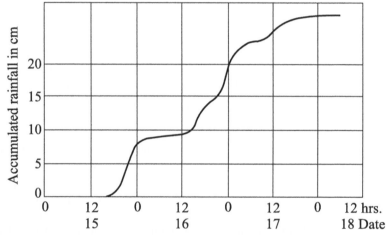

Figure 2.4 Rainfall mass curve.

Having obtained directly and by interpolation, the mass rainfall curves of all the raingauge stations available in the drainage basin, the next step is to determine the influence area of each, that is to say the area over which the mass rainfall curve obtained at a particular station may be considered to be applicable. A very good method of doing so is to divide up the area into what are known as Thiessen polygons (Fig. 2.5). These are obtained by joining raingauge stations by straight lines and drawing their perpendicular bisectors, thus dividing the area into a number of polygons. Each such polygon represents the influence area of the station which is located within that polygon. For more accurate analysis it would be necessary to prepare an 'isohyetal' map of the area for the storm under investigation (an 'isohyetal' is a line joining all the points of equal rainfall over the area) (Fig. 2.6). From the isohyetal map, the weighted average rainfall within each Thiessen polygon can be determined, and this would be a more accurate figure for average rainfall within the polygon area than the rainfall figure at the raingauge station of the polygon. This refinement, however, is usually not done since the error involved is small in relation to the general accuracy of the assumptions.

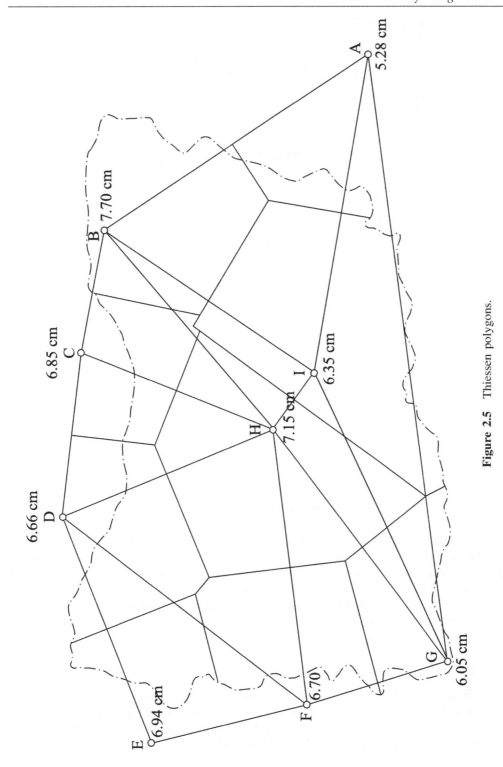

Figure 2.5 Thiessen polygons.

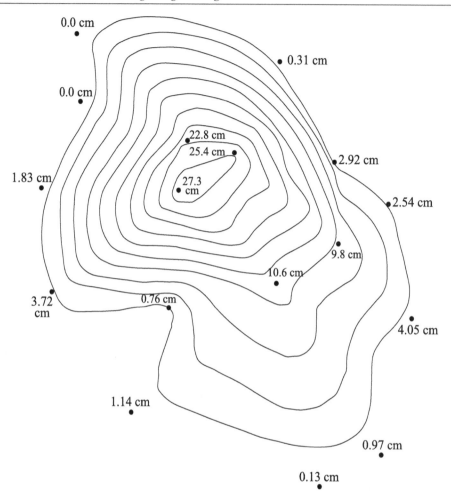

Figure 2.6 Isohyetal map.

It may further be noted that for making a flood forecast from a basin it is not enough to study the heavy storms which have occurred over it in the past but, it is also necessary to consider the heaviest precipitation that is possible in the future. It is possible that the basin may have escaped being the centre of unusually heavy storms in the past merely by chance but may nevertheless be subjected to such storms in future. A study of this nature can only be made by competent meteorologists and their help should be sought where the hydrograph of the worst possible flood is required for an important project.

2.2.4.2 Runoff analysis

General: Runoff is the balance of precipitation after the demands of evaporation and interception by trees, etc. have been met with. It consists of surface flow plus the interflow. Water infiltrating the soil surface and not retained as soil moisture either moves to the stream as interflow or penetrates to the water table and eventually reaches the stream as ground water or base flow. Therefore, flood in a river is the result of surface runoff plus interflow. The nature of runoff of stream is determined by two sets of factors, one depending on the nature of precipitation and the other on the physical characteristics of the drainage basins.

2.2.4.2.1 Infiltration approach

The most reliable method is the infiltration approach. As most of the losses are small during heavy rainstorms the determination of excess rainfall is essentially that of evaluation of the infiltration rates. Infiltration has been defined as the capacity of the soil cover to absorb water. Infiltration includes deep percolation which adds to the ground water-table, the moisture retained by soil and interflow. The definition of the infiltration capacity credited to Horton [4] is the maximum rate at which water can enter the soil at a particular point under a given set of conditions. The infiltration capacity formula can be expressed as:

$$I_c = I_a + (I_o - I_a)e^{-Kt_r} \tag{2.43}$$

where

I_c = infiltration capacity,
I_o = initial (original) infiltration,
I_a = asymptotic limit infiltration capacity or the constant rate of infiltration,
K = a constant known as infiltration constant,
t_r = time, i.e. the duration of rainfall (time from the beginning of rainfall),

which is valid for the case of immediate and continued ponding of the soil surface, i.e. the situation when the water availability at the surface is so great that the soil at the surface saturates instantaneously and remains saturated indefinitely.

The above equation can be used to determine the relationship of duration of rainfall and the rate of infiltration as follows:

By transposing equation and taking logarithms of both sides

$$\ln (I_c - I_a) = - Kt_r + \ln (I_o - I_a)$$

or

$$t_r = \frac{\ln (I_o - I_a) - \ln (I_c - I_a)}{K}$$

Since I_o, I_a and K are constant values for a given area and for a given storm, the above equation becomes:

$$t_r = \ln \frac{(I_o - I_a)}{K} - \ln \frac{(I_c - I_a)}{K}$$

which is of the form of the equation of a straight line, viz.

$$y = mx + C, \text{ where } y = t_r$$

$$m = -1/K \text{ and } x = \ln (I_o - I_a) \text{ and}$$

$$C = \ln \frac{(I_o - I_a)}{K}$$

the plot of which is a straight line on semilogarithm paper. The plot of infiltration rate with time on Cartesian coordinate will show a nature-like as shown in Fig. 2.7. Generally, it is cumulative

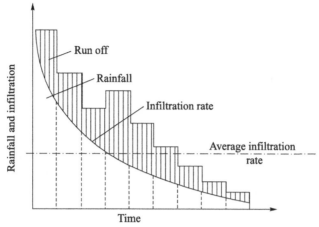

Figure 2.7   Infiltration rate with time (after Horton).

infiltration which is measured in the field and also by the infiltrometer experiment and not the infiltration rate. The cumulative infiltration with respect to time can be expressed in the form

$$y = at^\alpha + b$$

where y = accumulated infiltration in time, t from the beginning of experiment, a, b, α are constants for that area to be determined from relevant data. Infiltration index is an average value of infiltration rate such that the value of rainfall excess of the rate is the surface runoff. Generally two indices are used for signifying the infiltration rate (a) ϕ-index and (b) W-index. A 'ϕ'-index is defined as the average rainfall above which the volume of rainfall equals the volume of runoff. A 'W'-index is the average rate of infiltration during the time the rainfall intensity exceeds the infiltration capacity

$$\text{'}W\text{' index} = \left(\frac{P - R}{t_r}\right)\text{cm} \tag{2.44}$$

where

      P = rainfall in cm/hr.
      R = infiltration rate in cm/hr.
      t_r = the time period of rainfall in hr.

2.2.4.2.2   By runoff percentage

In this case it is assumed that runoff is a certain factor of the precipitation. The factor or runoff coefficient is the ratio of volume of stream flow above the base flow to the volume of rainfall. In estimating the volume of streamflow care should be taken not to include snowmelt in runoff. According to this

$$R = KP \tag{2.45}$$

where

      R = runoff in cm.
      P = rainfall in cm.
      K = runoff coefficient.

Different authors have given the value of K under different conditions. Some of them are discussed below:

Richard:

Richard gave the values of K for various types of catchments. These are given in Table 2.1. These values observed from natural conditions can be used as rough estimate at best.

Table 2.1   Richard's Runoff Coefficient

Type of catchment	Value of runoff coefficient
Rocky and impermeable	0.80 to 1.00
Slightly impermeable base	0.60 to 0.80
Slightly permeable, cultivated or covered with vegetation	0.40 to 0.60
Cultivated absorbent soil	0.30 to 0.40
Sandy absorbent soil	0.20 to 0.30
Heavy forest	0.10 to 0.20

Barlow:

Barlow on his experience in rivers in a province of India has recommended the following values of K given in Table 2.2.

Table 2.2   Barlow's Runoff Coefficient

Class	Type of catchment	Runoff coefficient
A	Flat, cultivated and black cotton soil	0.10
B	Flat, partly cultivated, porous soil	0.15
C	Average	0.20
D	Hills and plains with little cultivation	0.35
E	Very hilly and steep, with hardly any cultivation	0.45

The above values correspond to average monsoon condition. For variation of monsoon from the average condition, these values are modified by the coefficient given in Table 2.3:

Table 2.3   Barlow's Class of Catchment

Nature of season	A	B	C	D	E
Light rainy or no heavy shower	0.70	0.80	0.80	0.80	0.80
Average or varying rainfall, no continuous downpour	1.00	1.00	1.00	1.00	1.00
Continuous downpour	1.50	1.50	1.60	1.70	1.80

Binnie's percentage:

Binnie has suggested that the run-off from a catchment is dependent on rainfall and as the amount of rain increases, the runoff percentage also increases. This is given by the following values in Table 2.4:

Table 2.4 Binnie's Table.

Annual rainfall (cm)	Runoff (per cent)
50	15
60	21
70	25
80	29
90	34
100	38
110	40

2.2.4.3 Soil conservation service—US curve number method

The *SCS-CN* [37] method is based on the water balance equation and two fundamental hypotheses. The first hypothesis states that the ratio of the actual amount of surface runoff to the maximum potential runoff is equal to the ratio of the actual amount of actual infiltration to the amount of potential maximum retention. The second hypothesis states that the amount of initial abstraction is some fraction of the potential maximum retention. Mathematically the water balance equation and the hypothesis can be expressed as follows

$$P = I_a + F + Q \tag{2.46}$$

$$\frac{Q}{(P - I_a)} = \frac{F}{S} \tag{2.47}$$

$$I_a = 0.2S \tag{2.48}$$

The combination of Eqs (2.46), (2.47) and (2.48) leads to the following equation

$$\frac{(P - I_a)^2}{P - I_a + S} = \frac{(P - I_a)^2}{P + 0.8S} \tag{2.49}$$

where

P = total rainfall,
I_a = initial abstraction,
F = cumulative infiltration excluding I_a,
Q = direct runoff and
S = potential maximum retention.

The parameter S depends on characteristics of the soil vegetation land use complex (*SVL*) and antecedent soil moisture conditions (*AMC*) in a watershed. For each *SVL* complex a lower and upper limit of S exist. The *SCS* expressed S as a function of what is termed as curve number as:

$$S = \frac{1000}{CN} - 10 \tag{2.50}$$

where *CN* = curve number and is a relative measure of retention of water by a given *SVL* complex and takes on values from 0 to 100. The number is derived from soil characteristics, vegetation including crops, landuse of that soil as well as intensity of use. In the above equation the unit of S is inches. The equation when expressed in mm becomes

$$CN = \frac{25400}{S + 254} \tag{2.51}$$

In paved areas $S = 0$, and $CN = 100$ and in the area where no surface runoff occurs $S = \infty$ and $CN = 0$. The volume of direct runoff as a function of rainfall and curve number can be represented graphically in Fig. 2.8. The CN depends on the basin characteristics and soil moisture conditions at the time of occurrence of rainfall. It is normally evaluated with the help of various Tables 2.20 to 2.24 given by US *SCS* report 1972 as a function of hydrologic soil group, antecedent rainfall, land use pattern, density of plant cover and conservation practices followed in the area. (see Appendix II)

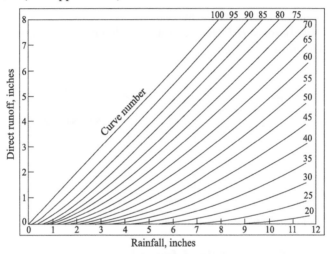

Figure 2.8   Volume of direct runoff as a function of rainfall and curve.

2.2.4.4   Rational method

The rational method is mostly used for determination of peak flow rate, for urban catchments and is based on the assumption that a constant intensity of rain uniformly spread over an area and the effective rain falling on the most remote part of the basin takes a certain period of time to arrive at the basin outlet. It is thus evident that the maximum rate of outflow will occur when the rainfall duration equals the time of concentration. Figure 2.9 shows the runoff hydrograph due to uniform rainfall over a catchment area with rainfall duration exceeding T_c.

The relationship for peak value of runoff rate

$$Q_p = \frac{1}{3.6} \, CIA \tag{2.52}$$

where

Q_p = peak discharge in cubic m³/s,
C = coefficient of runoff,
A = area of drainage basin in km²,
I = intensity of rainfall in mm/hr for a duration equal to the time of concentration.

The runoff coefficient encompasses the effects of other factors like vegetation type, slope of ground, and soil characteristics. The value of C for different condition are given in Appendix III. The formula for the intensity of rainfall is expressed as

$$I = \frac{KT_r^b}{(T_c + b)^n},$$
(2.53)

where

T_r = recurrence interval
T_c = time of concentration.
K, a, b, and n are parameters whose values are given in Appendix III.

$$T_c = 0.01947 L^{0.77} \, S^{-0.385} \text{ as per Kirpich (52)}$$
(2.54)

where

T_c in mins,
L = max^m length of water land in m, *and*
S = slope of the basin = H/L; H = elevation difference between the most remote point and outlet, in m.

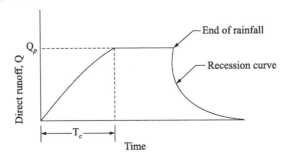

Figure 2.9 Runoff hydrograph due to uniform rainfall.

2.2.4.5 Derivation of regional flood formula

The form of regional flood frequency relationship for the General Extreme Value (GEV) distribution is

$$\frac{Q_T}{\overline{Q}} = u + \alpha \, y_T$$
(2.55)

where,

$$yT = \frac{[1 - \ln(1 - 1/T)K]}{k}$$
(2.56)

where

u, α and k, i.e. shape, location and scale parameters of the distribution,
Q_T = $T - yr$ return period flood estimate,
y_T = reduced variate corresponding to $T - yr$ return period flood.

The conventional Dicken's formula is:

$$Q = C A^{0.75}$$
(2.57)

The form of this formula may be generalized as

$$Q_T = C_T A^b$$
(2.58)

The form of regional relationship between mean annual peak flood and catchment area is

$$\overline{Q} = a A^b$$
(2.59)

Dividing Eq. (2.58) by Eq. (2.59), the following expression is obtained

$$\frac{Q_T}{\overline{Q}} = \frac{C_T}{a} \qquad (2.60)$$

Alternately it may be expressed as

$$C_T = \frac{Q_T}{\overline{Q}} a \qquad (2.61)$$

Or, Substituting the value of $\frac{Q_T}{\overline{Q}}$ from Eq. (2.55)

$$C_T = (u + \alpha \, y_T) \, a \qquad (2.62)$$

And after substituting the value of C_T in Eq. (2.58)

$$Q_T = (u + \alpha \, y_T) \, a \, A^b \qquad (2.63)$$

And again substituting the value of y_T from Eq. (2.56)

$$Q_T = [ua + a \, \alpha y_T] \, A^b$$

or,

$$Q_T = \left[ua + a\alpha \frac{\left[1 - \left\{ -\ln\left(1 - \frac{1}{T}\right)\right\}^k\right]}{k} \right] A^b \qquad (2.64)$$

$$Q_T = \left[ua + \frac{a\alpha}{k} - \frac{a\alpha}{k}\left\{ -\ln\left(1 - \frac{1}{T}\right)\right\}^k \right] A^b \qquad (2.65)$$

or,

$$Q_T = \left[a\left(\frac{\alpha}{k} + u\right) - \frac{a\alpha}{k}\left\{ -\ln\left(1 - \frac{1}{T}\right)\right\}^k \right] A^b$$

$$Q_T = \left[\beta + \gamma\left\{ -\ln\left(1 - \frac{1}{T}\right)\right\}^k \right] A^b \qquad (2.66)$$

where,

$$\beta = a \left(\frac{\alpha}{k} + u\right) \text{ and } \gamma = \frac{-\alpha}{k} a$$

Studies conducted by National Institute of Hydrology, Roorkee, India, gives the following regional flood formula for ungauged catchments of north Brahmaputra (river) system as

$$Q_T = \left[-51.05 + 54.6\left\{ -\ln\left(1 - \frac{1}{T}\right)\right\}^{-0.025} \right] A^{0.72} \qquad (2.67)$$

where Q_T = Flood estimate in m^3/s for T yr, return flood and A = catchment area in km^2.

2.2.4.6 Hydrograph and derivation of unit hydrograph

A hydrograph is a graphical representation of the discharge of a stream as a function of time. For the purpose of identifying general runoff characteristics and to provide a basis for analysis of its component parts a graphical presentation of stream-flow data is prepared in the form of hydrographs. The stream flow is generally composed of (a) precipitation falling directly into the streams, (b) true overland flow, (c) subsurface storm flow or interflow without having reached the main groundwater table, and (d) the groundwater flow or the water contributed as underground flow from the groundwater. The characteristics of direct and groundwater runoff differ greatly and there are no practical means of differentiating between the two after they have been intermixed in the stream. The typical hydrograph resulting from a single storm consists of a rising limb, peak and recession. The recession represents the withdrawal of water stored in the stream channel during the period of rise. Double peaks are sometimes caused by the geography of the basin but more often result from two or more periods of rainfall separated by a period of little or no rainfall.

The concept of unit hydrograph as a tool for the analysis of surface runoff was first introduced by Sherman [5] and later developed by different authors. It is essentially the hydrograph which is produced by a rainfall of given duration uniformly distributed over the catchment which produces a direct runoff volume of unit depth, i.e. say, 1 cm over the entire catchment area. For obtaining the unit graph, the rainfall records are to be searched for an ideal single interval isolated and uniform rainfall. From the corresponding resultant runoff hydrograph the base flow (inclusive of the ground water inflow if possible to estimate) is deducted and the distribution of the residual hydrograph in different time-intervals is worked out. This gives the distribution hydrograph which is preferred by Bernard [6] to the actual unit graph. The method of obtaining the distribution hydrograph and the corresponding unit hydrograph is shown in Table 2.5.

Table 2.5 Computation of Distribution and Unit Hydrograph Ordinates.

Time interval	*Observed discharge* (Q)	*Base flow* (Q')	*Residual discharge* $(F = Q - Q')$	*Percentage distribution* (p)	*Ordinates of unit hydrograph* (q) (m³/s)
1	Q_1	Q_1'	F_1	p_1	$q_1 = p_1 \Sigma q$
2	Q_2	Q_2'	F_2	p_2	q_2
3	Q_3	Q_3'	F_3	p_3	q_3
.	.	.	.	.	.
.	.	.	.	.	.
.	Q_n	Q_n'	F_n	p_n	q_n
Total		ΣF	$\Sigma 100$	Σq	

$$p_i = \frac{100 F_i}{\Sigma F}, \Sigma q = \frac{A \times 10^4}{100 t}$$

In the above Expression A is the catchment area in *ha* and t is the number of seconds in one time-interval.

i.e., accordingly
$$\left[\frac{\Sigma q \cdot t \cdot 100}{A(10^4)} \right] = 1$$

2.2.4.6.1 *Derivation of unit hydrograph*

The theory of the unit hydrograph is based on three fundamental assumptions, viz., that for a particular catchment, (i) the duration of surface runoff from a single interval storm is essentially constant whatever the amount of runoff, (ii) the surface runoff hydrographs resulting from different single-interval storms are proportional to each other, i.e. the distribution of surface runoff is the same for all such storms, and (iii) that the resultant hydrograph from a composite multiple-period storm is obtained by superimposition of the component hydrographs obtained from the precipitations at different intervals.

Unit hydrographs can be derived from hydrographs produced either by isolated uniform intensity storms (which seldom occur) or by complex storms. The hydrograph resulting from a known intense storm of some small unit duration is chosen and the groundwater contribution along with the base flow is separated to obtain the total volume of the storm runoff.

Groundwater flow is obtained by extension of the groundwater depletion curve beyond A, i.e. the time of beginning of storm runoff. This is shown in Fig. 2.10 which represents the estimate of groundwater discharge as it would have continued, had no rain occurred.

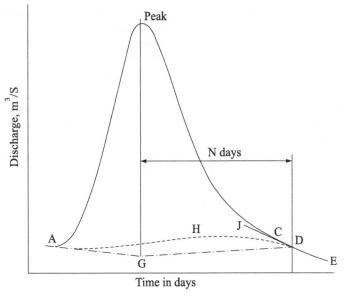

Figure 2.10 Analysis of hydrograph.

Line DE represents the groundwater depletion after the storm runoff has largely passed the point of measurement; it is obtained by shifting a template of the groundwater depletion curve JDE horizontally to coincide with the observed runoff. Point D, where the depletion curve extended backward departs from the recorded discharge, is considered to represent the end of storm water discharge. The point D, representing the end of surface runoff is taken at the moment, where the recession limb of the hydrograph has its points of greatest curvature. The period, N, from peak to point D may be estimated by inspection of several hydrographs from the basin. A rough guide to the selection of the time period, N is given by

$$N \text{ (in days)} = \left[\frac{A_1^{0.20}}{1.21} \right]$$

where

A_1 = drainage area in km^2.

The transition from A to D may be accomplished by AGD which is groundwater depletion extended from A to G, a point under the crest of the hydrograph, and then straight line to

D or by the curved line *AHD*, based on the fact that the transition should be a gradual one. When performing the segregation, the error due to the possible incorrectness can be reduced by applying a consistent procedure in analysis and synthesis. It is assumed that there is no deep seepage flow into or out of the drainage basin. With the groundwater contribution separated, the balance would represent the storm runoff.

The ordinates of the storm runoff are then divided by the total runoff volume over the drainage basin expressed as depth in cm above the point of measurement in order to reduce to the unit volume. The unit hydrograph thus obtained holds for storms the duration of which equals that of the analysed storm. It is a good practice to analyse several storms of the same duration and to average the peak discharges and the times to peak and then to sketch the unit hydrograph, analogous to the observed hyrographs, keeping in mind that the total volume should be unity. The simplest way to calculate volume of storm runoff is to sum up the product of the ordinates of the hydrograph and the corresponding value of time interval of the base period.

The derivation of unit hydrographs from complex storms is a more complicated procedure but results in a more accurate average hydrograph. Several procedures have been developed but the method of successive approximation suggested by Collins [7] is the simplest one. First, a unit hydrograph (derived from whatever storm records are available) is tentatively selected. With its aid the hydrograph of all the partial storms but excluding the largest portion of the adopted complex storm is constructed. The hydrograph thus obtained, subtracted from the actual observed hydrograph of the adopted complex storm, should after reduction to the unit volume yield the estimated unit hydrograph, the ordinates of which are made equal to the averages of the first estimated unit hydrograph, and the one obtained with it.

In practice one usually finds that the catchment responds in a manner sufficiently non-linear or time variant, and that great refinement in deriving unit hydrograph is not justified.

COMMENTS

The unit hydrograph should be derived from a fairly large flood, otherwise it might not be representative of the true conditions during large floods. The idea that a catchment has a linear time invariant response (i.e. that the superposition principle is applicable or that the unit hydrograph of duration *T*, is unique) is a pure assumption and it must not be stressed any further than necessary; it is obvious that the ground conditions at the time of storm affect the speed at which the water runs over or through the soil. The greater the discharge, the greater the velocity in the streams. Consequently, unit hydrographs derived from large floods must be expected to show shorter time elements than those derived from small floods. From the analysis of hydrologic data for minor and major floods recorded in a large number of basins in USA, it has been found that in most of the basins considered, the peak ordinates of unit hydrographs derived from major flood hydrographs, representing runoff volumes greater than approximately 13 cm in depth from the drainage area, were 25–50 per cent higher than values computed from records of minor floods, in which the runoff was from 3–5 cm. The differences were not proportional to the volumes of flood runoff but apparently were the results of differences in a real distribution of rainfall and in hydraulic relations.

2.2.4.6.2 *Duration of unit storm period*
To reduce the labour of the unit graph computations the largest period consistent with accuracy required to define the hydrograph should be chosen. A good thumb rule is to choose the unit

period equal to about 1/4 to 1/5th of time to peak so that at least 4-points on the rising limb are defined. Unit period should, however, be adopted in full hourly values such as 1, 2, 3, 4, etc. It can be 1/2 hr for very small catchments since the rainfall data is generally obtained at intervals of 1/2 hr.

Hydrographs which give appreciable direct surface runoff of, say 1.3–2.5 cm (i.e. 0.5–1 in) and above over the basin should be selected. The volume of surface runoff is computed and a loss rate (ϕ-index i.e. loss rate over effective rainfall period) is applied to the hyetograph such that the rainfall excess is equal to the volume of direct surface runoff. Rainfall prior to the starting of the flood hydrograph is taken as initial loss (i.e. filling up depression etc.) and is not included in the effective rainfall.

The duration of the unit hydrograph means the duration of effective rainfall. Short intense storms are advantageous for the derivation of unit hydrograph because the determination of the effective rainfall is rendered much easier. Variations in the distribution of intensity of effective rainfall are obviously much less serious if the rainfall is of short duration, than if a long storm is involved. On many catchments, however, it is not possible to obtain storms of sufficiently short duration containing sufficient rainfall to produce a reasonable flood; when this happens storms of greater duration must be used. Valley storage tends to eliminate the effects of minor variations in rainfall intensity and, therefore, longer unit rainfall durations are suitable for basins having large valley storage capacities. A 6-hour unit rainfall duration is suitable and convenient for most studies relating to drainage areas larger than approximately 300 km^2, while 12-hour duration is adopted for preliminary approximate studies. For drainage areas less than 300 km^2, a unit duration of 1/2–1/4 time of basin lag is generally satisfactory. For practical purposes the following table gives a rough outline of the relation between the size of area and unit duration.

Area in km^2	Unit duration
2	1/4 × time of basin lag (centre of mass of rainfall to peak)
20–200	2 hr
200–1,000	6–12 hr
1,000–2,500	12 hr
2,500–5,000	12–24 hr
5,000–10,000	24 hr

2.2.4.6.3 Limitations of the unit hydrograph theory

The unit hydrograph should not be applied to basins larger than 10,000 km^2 because of the non-uniform rainfall distribution unless reduced accuracy is acceptable; however, much depends on regional and climatic characteristics. For small basins the construction of a unit hydrograph necessitates the presence of recording raingauges and continuous stream discharge records. Besides, for smaller plots the major part of flow is laminar instead of turbulent and the base width will not be independent of rainfall intensity. In narrow elongated basins, uniform intensity storms will seldom occur. However, if rainfall pattern is reasonably invariant from storm to storm, its effect is neglected unless the non-uniformity in areal distribution of the effective rainfall is pronounced.

2.2.4.6.4 Concentration of runoff near peak

A study of unit hydrographs for a large number of drainage basins has revealed an approximate relationship between the peak discharge rate and the widths of unit hydrographs at ordinates exceeding approximately 50 per cent of the maximum. Curves *W-50* and *W-75* are usually drawn to envelope the majority of values of unit hydrograph widths measured at discharge ordinates equal to 50 and 75 per cent of peak ordinate respectively as obtained from drainage basins of various configurations and runoff characteristics. The curves may be used in determining conservative widths for synthetic unit hydrographs, alternately the following set of equations may be adopted

$$\text{Curve } W\text{-}50 \text{ (hr)} = \frac{5.6}{(q_{pR})1.08} \tag{2.68}$$

$$\text{Curve } W\text{-}75 \text{ (hr)} = \frac{3.21}{(q_{pR})1.89} \tag{2.69}$$

where

q_{pR} = peak discharge rate, per unit drainage area of unit hydrograph for duration t_R in cumec/km^2.

2.2.4.6.5   *Synthetic unit hydrograph*

In many important hydrologic studies, synthetic unit hydrographs are required either as a substitute for derivations from hydrologic records or as a means of correlating the supplementing observed data. Several methods of computing synthetic unit hydrographs have been presented in technical publications. In developing unit hydrographs for use in estimating critical flood hydrograph, conservative determinations of peak discharge, the degree of concentration of runoff near peak, and the lag time are of primary importance; the shape of the rising and recession sides and the base width of the unit hydrographs are usually of secondary importance. Snyder [8] based on his studies on the relationship between lag-to-peak of the area-time curve and that of the actual hydrograph presented the synthetic unit hydrograph relations in 1938. The basis of the method is an empirical relationship between the catchment characteristics; like area and shape (slope being secondary), and the shape of the unit hydrograph as characterised by

(a) Time of occurrence of the peak,
(b) Magnitude of the peak flow, and
(c) Base width.

Snyder presented three empirical equations:

(i) **Lag equation**:

$$t_p = C_t (L_c L)^{0.3} \tag{2.70}$$

where

t_p = lag time from mid-point of effective rainfall duration (t_r) to the peak of unit hydrograph; in hr,
L_c = river distance from the outlet to the centre of gravity of the drainage basin; in km,
L = river distance from the outlet to the upstream limit to the drainage area, in km,
C_t = coefficient depending upon units and drainage basin characteristics,
t_r = basic rainfall duration equal to $\dfrac{t_p}{5.5}$

(ii) *Peak flow equation*:

$$q_p = \frac{C_p}{t_p},$$ (2.71)

$$q_p R = \frac{C_p}{t_{pR}},$$ (2.72)

where

C_p = coefficient depending on units and basin characteristics,

q_p = peak flow (cumec/km^2) of the basin for standard rainfall duration t_r,

q_{pR} = peak flow (cumec/km^2) of the basin for duration t_R,

t_{pR} = lag time from mid-point of duration t_R to peak of unit hydrograph in hours, i.e.

$$t_{pR} = t_p + \frac{(t_R - t_r)}{4},$$ (2.73)

(iii) *Base length equation*: The base length of the unit hydrograph in days may be estimated from:

$$T = 3 + \frac{3t_p}{24}$$ (2.74)

These relationships have proved to be particularly useful in the study of runoff characteristics of drainage areas where stream-flow records are not available. However, in applying the method, the validity should be assured to start with by checking the similarity of the basin characteristics. Proper values of the coefficients are to be determined first for the region where the equations are to be applied.

2.2.4.6.6 *Changing the duration of a unit hydrograph*

If we have a unit hydrograph obtained from a storm of duration T we can use the superposition principle to obtain a unit hydrograph of duration equal to any integral multiple of T. The T-hour unit hydrographs are staggered by periods of T and the simultaneously occurring ordinates are added to reproduce flood hydrographs corresponding to n units of runoff from a storm duration of nT-hours. The ordinates of the hydrograph are divided by n to give the ordinates of nT-hour unit hydrograph.

If we require the unit hydrograph for a duration other than an integral multiple of the given duration, we must resort to 'S-curve hydrograph'. This is the hydrograph of storm runoff due to a continuous uniform effective rainfall of unit intensity lasting indefinitely. A more convenient procedure for its derivation is shown in Table 2.6.

Table 2.6 Application of S-curve to Derive 6-hr Unit Hydrograph.

Time, (hr)	12-hr *unit hydrograph ordinates* (m^3/s)	12-hr *S-curve hydrograph* (2) *ordinates origin time shifted* 12-hr *later*	12-hr *S-curve hydrograph* (1) *ordinates*	12-hr *S-curve hydrograph* (1) *ordinates shifted* 6-hr	*Ordinates of hydrograph from 0.5 cm runoff in* 6-hr	6-hr *unit hydrograph coordinates* (m^3/s)
0	0	—	0	—	0	0
6	900	—	900	0	900	1800
12	3400	0	3400	–900	2500	5000
18	6900	900	7800	–3400	4400	8800
24	10100	3400	13500	–7800	5700	11400
30	12300	7800	20100	–13500	6600	13200
(*Contd.*)	....	....	....	....	....	....

The process is continued until the S-curve hydrograph discharge rate is equal to the basic rate of rainfall excess.

When applying the S-curve principle for making the adjustments of the unit hydrograph, it may be convenient to work from the right end of the unit hydrograph where the correct S-curve value is known. Sometime it is found that S-curve hydrograph becomes wobbly towards the end; this means that either the basic unit hydrograph is incorrect or the catchment does not respond in an exactly linear manner.

2.2.4.7 Estimation of design storm and the design flood therefrom

The design flood discharge is the maximum flood that would occur under average physiographic conditions of the watershed due to a design storm of a given frequency. The design storm is defined as the storm which gives rise to the design flood for the particular catchment and has to be selected based on 'basin lag' time and the desired return period of flood for which the structure is designed. Based on analysis of a number of combinations of storms and basin lags it has been found that for a given watershed the critical storm has a duration equal to the basin lag.

Reliable precipitation data in many cases are available for a much longer period, say 50 or 100 years or even longer, and as a result more reliable estimates of the rare storms are possible than in the case of rare floods, whether by probability methods otherwise. As such the practice of estimating design storm first and then calculating the design flood therefrom is in frequent use by the designers. The design storm again may be estimated by any one of the following methods:

(i) probability method of estimation of maximum precipitation of duration of one, two, three or more days,
(ii) method of rainfall-depth-duration curves or rainfall-intensity-duration curves, and
(iii) method of storm transposition, i.e. transposition of the worst storm recorded in the neighbourhood of the catchment so that the catchment lies at the centre of the storm.

The probabilities again, may be based upon the annual highest one-day rainfall or on the frequency distribution of all the one-day rainfalls, etc. As in the case of flood peaks, the observed frequencies of rainfalls of different order may be calculated as before, and the frequency curve extended by methods of probability. In this case, the probability estimates would be more reliable because of the greater length of the basic data.

The maximum depth-duration curve for the size of the area involved may be computed from a comprehensive study and analysis of the major storms in order to determine the most critical combination of the meteorological conditions. The depth-duration area curves express graphically the relations between the average depths of precipitation in different areas around the storm centre and the corresponding area, for different durations of the rainfall.

For the transposition of the storm to neighbouring catchments the isohyetals for different durations of the storm may be drawn and the areas bounded by each closed isohyetal may be determined along with the average rainfall in this area in each case. These values would then give the critical depth-area relationships for different durations like one day, two days, three days, etc. For a particular catchment precipitations may be read off from these curves so as to constitute the design storm.

After the design storm is estimated, the design flood may be computed therefrom by the application of the unit hydrograph.

COMMENTS

To obtain a design storm a combination of all the worst conditions is chosen and as it is not likely what all the worst conditions would operate in the case of every big storm the factor of safety is thus increased. Thus, a 50-year storm, when combined with the worst set of other worst conditions, would give a 100-year flood. It would, however, be safe if a 66-year storm is taken for arriving at a 100-year flood. For small basins where the critical pattern of design storm will occur, often the factor of safety involved in assuming a combination of worst condition would be less. In such cases a 100-year storm should be taken to derive a 100-year flood. A basin is small when the time of concentration is less than one day and storm generally covers the entire area. The Central Water and Power Commission of India [9] recommends that in the case of major and medium projects with storage of more than 6000 hect met. (50,000 acft) the design flood to be adopted should have frequency of not less than once in 1000 years and in the case of minor dams with less than 6000 hect met. (50,000 acft) of storage the design flood should have a frequency of once in 100 years.

2.2.4.7.1 *Estimation of design flood*

The utility of the unit graph method has been in computing the runoff hydrograph when the storm rainfall is known and also in flood routing computations.

The computation of the runoff hydrograph from a storm with known rainfall is done by superimposition of the component runoff graph's for the precipitation in each interval. If the time interval is taken as one day and, the precipitations on successive days are R_1, R_2 and R_3 with estimated runoff ratios of a_1, a_2, a_3 respectively, the effective rainfall contributing to the surface runoff on different days would be $r_1 = a_1 R_1$, $r_2 = a_2 R_2$, $r_3 = a_3 R_3$ and the computation of the runoff in such a case is illustrated in Table 2.7.

Table 2.7 Application of Unit Hydrograph to Estimate Runoff Hydrograph.

Day	Effective rainfall	Ordinates of unit hydrograph	Computed runoff
1	r_1	p_1	$Q_1 = p_1 r_1$
2	r_2	p_2	$Q_2 = p_2 r_1 + p_1 r_2$
3	r_3	p_3	$Q_3 = p_3 r_1 + p_2 r_2 + p_1 + r_3$
.	.	.	.
.	.	.	.
.	.	.	.
n		p_n	$Q_n = p_n r_1 + p_{n-1} r_2 + p_{n-2} r_3$

It is observed in practice that the unit hydrograph peak obtained from heavier rainfall is about 25–50 per cent higher than that obtained from the smaller rainfalls. Therefore the unit hydrograph from the observed floods may have to be suitably maximized up to a limit of 50 per cent depending on the judgement of the hydrologist. In case the unit hydrograph is derived from very large floods then the increase may be of a very small order, if it is derived from low floods the increase may have to be substantial.

2.2.4.8 Design flood from IUH—Basic concepts of instantaneous unit hydrograph

It may be recalled that in the earlier pages the design flood corresponding to an effective precipitation over a catchment area has been based on unit hydrograph concept. If the unit of effective precipitation becomes infinitesimally small, the resulting unit hydrograph is called an instantaneous unit hydrograph or IUH which is expressed simply by the expression $u(t)$ (Fig. 2.11). In other words, for an IUH, the effective precipitation is applied to the drainage basin in zero time. By the principle of the superposition of the linear unit hydrograph theory where an effective rainfall function $I(\tau)$ of duration t_0 applied to each infinitesimal element, of this effective rainfall hydrograph (ERH) will produce a direct runoff hydrograph (DRH) equal to the product of $I(\tau)$ and IUH expressed by $u(0, t - \tau)$ or $u(t - \tau)$. Thus, the ordinate of DRH at time t is given by

$$Q(t) = \int_0^t u(t - \tau) I(\tau) d\tau \tag{2.75}$$

In the equation the function $u(t - \tau)$ known as the Kernel function [10] which translates the integrated effect of the excitation $I(\tau)$ occurring from the beginning of the rainfall ($\tau = 0$) into the response time t, into the response at time t. If in addition one assumes that the part of the response at time t, due to the excitation at time t, does not depend on the dates of t and τ but only on their difference (i.e. time) then the system is not only linear but in addition time invariant. As such the Kernel function depends on only the arguments $u(t - \tau)$ rather than both arguments t and τ. What the equation tells us is that the catchment response is entirely characterized by one function of one argument, i.e. the catchment Kernel. If this function is known then the direct runoff can be forecast after any excess rainfall pattern by equation Eq. (2.75). In the more usual terminology of hydrologists, the Kernel function is known as the instantaneous unit hydrograph.

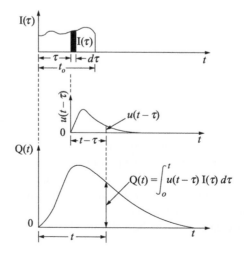

Figure 2.11 IUH and convolution of $I(\tau)$ with IUH.

2.2.4.8.1 *Derivation of IUH from conceptual model*

Various types of conceptual models have been proposed to delineate an IUH. The model proposed by Nash has been dealt with here. Nash considered a drainage basin as equivalent to n identical linear reservoir in series (Fig. 2.12). By routing a unit inflow through a fictitious reservoir where the storage S is directly proportional to outflow q, i.e.

$$S = Kq \tag{2.76}$$

K being the reservoir constant called as storage coefficient. As the difference between inflow p to the reservoir and outflow q is the rate of change of storage the continuity equation is

$$p - q = dS/dt \tag{2.77}$$

Substituting Eq. (2.76) in Eq. (2.77) results in a differential equation in q which when solved for the initial condition, i.e. $q = 0$ at $t = 0$ results in the following expression for outflow

$$q = p \, (1 - e^{-t/K}) \tag{2.78}$$

At $t = \alpha$, $q = p$, that means the outflow approaches an equilibrium condition. Suppose the inflow terminates at time t_0 after outflow began, the outflow at t in terms of discharge q_0 at t_0 can be expressed as

$$q = q_0 \, e^{-t/\tau} \tag{2.79}$$

where $\tau = (t - t_0)$ i.e. equal to the time since inflow is terminated. For an instantaneous inflow which fills the reservoir of storage S_0 in $t_0 = 0$, we have from Eq. (2.76) $q_0 = S_0/K$ and Eq. (2.77) gives the outflow as

$$q = p \, e^{-t/K} = S_0/K \, e^{-t/K} \tag{2.80}$$

For an unit input or $S_0 = 1$, the IUH of linear reservoir is

$$u(t) = \frac{1}{K} e^{-t/K} \tag{2.81}$$

This is represented as the outflow from the first reservoir (Fig. 2.12). In routing the flow this outflow is considered as the inflow to the second reservoir. Using Eq. (2.80) as the input function

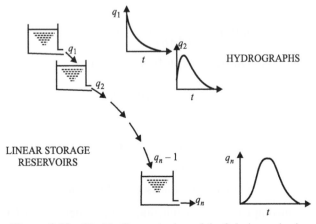

Figure 2.12 Nash's Conceptual model of drainage basin.

with τ being the variable and using the Kernel function with $(t - \tau)$ being the variable the convolution integral of Eq. (2.78) gives the outflow from the second reservoir as

$$q_2 = \int_0^t \frac{1}{K} e^{-t/K} \frac{1}{K} e^{-(t-\tau)/K} d\tau = \frac{1}{K^2} e^{-t/K} \tag{2.82}$$

Continuing the routing procedure up to the n-th reservoir the outflow q_n from it can be expressed as

$$u(t) = \frac{1}{K(n-1)!} \left(\frac{t}{K}\right)^{n-1} e^{-t/K}$$

or

$$\frac{1}{K\Gamma(n)} (t/K)^{n-1} e^{-t/K} \tag{2.83}$$

which is the IUH of the simulated drainage basin. The value of K and n can be found out by taking the first and second moments of IUH about the origin $t, = 0$, i.e.

$$M_1 = nK, \; M_2 = n (n + 1) K^2 \tag{2.84}$$

The first moment represents the lag time of the centroid of IUH. In the application of IUH in the convolution integral for relating ERH and DRH, the linearity principle requires that each infinitesimal element of ERH yield its corresponding DRH with the same lag time.

This means the difference in time between the centroids, of ERH and DRH should be equal to M_1. Hence,

$$(M_{DRH})_1 - (M_{ERH})_1 = nK \tag{2.85}$$

where $(M_{DRH})_1$ and $(M_{ERH})_1$ are the first moment arms respectively of ERH and DRH about the time origin. Further, it can be shown that

$$(M_{DRH})_2 - (M_{ERH})_2 = n (n + 1) K^2 + 2nK(M_{ERH})_1 \tag{2.86}$$

where $(M_{DRH})_2$ and $(M_{ERH})_2$ are the second moment arms respectively of DRH and ERH about the time origin. The values of n and K needed for defining IUH can thus be found out from the computed values of first and second moment of the given ERH and DRH. The general IUH equation is given in Eq. (2.83) from which the equation of unit hydrograph of any period T is found out as follows:

$$u(0, t) = \frac{1}{K\Gamma(n)} e^{-t/K} (t/K)^{n-1} \tag{2.87}$$

Now the equation of S-curve hydrograph can be expressed as

$$S(t) = \int_0^t u(0, t) \, dt$$

or

$$S(t) = \frac{1}{K\Gamma(n)} \int_0^{t/k} e^{-t/K} (t/K)^{n-1} d(t/K) \tag{2.88}$$

or

$$S(t) \text{ can be expressed as } I \, (n, t/K)$$

where $I \, (n, t/K)$ is the incomplete gamma function of the order n at (t/K).

The UH of period T accordingly can be expressed

$$u(T, t) = \frac{1}{T} [s(t) - S(t - T)]$$

or

$$u(T, t) = \frac{1}{T} [I(n, t/K) - I(n,(t - T)/K)] \tag{2.89}$$

which is the general equation of UH of period T. Tables of $I(n, t)$ are available for writing down the ordinates of $u(T, t)$. With the unit hydrograph thus known the desired flood can be computed for a given excessive rainfall following methods as outlined in section 2.2.4.7.1.

2.2.4.8.2 *Diskin's method of obtaining IUH from derivatives*

The method [11] is based on approximating the IUH by a polygon composed of a number of straight line segments. The second derivative of such a segmental unit hydrograph is a train of pulses located at the points of change of slope of the polygon. The magnitude and location of these pulses is evaluated from the shape of the rainfall hyetograph and that of the surface runoff hydrograph. Integration of the function represented by the train of pulses yields a block-shaped diagram which gives the slope of the IUH at all points; integration of this diagram leads to the polygon representing IUH.

The method is based on the theorem that the convolution of the first derivative of the input function with the second derivative of the impulse response function of the system yields the third derivative of the output function of the system—Impulse train functions are functions which have non-zero values only at a finite or infinite number of distinct points. The value of the function at the points when they are not zero is undefined except in the sense of the definition of a unit impulse $\delta(t_1)$. This definition, for a unit impulse at the point $t = t_1$ is summarised by the following equations

$$\delta_{t_1} = 0 \text{ for } t \neq t_1; \quad \delta_{t_1} \to \infty \text{ for } t = t_1$$

and

$$\int_{t_1 - \varepsilon}^{t_1 - \varepsilon} \delta_{t_1} \, dt = 1.0 \text{ for all } \varepsilon > 0 \tag{2.90}$$

The value of an impulse-train function $f(t_1)$ at a point t_1 where it is not zero is taken as the product of some positive or negative quantity A_1 and the unit impulse at that point.

$$f_{(t_1)} = A_1 \delta_{(t_1)} \tag{2.91}$$

The convolutions of two impulse train functions also results in an impulse train function. Consider the case of convolution of two functions each of which is composed of a single impulse. The two impulses are taken to be one having a value A_1 located at time t_1 and the second having a value B_1 located at time s_1 (Fig. 2.13). For any value of τ which is not equal to t_1 the first function in the convolution integral is equal to zero. Also for any value of the time t which is not equal to $(t_1 + s_1)$ the value of $(t_1 - \tau)$ will be such that the second function in the convolution operation will be zero.

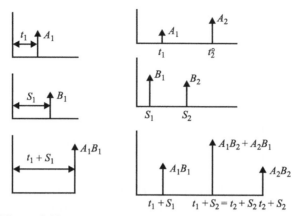

Figure 2.13 Convolutions of two impulse train functions.

It follows that the function resulting from the convolution operation will have a non-zero value at the point $t = t_1 + s_1$. The value of the resulting function at that point is seen to be equal to the product of the values of the two impulses $A_1 B_1$. The procedure can be extended to the convolution of two functions that are composed of impulse trains. It is clear that if there are n_1 impulses $(A_1, A_2, \dots A_{n1})$ located at $(t_1, t_2, \dots t_{n1})$ in the first impulse train and n_2 impulses $(B_1, B_2, \dots B_{n2})$ located at $(s_1, s_2, \dots s_{n2})$ in the second function, the number of impulses in the function obtained by their convolution will be $n_1 n_2$. Their values will be $A_i B_j$ and their location $(t_i + t_j)$ in which $i = 1, 2, \dots n_1$ and $j = 1, 2, \dots n_2$. The number of impulses in the result will be less than $n_1 n_2$ if there are two or more pairs of impulses in the two functions for which the sum of $(t_i + t_j)$ is equal to a given value of t. In such a case the value of the function obtained by the convolution at that point will be the sum of the products of the corresponding pairs of pulses. The procedure is repeated in Fig. 2.13 where two functions each containing two impulses are convoluted. In the resulting function three impulses are obtained at the points:

$$(t_1 + s_1); (t_1 + s_2) = (t_2 + s_1) \text{ and } (t_2 + s_2)$$

the corresponding values of the impulses at these three points are respectively $A_1 B_1$, $(A_1 B_2 + A_2 B_2)$ and $A_2 B_2$ (Fig. 2.13).

The problem of deriving the IUH, $h_{(t)}$ from a given pair of functions, viz., the direct surface runoff hydrograph y_t and the rainfall excess hyetograph x_t is now described with the help of an example after Diskin. Consider the direct surface hydrograph as shown in Fig. 2.14. The hydrograph is composed of straight line segments (S) joined by parabolic arcs (C) spanning a duration equal to the duration of the rainfall excess. The straight line sections may be of varying length but the curved sections must all be of equal length.

The derivative (dy/dt) of the diagram y_t, which approximates the given direct surface runoff hydrograph will be a polygon composed of sloping and horizontal sections. The slopes of these segments should be proportional to the heights of the blocks in the hyetograph and their total height equal to the change in slope between adjacent straight lines in the y_t diagram. Let $x_1, x_2 \dots x_m$ denote the heights of the blocks in the rainfall excess hyetograph and let $H_1, H_2, \dots H_n$ denote the heights of the horizontal sections of the (dy/dt) diagram above, the base-line and let B_0, B_1, B_n denote the unknown values of the impulses in the impulse train function representing the second derivatives (d^2h/dt^2) of the IUH. It follows from the method

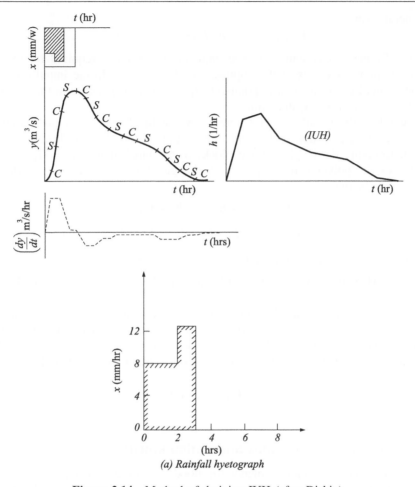

Figure 2.14 Method of deriving IUH (after Diskin).

of computations of the curves in Fig. 2.14 that the height of the first horizontal section in the (dy/dt) diagram is given by

$$H_1 = B_0 \, (x_1 + x_2 + \ldots x_m) \, T \qquad (2.92)$$

where T = width of each block in the rainfall excess hyetograph assuming it is composed of m blocks of equal width T. The change in height between the first horizontal section and the second horizontal section in the (dy/dt) diagram is, given by $(H_2 - H_1) = B_1 \, (x_1 + x_2 + \ldots + x_m) \, T$ and it will be apparent that the expression

$$E = (x_1 + x_2 + \ldots + x_m) \, T \qquad (2.93)$$

is the total rainfall excess represented by the area under the hydrograph

$$H_1 = B_0 \, E \qquad (2.94)$$

$$H_2 - H_1 = B_1 \, E \qquad (2.95)$$

and the general term

$$H_{k+1} - H_k = B_k E \quad (k = 0, 1, \dots n) \tag{2.96}$$

This set of equations contain the exact number of equations required for the evaluation of the $(n + 1)$ unknown values of the impulses $B_0, B_1, B_2 \dots B_n$ in the impulse train function (d^2h/dt^2). The impulses are located at points $(t_0, t_1, \dots t_n)$ corresponding to the times when the curved sections begin in the y_t diagram.

The integral of the impulse train function gives a block diagram representing the slope (d_h/d_t) of the instantaneous unit hydrograph. Integration of this diagram leads to the polygon-shaped IUH. Denoting the heights of the blocks in the block diagram by $G_1, G_2, \dots G_n$ and the ordinates of the polygon at the points of change of slope $(t_1, t_2, \dots t_n)$. The following relationships are obtained:

$$G_1 = B_0 = H_1/E; \quad h_1 = G_1 t_1 \tag{2.97}$$

$$G_2 = B_1 + G_1 = H_2/E; \quad h_2 = h_1 = h_1 + G_2 (t_2 - t_1) \tag{2.98}$$

and in general
$$G_k = H_k/E \text{ and } h_k = h_{k-1} + G_k (t_k - t_{k-1}) \tag{2.99}$$

The resulting IUH defined by the set of ordinates $(h_1, h_2, \dots h_n)$ for the example is given in Fig. 2.14.

It should be noted that some trial and error procedure is necessary in fixing the location of the points separating the straight line sections and the curved sections in the original y_t curves. This adjustment becomes necessary because the values of the ordinates of the IUH derived by Eqs (2.96) and (2.99) must satisfy the condition that the area under the h_t polygon is unity, i.e.

$$\int_0^t h_1 \, dt = 1.00 \tag{2.100}$$

2.2.4.9   Design criteria for dams and embankments

In the design of dams and embankments for flood control it is usually not feasible from the economic points of view to provide complete protection against severe flood occurrence. The normal practice is to provide detention storage sufficient to absorb the standard project flood. The standard project flood is defined as the flood resulting from most severe storm or meteorological conditions considered reasonably to be the characteristics of the region. All the flood producing storms in the region are tabulated and about 10 per cent of highest storms of record are omitted. The next lower one is taken as the storm for the standard project flood. For the safety of the dams with storage over 60 million m^3, the spillways capacity provided should be sufficient to pass the maximum probable flood. This is to be determined by unit hydrograph method for maximum portable storm. The latter is obtained from the storm studies of all the storms that occurred in the region, maximised for all conditions including moisture content. Flood embankments may be designed to pass a flood with a return period of 100 years depending on its importance.

Problem 2.1

The analysis of annual flood peaks on a stream with 60 years of record indicates that the mean value of the annual peaks is 2264 m^3/s. The standard deviation of the peaks is 340 m^3/s.

(i) Find the probability of having a flood next year magnitude equal to or greater than 3170 m³/s.
(ii) What is the probability that at least one flood of such magnitude will occur during the next 25 years?
(iii) What is the magnitude of a flood with recurrence interval of 20 years?

Solution

Mean annual peaks = 2264 m³/s = $\overline{X}$
Standard deviation σ = 340.0 m³/s.

(i) As per Gumbel's method

$$P = 1 - e^{-e^{-b}} \text{ where}$$

$$b = \frac{1}{0.78\sigma}[X - \overline{X} + 0.45\sigma]$$

X = 3170 m³/s or

$$b = \frac{1}{0.78 \times 340.0}[3170 - 2264 + 0.45 \times 340.0] = 3.998$$

Hence

$$P = 1 - e^{-e^{-3.998}} = 1 - e^{-2.718^{-3.998}} = 1 - e^{-0.0183}$$

$$= 1 - 0.98 = 0.02 \text{ i.e. probability equals 2 per cent.}$$

(ii) The probability is given by

$$P(E) = 1 - \left[1 - \frac{1}{T(X)}\right]^{L}$$

In the present problem $T(X)$ = 25 years, L = 60 years
Hence

$$P(E) = 1 - \left(1 - \frac{1}{25}\right)^{60} = 1 - (1 - 0.4)^{60}$$

$$= 1 - 0.86 = 0.914, \text{ i.e. probability equals 91.4 per cent.}$$

(iii) The magnitude of a flood with a given recurrence, interval or probability P is

$$X = 2264 - 340 [0.78 \ln \{ - \ln (1 - 1/20) \} + 0.45]$$
$$= 2898.7 \text{ m}^3/\text{s}.$$

Problem 2.2

Flood frequency computations for a flashy river at a point 50 km upstream of a bund site indicated the following:

Return period	T yr	50	100
Peak flood	m³/s	20600	22150

Estimate the flood magnitude in the river with a return period of 500 yrs through use of Gumbel's method.

Solution

$$X_{100} = \overline{X} + K_{100}\,\sigma_{n-1},\ X_{50} = \overline{X} + K_{50}\,\sigma_{n-1}$$

Subtracting

$$(K_{100} - K_{50})\,\sigma_{n-1} = X_{100} - X_{50} = 22150 - 20600 = 1550$$

but

$$K_T = \frac{b_T}{\sigma_n} - \frac{\overline{b}_n}{\sigma_n} \text{ where } \sigma_n \text{ and } \overline{b}_n \text{ are constants for the given data}$$

$$(b_{100} - b_{50})\,\frac{\sigma_{n-1}}{\sigma_n} = 1550$$

where

$$b_{100} = -(\ln \ln 100/99) = 4.60015 \text{ and } b_{50} = -(\ln \ln 50/49) = 3.90194$$

$$\frac{\sigma_{n-1}}{\sigma_n} = \frac{1550}{4.60015 - 3.90194} = 2220, \text{ for } T = 500 \text{ yrs}$$

$$b_{500} = -(\ln \ln 500/499) = 6.21361,$$

or

$$(b_{500} - b_{100})\,\frac{\sigma_{n-1}}{\sigma_n} = X_{500} - X_{100}$$

$$(6.21361 - 4.60015)\,2220 = X_{500} - 22150$$

or

$$X_{500} = 22150 + 3582 = 25732 \text{ m}^3/\text{s}$$

Problem 2.3

The following gives the rainfall of 6-hr duration occurring over a catchment area of 3000 sq. km.

July	5, 81	time	9 a.m.	to	3 p.m. — 2.75 cm
,,	6, 81	,,	9 a.m.	to	3 p.m. — 1.75 cm
,,	7, 81	,,	3 a.m.	to	9 a.m. — 5.75 cm
,,	7, 81	,,	9 p.m.	to	3 a.m. — 3.75 cm

The ordinate of the 6-hr unit hydrograph for the catchment area is given in tabular form (column 2) of Table 2.8. Consider infiltration and other losses as 0.75 cm and assume a constant base flow of 100 m^3/s. Obtain the flood hydrograph corresponding to the above rainfall.

Solution

Table **2.8** Calculation of Ordinates of Flood Hydrograph Solution of the Problem.

Time (hr)		Ordinate of 6-hr, unit hydrograph (m³/s)	Rainfall at 6-hr, interval (cm)	Surface runoff from the rainfall (m³/s) Rainfall excess				Subtotal (m³/s)	Base flow (m³/s)	Runoff (m³/s)
				2.0 cm	1.0 cm	5.0 cm	3.0 cm			
July 5th	0	0						0		100
	6	13						0		100
	12	130	2.75	26				26	100	126
	18	325		260				260	100	360
	24	340		650				650	100	750
July 6th	30	275		680				680	100	780
	36	190	1.75	550	13			563	100	663
	42	100		380	130			510	100	610
	48	80		200	325			525	100	625
July 7th	54	50	5.75	160	340	65		565	100	665
	60	30		100	275	650		1025	100	1125
	66	25		60	190	1625		1875	100	1975
July 8th	72	9	3.75	50	100	1700	39	1889	100	1989
	78	4		18	80	1375	390	1863	100	1963
	84	1		8	50	950	975	1983	100	2083
	90	0		2	30	500	1020	1552	100	1625
	96				25	400	825	1250	100	1350
	102				9	250	570	829	100	929
	108				4	150	300	454	100	554
	114				1	125	240	366	100	466
	120					45	150	195	100	295
	126					20	90	110	100	210
	132					5	75	80	100	180
	138						27	27	100	127
	144						12	12	100	112
	150						3	3	100	103

Problem 2.4

Develop a synthetic unit hydrograph of 6 hr duration for a basin of area 300 sq. km. The length of the water course as estimated from the topography is 30 km and the distance from the centroid to the stream outlet has been found to be equal to 25 km. The value of C_t, and C_p for the catchment being 0.92 and 5 respectively.

Solution

The following calculation is made using the parameters as adopted by Snyder

Basin lag,

$$t_p = C_t (L_c L)^{0.3}$$
$$= 0.92 (25 \times 30)^{0.3}$$
$$= 6.56 \text{ hrs}$$

$$t_r = \frac{t_p}{5.5} = 1.19 \text{ hr}$$

$$t_{pR} = t_p + \left(\frac{t_R - t_r}{4}\right) = 6.56 + \frac{(6 - 1.19)}{4}$$

$$= 7.763 \text{ hr}$$

In the problem $t_R = 6$ hr

$$q_{pR} = \frac{C_p}{t_{pR}} \text{ or } Q_p = \frac{C_p A}{t_{pR}} = 193 \text{ cumec.}$$

$$C_p = 5.0$$

The length of the base period

$$T = 3 + \frac{3t_p}{24} \text{ days} = 3.97 \text{ days}$$

$$= 95 \text{ hr}$$

Time from the beginning of rising limb, to the peak

$$= t_{pR} + \frac{t_R}{2} = 10.76 \text{ hr}$$

$$q_{pR} = 193/300 = 0.643 \text{ cumec/km}^2$$

To draw the synthetic unit hydrograph the W_{75} and W_{50} has to be determined

$$W_{75} = \frac{3.21}{(q_{pR})^{1.08}} = 5.17 \text{ hr}$$

$$W_{50} = \frac{5.6}{(q_{pR})^{1.08}} = 9.02 \text{ hr}$$

Thus we know peak of unit hydrograph is 193 cumec, time of peak 10.76 hr time base 95 hr, W_{75}, (5.17 hr) and W_{50} (9.02 hr). Accordingly, the hydrograph is sketched so that the runoff over the whole catchment comes to 1 cm (Fig. 2.15).

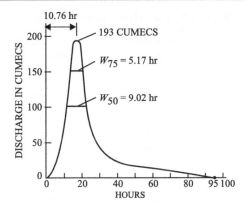

Figure 2.15 Synthetic unit hydrograph.

Problem 2.5

The hydrograph resulting from a 10-hr rainfall over a catchment area of 6000 sq. km is given in table below. The hourly variation of precipitation is given below.

Time in hrs	0	1	2	3	4	5	6	7	8	9	10
Rainfall in cm	1.2	0.8	4.2	5.1	6.3	7.5	6.0	4.0	1.5	0.5	0

Find out the effective duration of the storm and obtain the unit hydrograph.

Solution

The solution of the above problem is carried out in tabular form Table 2.9. The steps involved are as follows:

(i) Column (1) shows the time from the beginning of the storm and column (2) the discharge observed from the rainstorm.

(ii) The hydrograph is plotted in Fig. 2.16 and the base flow is separated. The estimated base flow is entered in column (3).

(iii) The direct runoff is calculated by subtracting the base flow from the hydrograph and entered in column (4).

(iv) The ordinates of surface runoffs are added and converted to volume. The volume divided by the catchment area gives the rainfall excess which is 5.38 cm.

(v) The unit hydrograph is then obtained by dividing the ordinate of the hydrograph by 5.38 and this is also shown in tabular and graphical form (Fig. 2.16).

(vi) The effective rainfall duration has been found out by assuming an infiltration volume of x cm. The total rainfall excess of 5.38 cm is obtained as follows:

$$(1.2 - x) + (0.8 - x) + (4.2 - x)$$
$$+ (5.1 - x) + (6.3 - x) + (7.5 - x)$$
$$+ (6.0 - x) + (4 - x) + (1.5 - x) + (0.5 - x) = 32.28$$

or
$$10x = 37.10 - 32.28$$

or
$$x = 0.482 \text{ cm}$$
$$= 0.5 \text{ cm.}$$

(vii) A line is drawn on the rainfall hydrograph at 0.5 cm above the base, which gives an effective storm runoff duration equal to 9 hr, (Fig. 2.16).

Table 2.9 Unit Hydrograph Calculation.

Time (hr)	Discharge (m³/s)	Base flow (m³/s)	Direct runoff (m³/s)	Ordinate of unit hydrograph (m³/s)
1	2	3	4	5
0	150	150	0	0
6	190	160	30	5.58
12	600	180	420	78.07
18	2000	190	1810	336.43
24	2900	210	2690	500.00
30	2750	220	2525	469.33
36	2400	240	2160	401.49
42	1950	260	1690	314.13
48	1500	280	1220	226.77
54	1100	295	805	149.63
60	900	305	595	110.63
66	750	325	425	79.00
72	640	340	300	55.76
78	540	360	180	33.46
84	450	380	70	13.01
90	425	390	35	6.51
96	400	400	0	0

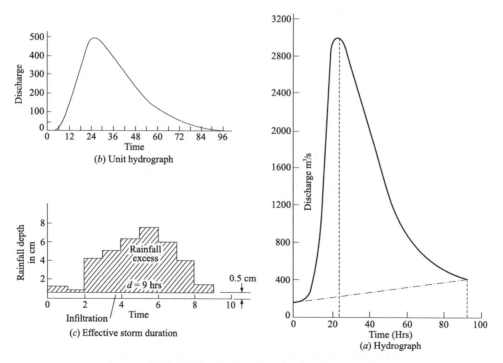

Figure 2.16 Hydrograph and derived unit hydrograph.

Calculation

Runoff in m^3 = $\dfrac{6\times60\times60}{3}$[0 + 0 + 2 (30 + 1810 + 2525 + 1690 + 805 + 425 + 180

+ 35) + 4 (420 + 2690 + 2160 + 1220 + 595 + 300 + 70)]

$$= 3.22704 \times 10^8 \text{ m}^3$$

∴ Uniform depth of runoff over the catchment area = $\dfrac{3.22704\times10^8}{6000\times10^6}$ m

$$= 0.0538 \text{ m}$$

$$= 53.8 \text{ mm} = 5.38 \text{ cm.}$$

Since a unit hydrograph represents, 1 cm of runoff, hence each ordinate of this isolated hydrograph should be divided by $\left[\dfrac{53.8}{10}\right]$ = 5.38 unit.

CASE STUDY 1

Steps to be followed for computing infiltration indices to be used in estimating runoff from major storms in comparatively large drainage areas.

(i) Select known hydrographs of major floods in the basin from the records for study and compute for each the volume of surface runoff. This is obtained by deducing 'base flow' (or flow which would have continued during the period of flood had there been no storm) and runoff due to rainfall earlier than or subsequent to the storm causing the flood under investigation, from the total flood volume.

(ii) Prepare mass rainfall curves for the raingauge stations within the area. Divide the basin into a series of Thiessen polygons. Determine the area of each Thiessen polygon within the basin.

(iii) From the mass rainfall curves, the rainfall within successive units of time intervals is measured and tabulated for each station. In selecting unit of time, the accuracy of the rainfall record and the density of raingauge stations in the area should be considered. An interval of three hours is usually satisfactory for studies pertaining to large drainage basins.

(iv) Compute the volume of rainfall within the successive unit time intervals in each polygon by multiplying the area of the polygon with the rainfall quantities in unit intervals obtained in step (iii).

(v) Assume a value for initial loss in mms and convert this into million cubic metres for each polygon area. The total rainfall must exceed the initial loss before a steady infiltration and corresponding surface runoff may be deemed to begin.

(vi) Assume a trial value for infiltration index, φ in mm per hour. For each polygon area determine the corresponding infiltration loss in millions of cubic metres during the selected unit time interval, which is given by φ (mm per hour) × polygon area Ap (sq. km) × unit interval (hours) × (1/1000). After the satisfaction of the initial loss, determine the rainfall volumes in excess of infiltration loss during successive time intervals as calculated above. Total up the accumulated excess of rainfall over infiltration for the entire storm period and compare with the observed volume of surface runoff computed in step (i). If the assumed value of φ is correct the two must tally with each other. Otherwise a trial must be made with a different value of φ till the results are equal.

The same procedure may be repeated for a number of known major flood hydrographs to arrive at a dependable average value of φ. The procedure is illustrated by Tables 2.10 and 2.11 which determine the infiltration index for a flood hydrograph shown in Fig. 2.17 for the drainage basin shown in Fig. 2.18. The mass rainfall curves used are also shown in Fig. 2.19.

The rainfall and rainfall excess data for representative zones are plotted as 'hyetographs' above the runoff hydrograph. The term hyetograph is used to refer to a graphical representation of average rainfall and rainfall excess rates, or volumes, over specified areas during successive unit time intervals during a storm.

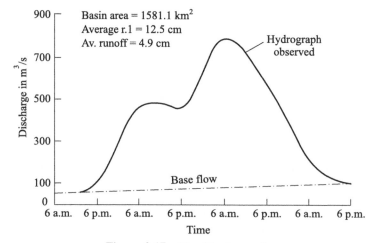

Figure 2.17 Flood hydrograph.

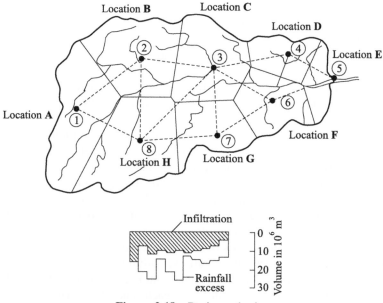

Figure 2.18 Drainage basin.

Table 2.10 Rainfall Volumes.

					Volumes of rainfall within area Ap = Ap × three hour rainfall increment × 1/1000 (millions of m³).												
					Time at end of 3-hour periods												
Sl. no.	*Name of drainage basin*	*Raingauge station*	*Area of polygon Ap in sq. km.*	*Precipitation at the station in mm*	*1st 6 a.m.*	*1st 9 a.m.*	*1st 12.00 noon*	*1st 3 p.m.*	*1st 6 p.m.*	*1st 9 p.m.*	*1st 12.00 midnight*	*2nd 3 a.m.*	*2nd 6 a.m.*	*2nd 9 a.m.*	*2nd 12.00 noon*	*2nd 3 p.m.*	*Total*
1	2	3	4	5	6	7	8	9	10	11	12	13	14	15	16	17	18
1		Location **A**	264.0	91	0	1.7	2.01	5.35	1.34	2.01	5.35	2.01	0.67	2.68	1.34	0.00	24.46
2		Location **B**	297.2	122	0	3.78	3.78	9.05	2.27	3.78	6.05	3.02	2.27	1.51	0.75	0.00	36.26
3		Location **C**	231.0	102	0	0	4.41	4.41	1.76	2.35	2.94	2.35	2.35	0.59	1.18	1.18	23.52
4	MAYURKASHI	Location **D**	105.4	132	0	2.14	2.94	2.38	0.54	2.14	1.34	0.54	0.54	0.81	0.54	0.00	13.91
5		Location **E**	63.3	162	0	0.84	1.18	0.34	0.34	1.02	2.19	0.50	0.85	1.18	0.85	1.52	10.81
6		Location **F**	165.2	81	0	0	1.05	1.28	0.84	1.26	1.26	1.68	1.68	0.84	2.09	0.84	12.82
7		Location **G**	156.0	122	0	0.99	1.39	1.19	1.58	1.58	1.58	0.79	2.73	3.16	2.37	1.58	18.94
8		Location **H**	298.0	193	0	6.04	5.29	1.51	6.04	7.55	5.29	2.26	3.78	5.29	6.04	8.30	57.39
		Total	1581.1	Average = 125		15.49	22.05	25.51	14.71	21.69	26.00	13.15	14.87	16.06	15.16	13.42	198.11

Table 2.11 Rainfall Excess Volume.

Sl. no.	Name of drainage basin	Raingauge station	Area of polygon Ap. in sq. km.	mm, initial loss	$m^3 \times 10^6$	mm/hour, φ	$m^3 \times 10^6$, φ 3 hrs.	1st 6 a.m.	1st 9 a.m.	1st 12.00 noon	1st 3 p.m.	1st 6 p.m.	1st 9 p.m.	1st 12.00 midnight	2nd 3 a.m.	2nd 6 a.m.	2nd 9 a.m.	2nd 12.00 noon	2nd 3 p.m.	Total
1	2	3	4	5	6	7	8	9	10	11	12	13	14	15	16	17	18	19	20	21
1		Location A	264.0	10	2.64	2.5	1.98	0	0	0	3.37	0	0.03	3.37	0.03	0	0.70	0	0	7.50
2		Location B	297.2	10	2.97	2.5	2.23	0	0	1.55	6.82	0.04	1.55	3.82	0.79	0.04	0	0	0	14.61
3		Location C	231.0	10	2.31	2.5	1.73	0	0	0.37	2.68	0.03	0.62	1.21	0.62	0.62	0	0	0	6.15
4		Location D	105.4	10	1.05	2.5	0.79	0	0.30	2.15	1.58	0	1.35	0.55	0	0	0.02	0	0	5.95
5		Location E	66.3	10	0.66	2.5	0.50	0	0.00	0.68	0	0	0.52	1.69	0	0.35	0.68	0.35	1.02	5.29
6		Location F	165.2	10	1.65	2.5	1.24	0	0.00	0	0.04	0	0.02	0.02	0.44	0.44	0.00	0.85	0	1.81
7		Location G	156.0	10	1.56	2.5	1.17	0	0	0	0.02	0.41	0.41	0.41	0	1.56	1.99	1.20	0.41	6.41
8		Location H	298.0	10	2.98	2.5	2.24	0	0.82	3.05	0	3.80	5.31	3.05	0.02	1.54	3.05	3.80	6.06	30.50
		Total	1581.1		15.82			0	1.12	7.80	14.51	4.28	9.91	14.12	1.90	4.55	6.44	6.20	7.49	78.22

Note: Observed runoff $= 76 \times 10^6$ m³. Calculated runoff 78.22×10^6 m³ based on φ = 2.5 mm/hr. This example is taken from [7] of Bibliography.

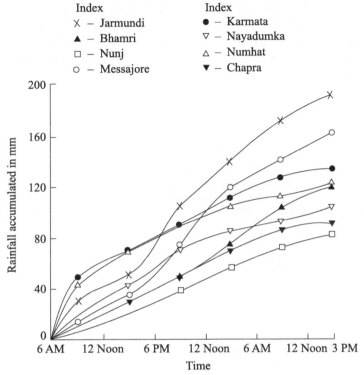

Index
X – Jarmundi
▲ – Bhamri
□ – Nunj
○ – Messajore

Index
● – Karmata
▽ – Nayadumka
△ – Numhat
▼ – Chapra

Figure 2.19 Rainfall mass curves.

The procedure explained above takes into account the variation in rainfall over different areas of the basin and change in intensity of rainfall over successive time intervals.

Figures in column 6 onwards in Table 2.10 are to be accumulated till they are more than the initial loss entered in column 6, Table 2.11. Rainfall excess up to such interval will be zero in Table 2.11. For subsequent time intervals deduct φ (column 8, Table 2.11) from the rainfall volume during the interval if the latter is more than the former and enter the difference in the rainfall excess column for the same time interval in Table 2.11. If the rainfall volume during the interval is less than φ, there will be no rainfall excess and a zero should be entered against that interval in the appropriate column of Table 2.11. This procedure is to be followed separately for each area. Thus for area 1, initial loss equals 2.64×10^6 m^3. The accumulated rainfall volume in this area exceeds the initial loss at 3 p.m. on 1st, column 9, Table 2.10. From column 9 to 11, Table 2.11, the rainfall excess is, therefore, zero. After that, infiltration is deducted from rainfall volume and if there is a positive remainder it is entered as rainfall excess, e.g. in column 9, Table 2.11 rainfall volume = 5.35×10^6 m^3. Deducting infiltration in this area which is 1.98×10^6 m^3 per 3 hours, the rainfall excess is 3.37×10^6 m^3 and is entered in col. 12, Table 2.11.

COMMENTS

The infiltration index having been determined for the basin from known flood hydrographs resulting from known storms in the past, it can then be applied to design storms of the future, adopted for working out design flood hydrograph for the basin. From the assumed value of initial loss and the calculated value of infiltration index, the rainfall excess during successive time intervals resulting from the design storm can be determined. The values of rainfall excess can then be utilised to work out the design flood hydrograph with the help of the unit hydrograph as explained in the text.

Problem 2.6

The infiltration capacity of an area at different intervals of time is indicated below in table. Find the equation of infiltration capacity in the exponential form

Time in hours	0	0.25	0.50	0.75	1.00	1.25	1.50	1.75	2.00
Infiltration capacity I_C (cm/hr)	10.4	5.6	3.2	2.1	1.5	1.2	1.1	1.0	1.0

Solution

We have I.C. curve equation as

$$I_c - I_a + (I_0 - I_a) \, e^{-kt_r}$$

where

I_a = Infiltration after reaching a constant value
I_0 = Infiltration capacity at beginning
I_c = Infiltration at time t
k = Constant
t_r = Time from the starting of precipitation

Hence, $I_a = 1$.

t (hr)	0	0.25	0.50	0.75	1.00	1.25	1.50	1.75	2.00
I_c (cm/hr)	10.4	5.6	3.2	2.1	1.5	1.2	1.1	1.0	1.0
$\log_{10} (I_c - I_a)$	0.973	0.663	0.342	0.041	−0.341	−0.699	−1.0		

Now graph is plotted between t and $\log_{10} (I_c - I_a)$ and from the graph slope of the line comes to (Fig. 2.20)

$$\frac{-0.730}{1} = -\frac{1}{1.37}$$

Again, from equation $I_c = I_a + (I_0 - I_a) \, e^{-kt_r}$, the slope of the line represented by

$$\left[-\frac{1}{k \, \log_{10} e} \right] = -\frac{1}{1.37}$$

but $e = 2.71828$. Therefore, $k = 3.155$.

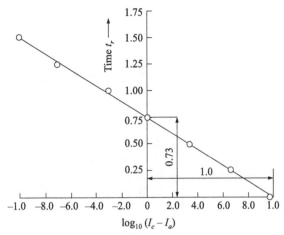

Figure 2.20 Infiltration capacity curve.

Hence, the equation of infiltration capacity curve is given by

$$I_c = 1 + (10.4 - 1) \, e^{-3155t_r}.$$

Therefore, $I_c = 1 + 9.4 \, e^{-3155t_r}$

Since $(I_0 = I_c)$ at 0 time = 10.4 cm/hr.

CASE STUDY 2

Steps to be followed for depth duration analysis of design storm.

In India standard duration of rainfall depths are in terms of maximum of 1-day, 2-day depths. This is because rainfall data is maintained at raingauge stations for 24-hour period. With the above background in view the steps are:

(i) The selected storm is assigned a definite beginning and ending from the rainfall records of stations. Ideally the duration is from a period of no rain to the next period of no rain.

(ii) For each day the total period of storm rainfall amounts for each raingauge station are noted down.

(iii) Separate isohyetal maps for each of the days are drawn based on rainfalls obtained as per step (ii).

(iv) Next isohyetal maps are constructed for the maximum of one day, two days, etc. rainfall. For this the isohyetal maps obtained in item (iii) are examined and the isohyetal map of the day that will provide maximum values at the storm centres and also over substantial areas around that is selected for computing the maximum one-day depth area contour (Table 2.12 and Fig. 2.21).

The maximum 2-day period is then found out by inspecting incremental isohyetal maps of 2-successive days that added together will provide maximum values for various areas. The isohyetal map for the maximum 2-day period thus determined will be prepared by adding the station amount of the 2-day period plotting individual storms and then by constructing new isohyetals.

Table 2.12 Maximum one day Depth Duration Curve from Isohyetal.

Centre	Isohyetal interval, (cm)	Average depth, (cm)	Area enclosed, (km²)	Area between two isohyetals, (km²)	Volume col 2 × col 4 (cm km²)	Total volume, (cm km²)	Average depth col 6/3, (cm)
1	2	3 4	5	6	7		
X	17.00 – 18.10	17.550	87	87	1.537×10^3	1.527×10^3	17.55
	15.00 – 17.50	16.250	785	698	11.343×10^3	12.870×10^3	16.39
	12.00 – 15.55	13.750	3220	2435	33.481×10^3	46.351×10^3	14.39
	12.25 – 12.50	11.875	5825	2605	30.934×10^3	77.285×10^3	13.27
	10.00 – 11.25	10.625	8100	2275	24.172×10^3	101.457×10^3	12.53
Y	12.50 – 12.80	12.650	21	21	0.266×10^3	0.266×10^3	9.54
	11.25 – 12.50	11.875	720	699	8.301×10^3	8.567×10^3	11.89
	10.06 – 11.25	10.625	1940	1220	12.936×10^3	21.530×10^3	11.10
XY	8.75 – 10.00	8.375	13.6×10^3	3.56×10^3	33.375×10^3	15.640×10^4	11.50
	7.50 – 8.75	8.125	1.88×10^3	5.90×10^3	42.250×10^3	19.870×10^4	10.57
	6.25 – 7.50	6.875	24.6×10^3	5.80×10^3	39.875×10^3	23.860×10^4	9.70

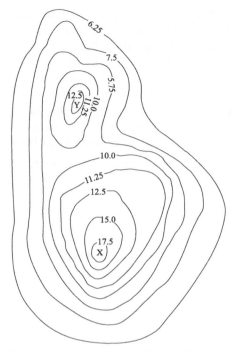

Figure 2.21 Isohyetal map for maximum of one day rainfall.

(v) Depth area computations are then made from each of the maximum isohyetal maps (Fig. 2.22).

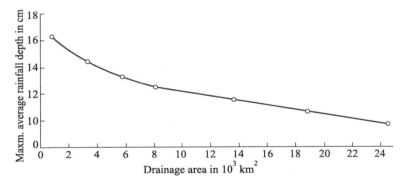

Figure 2.22 Depth duration curve for maximum one day storm.

CASE STUDY 3

Method for estimating the critical rate of inflow into a full reservoir corresponding to the spillway design storm:

The formation of a reservoir in a natural drainage basin materially alter the flood runoff regime. This is because of the synchronising effect of the high runoff rates originating above the head of the reservoir with maximum rates from areas contributing laterally to the reservoir. Under natural river conditions the runoff from the upper catchment area is retarded as a result of valley storage and frictional resistance as it passes along the reservoir reach. With the formation of a deep reservoir inflow near the upper and end of the long reservoir moves with a velocity equal to $\sqrt{gy}$, where y = flow depth. This means the time required for flood wave to traverse reservoir length is much smaller than what it used to be before. This may cause synchronisation effect of all contributing runoff resulting in an appreciably higher rates of inflow in a full reservoir although, in some reservoirs the differences may not be much.

In such case the steps that are followed are summarised below:

(i) The drainage area contributing to the reservoir is divided into subareas as shown in Fig. 2.23.

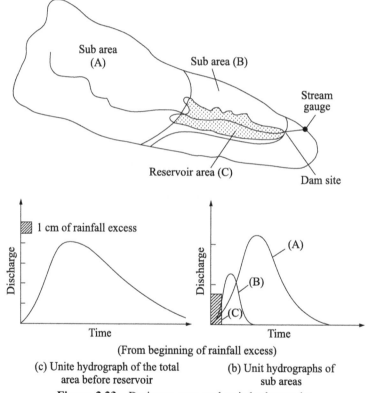

Figure 2.23   Drainage area and unit hydrographs.

(ii) Unit hydrographs are derived for the respective subareas using hydrologic records or by synthetic unit hydrograph computations.

(iii) For the reservoir surface the rate of runoff is taken as equal to the rate of rainfall.

(iv) The time required for flood waters entering the upper end of the reservoir to become effective in raising the reservoir level at the spillway site is found out by dividing the reservoir distance with the velocity $\sqrt{gy}$.

(v) Unit hydrographs for the subareas are then plotted in the same sheet in proper time sequence (Fig. 2.23).

(vi) Hydrographs of runoff from various subareas corresponding to design-storm rainfall excess are computed and combined in proper time sequence to obtain a composite hydrograph of reservoir inflow.

Problem 2.7

The effective rainfall and runoff hydrograph of a catchment is shown in Fig. 2.24. Find out the general IUH equation following Nash conceptual model and hence find out the equation of 1. hr U.H. Find out also the calculated discharge hydrograph.

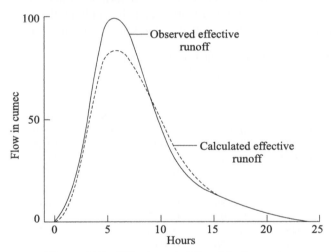

Figure 2.24 Effective rainfall and hydrograph.

Solution

The general IUH equation can be written as

$$u(0, t) = \frac{1}{K\Gamma(n)} (t/K)^{n-1} e^{t/K}$$

It is now necessary to find out the values of n and K from the moments of the effective rainfall and effective runoff by the following sets of equations: $nK = u_1$, $n = u_1^2/u_2$, where $u_1 = (Q_1 - I_1)$, Q_1 and I_1 being the centre of gravity of the effective runoff hydrograph and effective rainfall hyetograph with respect to ordinate at origin. Again $u_2 = (Q_2 - I_2)$, where Q_2 and I_2 being the moment of the respective runoff hydrograph and rainfall hyetograph about the ordinate passing through the origin. In this problem by numerical integration the following results were obtained $Q_1 = 7.8$ hrs, $I_1 = 0.6$ hrs, $Q_2 = 15.0$ hr^2, $I_2 = 0.1$ hr^2. Hence, $nK = 7.2$ hr and $n(n + 1) K^2 + 2nK (0.6) = 14.9$.

The solution of the above two equations give $n = 3.5$ and $K = 2.06$ hr. The equation of the IUH then

$$u(0, t) = \frac{1}{2.06\Gamma(3.5)}(t/2.06)^{2.5}\, e^{-t/2.06}$$

The equation of the 1 hour UH, therefore, can be obtained from

$$u(T, t) = \frac{1}{T}\, [I(n, t/K) - I(n, (t - T)/K)]$$

By substituting $T = 1$ hr, $n = 3.5$, $K = 2.06$

or

$$u(1, t) = \frac{1}{1}\left[I\!\left(3.5, \frac{t}{2.06}\right) - I\!\left(3.5, \frac{t-1}{2.06}\right)\right]$$

The solution of the above equation is shown in tabular form for values of t from 1 to 25 hours. Table 2.13 also shows the computation of storm runoff by applying the 1 hr UH to the effective rainfall. The computed hydrograph is plotted in the same figure for comparison.

Table 2.13 Computation of 1-hr Unit Hydrograph and Storm Runoff from IUH.

t (hr)	t/K	$I(n, t/K)$	$I\!\left(n, \dfrac{t-1}{K}\right)$	$u(T, t)$	$i(t)$	$u(T, t)\, i_1$	$U(T, t)\, i_2$	$q(t)$
0	0	0	0	0	608	0	0	
1	0.485	0.0052	0.0052	0.0052	107	3.16	0	3.16
2	0.971	0.0375	0.0375	0.0323		19.64	0.56	20.20
3	1.456	0.1073	0.1073	0.0693		42.44	3.46	45.90
4	1.942	0.2072	0.2072	0.0999		60.73	7.47	68.20
5	2.427	0.3221	0.3221	0.1149		69.86	10.69	80.55
6	2.913	0.4396	0.4396	0.1175		71.44	12.29	83.73
7	3.398	0.5493	0.5493	0.1097		66.90	12.57	79.27
8	3.883	0.6459	0.6459	0.0966		58.73	11.74	70.47
9	4.369	0.7276	0.7276	0.0817		49.67	10.34	60.01
10	4.854	0.7941	0.7941	0.0665		40.43	8.74	49.17
11	5.340	0.8465	0.8465	0.0524		31.86	7.12	38.98
12	5.825	0.8872	0.8872	0.0407		24.75	5.61	30.36
13	6.311	0.9180	0.9180	0.0308		18.73	4.35	23.09
14	6.796	0.9419	0.9419	0.0239		14.53	3.30	17.83
15	7.282	0.9579	0.9579	0.0160		9.72	2.56	12.28
16	7.767	0.9702	0.9702	0.0123		7.48	1.71	9.19
17	8.252	0.9791	0.9791	0.0089		5.41	1.32	6.73
18	8.738	0.9854	0.9854	0.0063		3.83	0.95	4.78
19	9.223	0.9899	0.9899	0.0045		2.74	0.67	3.41
20	9.709	0.9930	0.9930	0.0032		1.95	0.48	2.43
21	10.194	0.9952	0.9952	0.0022		1.34	0.34	1.68
22	10.680	0.9967	0.9967	0.0015		0.91	0.24	1.15
23	11.165	0.9978	0.9978	0.0011		0.67	0.16	0.83
24	11.650	0.9985	0.9985	0.0007		0.43	0.12	0.55
25	12.136	0.9989		0.0004		0.24	0.07	0.31

Note: This problem is taken from Proc. I.C.E, (Lond.), Nov. 1960.

Problem 2.8

For the rainfall excess hyetograph as shown in Fig. 2.14 occurring over a catchment area of 500 km^2 the values of direct surface runoff obtained are given in tabular form. The type of curve corresponding to the various ordinate values are also indicated in Table 2.14. Derive the IUH for the basin.

Table 2.14

t, (hr)	y, (m^3/s)		t, (hr)	y, (m^3/s)		t, (hr)	y, (m^3/s)	
0	0		19	104	C	38	56	
1	4	C	20	100		39	52	
2	12		21	96	S	40	50	C
3	36		11	90		41	48	
4	60	S	23	88		42	44	
5	100		24	84	C	43	36	
6	132		25	82		44	32	
7	148	C	26	80		45	26	S
8	154		27	78		46	20	
9	156		28	76		47	18	
10	160	S	29	74		48	12	
11	162		30	72		49	10	C
12	160		31	70		50	8	
13	156	C	32	66		51	6	
14	150		33	64		52	5	S
15	136		34	62		53	4	
16	128	S	35	60		54	2	
17	116		36	60	S	55	1	C
18	110		37	58		56	0	

Solution

Values of direct surface runoff at division points between straight and curved section of the runoff hydrograph and derivation of the IUH ordinate for the example are shown in tabular form Table 2.15 as given on next page:

Computation of column (8): We know vide Eq. (2.94) that $dy/dt = H_1 = B_0 E$, where $B_0 = d_2 h dt^2$ of the IUH and E = total depth of rainfall excess represented by the area under the hyetograph, i.e. vide Fig. 2.14 or $E = (2 \times 8 + 1 \times 13 = 29$ mm), or $B_0 = H_1 E = 32/E$ (m^3/s per hr). Now 29 mm rainfall excess over an area equal to 500 km^2 amounts to

$$\frac{29 \times 500}{3.6} \text{ (m}^3\text{/s)} = 8.05 \times 500$$

$$= 4025 \text{ m}^3\text{/s}$$

Note 3.6 mm per hr = 1 m^3/s per km^2

or

$$B_0 = \frac{d^2 h}{dt^2} = \frac{32}{4025} = 7.95 \times 10^{-3} \text{ hr}^{-2}$$

Table 2.15 Computation of IUH following Diskin's Approach.

t (hr)	y (hr)	Type curve	dy (m³/s)	dt (hr)	dy/dt (m³/s per hr)	(hr)	(hr⁻²)	(hr⁻²)	(hr⁻¹)
1	2	3	4	5	6	7	8	9	10
0	0	C							
3	36	S	64	2	32.00	5	7.95×10^{-3}	7.95×10^{-3}	
5	100	C				6	7.29×10^{-3}	$+0.66 \times 10^{-3}$	39.75×10^{-3}
8	154	S	8	3	2.67				
11	162	C S	−34	3	−11.33	6	3.48×10^{-3}	-2.82×10^{-3}	43.71×10^{-3}
17	116	C				5	$+1.57 \times 10^{-3}$		26.79×10^{-3}
20	100	S	−10	2	−5.00			-1.25×10^{-3}	
22	90	C					$+0.75 \times 10^{-3}$	0.50×10^{-3}	20.50×10^{-3}
25	81	S	−26	13	−2.00	16			
38	56	C				9	-0.75×10^{-3}		12.54×10^{-3}
41	48	S	−30	6	−5.00			-1.25×10^{-3}	
47	18	C				6	$+0.92$		1.29×10^{-3}
50	8	S	−4	3	−1.33			-0.36	
53	4	C					$+0.33$		
56	0								0.0

Similarly, for other values B,

$$\text{viz., } B_1 - (H_2 - H_1)/E = (2.67 - 32)/4025$$
$$= -7.29 \times 10^{-3} \text{ hr}^{-2}$$

Computation of column (9)

$$G_1 = B_0; \quad G_2 = G_1 + B_1 \text{ and so on}$$

Computation of column (10)

$$h_1 = G_1 \times h = 7.95 \times 10^{-3} \times 5 = 39.75 \times 10^{-3}$$
$$h_2 = h_1 + G_2 (t_2) = 39.75 \times 10^{-3} + 0.66 \times 6 \times 10^{-3}$$
$$= 43.71 \times 10^{-3} \text{ etc.}$$

Problem 2.9

The IUH and the rainfall excess hyetograph for a catchment area of 500 km^2 is given in (Fig. 2.25). Derive the direct surface runoff hydrograph.

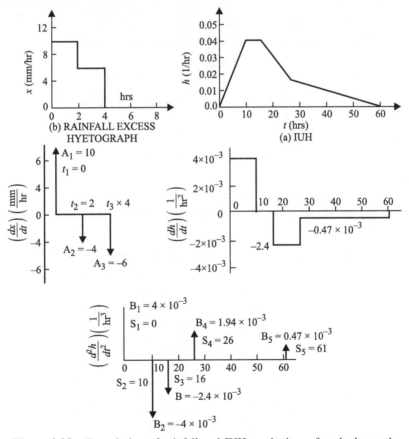

Figure 2.25 Convolution of rainfall and IUH to obtain surface hydrograph.

Solution

Convolution of impulse train function (Fig. 2.25). Calculations are shown in Tables 2.16–2.18.

Table 2.16

Location: Time (hr)	Ordinates $A_i B_j$ (mm/hr^4) = (d^3y/dt^3)
$(t_1 + s_1) = 0$	$A_1 B_1 = 10 \times 4 \times 10^{-3} = 40 \times 10^{-3}$
$(t_2 + s_1) = 2$	$A_2 B_1 = -4 \times 4 \times 10^{-3} = -16 \times 10^{-3}$
$(t_3 + s_1) = 4$	$A_3 B_1 = -6 \times 4 \times 10^{-3} = -24 \times 10^{-3}$
$(t_1 + s_2) = 10$	$A_1B_2 = 10 \times -4 \times 10^{-3} = -40 \times 10^{-3}$
$(t_2 + s_2) = 12$	$A_2B_2 = -4 \times -4 \times 10^{-3} = 16 \times 10^{-3}$
$(t_3 + s_2) = 14$	$A_3B_2 = -6 \times -4 \times 10^{-3} = 24 \times 10^{-3}$
$(t_1 + s_3) = 16$	$A_1B_3 = 10 \times -2.4 \times 10^{-3} = 24 \times 10^{-3}$
$(t_2 + s_3) = 18$	$A_2B_3 = -4 \times -2.4 \times 10^{-3} = 9.6 \times 10^{-3}$
$(t_3 + s_3) = 20$	$A_3B_3 = -6 \times -2.4 \times 10^{-3} = 14.4 \times 10^{-3}$
$(t_1 + s_4) = 26$	$A_1B_4 = 10 \times 1.93 \times 10^{-3} = 19.3 \times 10^{-3}$
$(t_2 + s_4) = 28$	$A_2B_4 = -4 \times 1.93 \times 10^{-3} = 7.7 \times 10^{-3}$
$(t_3 + s_4) = 30$	$A_3B_4 = -6 \times 1.93 \times 10^{-3} = 11.58 \times 10^{-3}$
$(t_1 + s_5) = 61$	$A_1B_5 = 10 \times 0.47 \times 10^{-3} = 4.7 \times 10^{-3}$
$(t_2 + s_5) = 63$	$A_2B_5 = -4 \times 0.47 \times 10^{-3} = -1.88 \times 10^{-3}$
$(t_3 + s_5) = 63$	$A_3B_5 = -4 \times 0.47 \times 10^{-3} = -1.88 \times 10^{-3}$

Values of (d^2y/dt^2) obtained by integration of the impulse train function.

Table 2.17

t (hr)	d^2y/dt^2 (mm/hr^3)	t (hr)	d^2y/dt^2 (mm/hr^3)
0 – 2	40×10^{-3}	18 – 20	-14.4×10^{-3}
2 – 4	24×10^{-3}	20 – 26	0
4 – 10	0	26 – 28	19.3×10^{-3}
10 – 12	-40×10^{-3}	28 – 30	11.58×10^{-3}
12 – 14	-24×10^{-3}	30 – 61	0
14 – 16	0	61 – 63	4.7×10^{-3}
16 – 18	-24×10^{-3}	63 – 65	2.82×10^{-3}

Values of the ordinates of the polygon (dy/dt) and the direct surface runoff function y expressed in m^3/s:

Table 2.18

t (hrs)	dt/dt) (mm/hr^2)	y (mm/hr^2)	y (m^3/s)
0	0	0	0
2	80×10^{-3}	80×10^{-3}	1.112
4	128×10^{-3}	288×10^{-3}	4.0
10	128×10^{-3}	1056×10^{-3}	14.7
12	48×10^{-3}	1232×10^{-3}	17.1
14	0	1280×10^{-3}	17.8
16	0	1280×10^{-3}	17.8
18	-48×10^{-3}	1232×10^{-3}	17.1
20	-76.8×10^{-3}	1107.2×10^{-3}	15.4
26	-76.8×10^{-3}	646.8×10^{-3}	9.0
28	-38.2×10^{-3}	570×10^{-3}	7.9
30	-15.04×10^{-3}	493.62×10^{-3}	6.85
61	-15.04×10^{-3}	0.0227×10^{-3}	3.76×10^{-4}
63	-9.40×10^{-3}	0.0096×10^{-3}	1.32×10^{-4}
65	0	0	0

Note: This example is taken from Ref. [11].

Problem 2.10

In a river diversion scheme for construction of a dam the probability of design flood being exceeded or equaled is 0.20. Assuming that there is likely to be one maximum flood in a year that would equal or exceed the design flood the series of annual floods can be considered as a Bernoulli sequence. Find the probability of design flood being exceeded exactly twice in 5 years and the probability of the design flood equaled or exceeded not more than twice in the next year.

Solution

The probability of the design flood being equaled or exceeded exactly twice in 5 years is given by

$$P(r = 2) = 5_{C_2}(0.2)^2(1 - 0.2)^{5-2} = 0.2$$

The probability of the design flood being equaled or exceeded not more than twice in the next 5 years is given by

$$P(r \leq 2) = P(r = 0) + P(r = 1) + P(r = 2)$$

$$= 0.25 + 0.4 + 0.2 = 0.85$$

APPENDIX I

Discharge Estimation at a Dam

The flow of a river at a dam site is generally diverted in many ways. Such as: (i) head regulator for feeding the water to an irrigation canal; (ii) penstocks leading water to the powerhouse for hydropower generation; (iii) intake tower for urban water supply; (iv) navigation lock for supplying water for navigation; and (v) water flowing through fish ladders, log passes, etc. besides the discharge over the spillway which is primarily meant for allowing flood discharge to escape downstream.

In case the river discharge is to be gauged at a dam spillway, all the above-mentioned additional withdrawals must be assessed and added to the spillway discharge. For dependable measurement of discharge it is desirable to have the spillway normal to the direction of flow with the same crest level and cross-section all along. It is always preferable to have free flow condition of flow over the spillway as submergence affects the coefficient of discharge. The formula for estimating discharge over a spillway can be expressed as

$$Q = CL\,H^{3/2} \tag{2.101A}$$

where

Q = discharge
C = coefficient of discharge
L = length of the crest
H = total head = $(h + \alpha\,V^2/2g)$,
h = depth of flow measured at a distance $4h$ behind the spillway crest where drawdown effect is absent
α = velocity correction coefficient depending on velocity distribution. In case the crest alignment is at an angle to the velocity of approach, component of the velocity normal to the crest alignment should be considered for estimating the velocity head. Values of C may be obtained from model experiments and verified by actual stream gauging downstream.

Approximate Discharge Estimation at Bridges

Measurement of discharge from the observed afflux at bridges provides a means of estimation of stream discharge. The method is developed by Kindsvater and Carter of Geological Survey of USA. The formula developed can be expressed as

$$Q = \sqrt{2g}\ CA_3 \left[\frac{\Delta h}{1 - \alpha_1\ C^2 \left(\dfrac{A_3}{A_1}\right)^2 + 2g\ C^2 \left(\dfrac{A_3}{K_3}\right)^2 \left(L + L_\omega \dfrac{K_3}{K_1}\right)} \right]^{\frac{1}{2}} \qquad (2.102A)$$

where
$\quad Q$ = discharge in m³/s,
$\quad C$ = coefficient of discharge,
$\quad A_3$ = gross area of Section 3 (Fig. 2.27) in m² (this is the minimum Section parallel to the constriction between abutment),
$\quad \Delta h$ = Difference in elevation of the water surface between Sections 1 and 3. Section 1 is located at least one bridge opening width upstream from the constriction,
$\quad A_1$ = Gross area of Section 1 in m²,
$\quad L$ = length of the abutment;
$\quad L_\omega$ = Length of the approach channel,
$\quad K_3$ = conveyance factor of Section 3;
$\quad K_1$ = Conveyance factor of Section 1 (Fig. 2.27).

$$\left(K = \text{Conveyance factor defined as } \frac{AV}{S^{\frac{1}{2}}} = \frac{AR^{2/3}}{n} \right)$$

α_1 = correction factor for velocity distribution at Section 1. For obtaining the value of α_1 the section is first divided into subsections $a, b, c \ldots n$ of approximately constant hydraulic properties so that $A_1 = A_a + A_b + \ldots + A_n$. Conveyances of the component sections are designated as K_a, K_b, etc., so that $K_1 = K_a + K_b + \ldots K_n$, α_1 is then given by

$$\alpha_1 = \frac{V^3\ dA}{A_1\ V_1^3} \qquad (2.103A)$$

Assuming discharge of each subsection to be proportional to its conveyance.

$$\alpha_1 = \left[\frac{\alpha_a\ K_a^3}{A_a^2} + \frac{\alpha_b\ K_b^3}{A_b^2} + \ldots \frac{\alpha_n\ K_n^3}{A_n^2} \right] \bigg/ \frac{K_1^3}{A_1^2} \qquad (2.104A)$$

where α_a, α_b, α_n are the correction factors for velocity distribution. Values of all the terms on the R.H.S. of the Eq. (2.102A) can be obtained by direct measurement and computation except for C, i.e. the discharge coefficient can be carried out.

Experiments show that C is a function of: (a) degree of channel contraction, (b) geometry of construction and (c) Froude number. Of these, the channel contraction and the length width ratio of the constriction (L/b) were found to be most significant, and the channel contraction ratio m is defined by:

$$m = \left(\frac{Q - q}{Q}\right) = \left(1 - \frac{q}{Q}\right) = \left(\frac{K_a + K_c}{K_{Q,}}\right) = \left(\frac{K_a + K_c}{K_1}\right)$$

In the above Q = total discharge, q = discharge that could pass through the opening without undergoing contraction. $K_Q = K_1$ conveyance of the total section while K_a and K_c = conveyance of that portion of the channel carrying the discharge ($Q - q$), see Fig. 2.26. The graphical relationship of C for various types of contraction are furnished in books on open channel flow such as Chow [27].

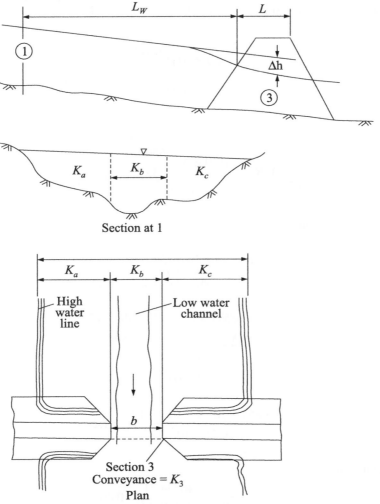

Figure 2.26 Definition sketch for river contraction at a bridge.

Problem 2.11

Estimate the flood discharge through a bridge opening having vertical abutment from the following data:

$$L\omega = 100 \text{ m}; \ (L = 10 \text{ m};) \ b = 60 \text{ m}, \ h = 2.5 \text{ m}.$$

The dimension of the river section is as follows:

Main river 60 m wide by 10 m deep, the side channel equals 130 m wide by 5 m deep. At the contracted section the bridge spans over the main channel only and the side channel is blocked by embankment. The values of Manning's roughness coefficient n and energy correction factor α are given as 0.035, 1.10 and 0.040, 1.11 for the main river and side channel respectively.

Solution

The conveyance for the river section at the uncontracted section $K_1 = K_s + K_m$, where the subscripts s and m stands for side and main channel respectively.

$$K_m = \frac{A_m R_m^{\frac{2}{3}}}{n_m} = \frac{(60 \times 10)\left(\dfrac{600}{75}\right)^{\frac{2}{3}}}{0.035} = 68571.43$$

and

$$K_s = \frac{A_s R_s^{\frac{2}{3}}}{n_s} = \frac{(130 \times 5)\left(\dfrac{650}{135}\right)^{\frac{2}{3}}}{0.040} = 46334.71$$

The energy correction factor α_1 for the section

$$\left[\frac{\alpha_m K_m^3}{A_m^2} + \frac{\alpha_s K_s^3}{A_s^2}\right] \bigg/ \frac{K_1^3}{(A_m + A_s)^2}$$

$$= \frac{\dfrac{1.10 \times 68571.43^3}{600^2} + \dfrac{1.11 \times 46334.71^3}{650^2}}{\dfrac{114906.14^3}{(1250)^2}}$$

$$= \frac{9.852 \times 10^8 + 2.613 \times 10^8}{9.71 \times 10^9} = 1.28$$

The contraction ratio

$$\frac{K_s}{K_m + K_s} = \frac{46334.71}{114906.14} = 0.40$$

The $L/b = \dfrac{10}{60} = 0.167$. Hence, for m = 0.40, $L/b = 0.165$ with $F_3 = 0.5$, the standard value of C for vertical type of constriction comes to 0.76 (vide ref. : 27).

For the contracted section: $\quad A_3 = 60 \times (10 - 1.5) = 510$ m^2

$$P_3 - 60 + 2 \times 8.5 = 77 \text{ m. } R_3, = \frac{510}{77} \text{ m.}$$

Hence $\qquad K_3 = \dfrac{A_3 R_3^{2/3}}{n_m} = \dfrac{510 \, (510/77)^{2/3}}{0.035} = 51716.05$

On substitution of the relevant values in the discharge Eq. (2.102A)

$$Q = 4.43 \times 0.76 \times 510 \left[\frac{1.5}{1 - 1.28 \times 76^2 \left(\dfrac{510}{1250}\right)^2 + 2 \times 9.81 \times 76^2 \left(\dfrac{510}{51716.05}\right)^2} \left(\frac{10 + 100 \times 51716.05}{115626.47}\right) \right]^{\frac{1}{2}}$$

$$= 4.43 \times 0.76 \times 510 \left[\frac{1.5}{1 - 0.1231 + 0.0602} \right]$$

$$= 4.43 \times 0.76 \times 510 \times 1.2652 = 2171.91 \text{ m}^3/\text{s}$$

Check for F_3

$$F_3 = \frac{Q}{A_3 \, (gy_3)^{1/2}} = \frac{2171.91}{510 \, (9.81 \times 8.5)^{1/2}} = 0.47$$

F_3 is slightly less than the assumed value of 0.5. So the discharge value has to be slightly adjusted after necessary trials.

Estimation of Discharge by Slope Area Method

Steps: (i) A suitable uniform straight reach of length not less than 304.8 m and preferably of the order of 10 times the stream width is selected. (ii) It should also provide a fall of at least 12 cm. (iii) A minimum of 3 gauges, two at each end and one at the centre should be installed. (iv) From the gauges, the hydraulic grade line should be ascertained and care should be taken so as not to have rapid fluctuations in the stage of the river. (v) When water surface slope is determined from flood marks this should be done with least possible delay after the occurrence of the flood. (vi) If the sections vary, velocity heads will be different and hence the energy and the hydraulic gradients will not be the same in such cases. The energy gradient has then to be computed by successive trials. (vii) Mean areas of the section is arrived at by averaging different sections in the reach. Average hydraulic mean depth in different segments are likewise computed from the average section. (viii) Velocity is then computed by Manning's formula, viz.,

$$V = \frac{1}{n} R^{2/3} S_e^{\frac{1}{2}} \qquad (2.105A)$$

where S_e = energy slope.

(ix) Velocity computation is usually done by successive trials starting with discharge computed by using surface slope in Manning's equation. (x) With this discharge and knowing the sectional area of upper and lower cross-sections, velocities and velocity heads are computed giving velocity heads and energy gradients S_e. In the next trial e substituted for surface slope in Manning's formula and, the new Q is estimated. The process continues till the difference in energy gradients computed in any two successive trials is narrowed down to within five per cent. (xi) When overflooding of banks, occurs discharge and velocity are worked out separately for different compartment using appropriate values of R, S_e and n. The value of n is obtained from actual field observations using a current meter, and otherwise reference may be made to the table compiled from observations in the field. (xii) When the n values over different parts of the sections are different, the velocities and discharges need to be calculated separately and they are then finally added to arrive at the total discharge.

CASE STUDY 4

Comparison of flood formulae in general use and world's maximum observed.

1. In many drainage basins both raingauge networks and river gauging are inadequate so planning for flood prediction and management must be based on the study of catchment characteristics. The various 'flood formulae development in various countries to relate the maximum peak flow at a site to the catchment area with a co-efficient and exponent being fixed for the area in question is furnished Figs. 2.27 a and b.

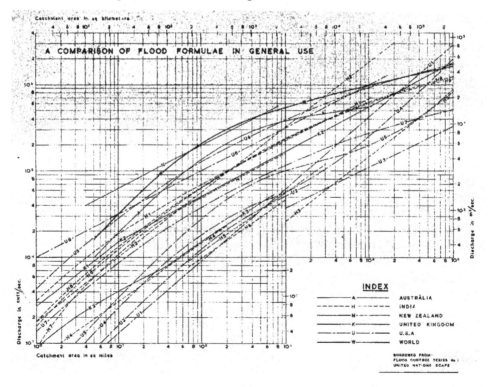

(a)

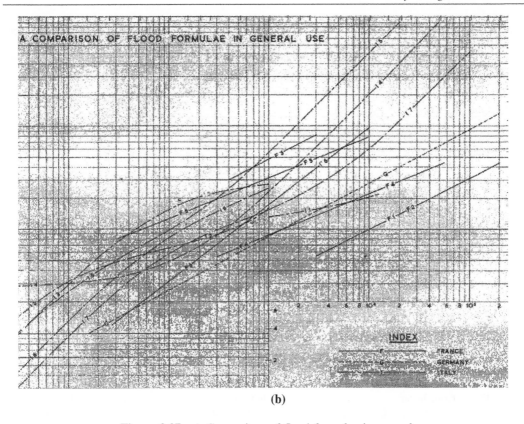

(b)

Figure 2.27 A Comparison of flood formulae in general use.

2. The knowledge of exceptionally large floods is essential to the hydrological and civil engineering solution to many problems in water resources management. In 1984 Rodier and Roche published their world catalogue of maximum observed floods and it involves the collection of flood information from 95 countries. The catalogue contains 1400 large floods. Of these 38 of world's maximum floods in relation to catchment indicates an equation, Fig. 2.28.

$$Q = 500A^{0.43} \qquad\qquad (2.106A)$$

where
   A = Area in km^2.
   Q = Discharge in m^3/s

3. The largest rainfall runoff floods in the world is shown below with dates & year of occurrence.

Flood study for the British Isles floods in the world (Roider and Ruchi, 1964)

Table 2.19   Maximum rainfall-runoff

Flood no	Stream	Date	Draining area (km^2)	Discharge (m^3/s^{-1})
1	San Rafael, California, USA	16 January, 1973	3.2	250
2	Little San Gorgonia, California, USA	25 February, 1969	4.5	311
3	Halawa, Hawaii, USA	4 February, 1965	12	762
4	s.F. Wailus, Hawaii, USA	15 April, 1963	58	2470
5	Buey, Cuba	7 October, 1963	73	2060
6	Papenoo, Tahiti, France	12 April, 1983	78	2200
7	San Bartolo, Mexico	30 September, 1976	81	3000
8	Quinne, New Caledonia, France	8 March, 1985	143	4000
9	Quaieme, New Caledonia, France	24 December, 1981	330	10400
10	Yatae, New Caledonia, France	25 December, 1981	435	5700
11	Little Nemaha, Nebraska, USA	9 May, 1950	549	6370
12	Haast, New Zealand	2 December, 1979	1020	7690
13	Mid fork American, California, USA	23 December, 1964	1360	8780
14	Cithuarlan, Maxico	27 October, 1959	1370	13500
15	Pioneer, Australia	23 January, 1918	1490	9800
16	Hualien, Taiwan, China	1973	1500	11900
17	Niyodo, Japan	9 August, 1963	1560	13500
18	Kiso, Japan	June, 1961	1680	11150
19	West Nueces, Texas USA	14 June, 1935	1800	15600
20	Machhu, India	11 August, 1979	1800	15600
21	Tamshui, Taiwan, China	11 September, 1963	1900	15600
22	Shingu, Japan	26 September, 1959	2110	16700
23	Pedernales, Texas, USA	11 September, 1963	2350	19025
24	Daeryorggang, North Korea	12 August, 1975	3020	13500
25	Yoshino, Japan	9 September, 1974	3750	14470
26	Cagayan, Phillipines	1959	4245	17550
27	Tone, Japan	15 September, 1947	5110	16900
28	Nueces, Texas, USA	14 June, 1935	5504	17400
29	Eel, California, USA	23 December, 1964	8060	21300
30	Pecos, Texas, USA	28 June, 1954	9100	26800
31	Betsiboka, Madgascar	4 March, 1927	11800	22000
32	Toedonggang, North Korea	29 August, 1967	12175	29000
33	Han, South Korea	18 July, 1925	23800	37000
34	Jhelum, Pakistan	1929	29000	31100
35	Hanjiang, China	1983	41400	40000
36	Mangoky, Madgascar	5 February, 1933	50000	38000
37	Narmada, India	6 September, 1970	88000	69400
38	Changing, China	20 July, 1870	1010000	110000
39	Lena, Russia	8 June, 1967	2430000	189000
40	Amazonas, Brazil	June, 1963	4640000	370000

Comparison of formulae for estimating flood discharge in general use
Compiled from papers presented at the fourth congress on large Dams, 1951
Q_m = Maximum Flood, Q_n = Average annual/flood, A = Discharge area,
Units: M = Metric system (one for Q, and sq·km for A)
E = Foot Pound system (of for Q, and sq-mi. for A)

Sl no.	Country	Equation	Units	Particular about equation	Author	Designation of curve	Source
1	The World	$Q_m = \dfrac{13100\,A}{(107+A)^{0.78}}$	E	Max. recorded flood throughout the world	Baird and Millwright	W	Baird and Millwright,
2	Australia	$Q_m = \dfrac{222000\,A}{(185+A)^{0.9}}$	E		Baird and Millwright	A	Baird and Millwright,
3	France	$Q_m = (10 \text{ to } 70)\,A^{0.5}$	M	Mild rain, A between 3000 and 160000 km².		F_1-F_2	A Coutagne
		$Q_m = 150\,A^{0.5}$	M	Violent rain, A between 400 and 3000 km².		F_3	A Coutagne,
		$Q_m = 54.6\,A^{0.4}$	M	River Garenne, A between 300 and 55000 km².		F_4	A Coutagne,
		$Q_m = 200\,A^{0.4}$	M	A between 30 and 10000 km².		F_5	A Coutagne,
		$Q_m = 10.76\,A^{0.737}$	M	Existing dams of "Massif Ceentral"		F_6	A Coutagne,
4	Germany	$Q_m = 2412\,A^{0.536}$	M	A between 15 and 200000 km².		0	A Coutagne,
5	India	$Q = \dfrac{7000\,A}{\sqrt{1+4}}$	E	For fan shape area	Inglie	M_1	Hunter & Wilmot,
		$Q = 1.795\,A^{0.75}$	E	Rain approximately 100 in.	Diekene	M_2	Hunter & Wilmot,
		$Q = 149\,A^{0.75}$	E	Rainfall 30 to 40 in.	Diekene	M_3	Hunter & Wilmot,
		$Q = 675\,A^{0.67}$	E	Maximum flood	Ryree	M_4	Hunter & Wilmot,

Sl no.	Country	Equation	Units	Particular about equation	Author	Designation of curve	Source
		$Q = 560\,A^{0.67}$	E	Average flood	Ryree	M_5	Hunter & Wilmot,
		$Q = 450\,A^{0.67}$	E	Minimum flood	Ryree	M_6	Hunter & Wilmot,
		Curve only	E	Bombay area	Whitting	M_7	Hunter & Wilmot
		$Q = 2000\,A^{\left(0.89 - \frac{1}{15}\log A\right)}$	E	Tungbhadra River	Madras Formula	M_8	Rao-K.L.,
		$Q = 1750\,A^{\left(0.92 - \frac{1}{14}\log A\right)}$	E	Tungbhadra River	Hyderabad Formula	M_9	Rao-K.L.,
6	Italy	$Q_m = \left(\dfrac{1.538}{A + 259} + 0.054\right)A$	M	A between 1000 and 12000 km².	Whistler	r_1	Tonini,
		$Q_m = \left(\dfrac{600}{A + 10} + 1\right)A$	M	A smaller than 1000 km².	Scimeni	r_2	Tonini,
		$Q_m = \left(\dfrac{2900}{A + 90}\right)A$	M	A between 1000 and 12000 km².	Pagliare	r_3	Tonini,
		$Q_m = \left(\dfrac{280}{A} + 2\right)A$	M	Mountain Basins	Baratta	r_4	Tonini,
		$Q_m = \left(\dfrac{532.5}{A + 16.2} + 5\right)A$	M	Mountain Basins	Oiandotti	r_5	Tonini,
		$Q_m = \left(3.25\dfrac{500}{A + 125} + 1\right)A$	M	A smaller than 1000 km². Max. rainfall 400 mm in 24 hr	Forti	r_6	Tonini,

Sl no.	Country	Equation	Units	Particular about equation	Author	Designation of curve	Source
		$Q_m = \left(2.35\dfrac{500}{A+125} + 0.5\right)A$	M	Maximum Rainfall 200mm in 24 hr	Forti	r_7	Tonini.,
7	New Zealand	$Q_m = 20000\,A^{0.5}$	E			N	Coutagne,
8	United Kingdom	$Q_m = 2700\,A^{0.75}$	EA	A smaller than 10 sq. mi.	Branaby Williams	K_1	Hunter & Eilmot,
		$Q_m = 4600\,A^{0.52}$	E	A greater than 10 sq. mi.	Branaby Williams	K_2	Hunter & Eilmot,
		Curve only	E		Institute of Civil Eng. 1933	K_3	Hunter & Eilmot,
9	USA	$Q = 200\,A^{5/6}$	E		Fanning	U_1	K.L. Rac,
		$Q = \left(\dfrac{46790}{A+320} + 15\right)A$	E	A between 5.5 and 2000 sq. mi	Marphy	U_2	K.L. Rac,
		$Q = 1400\,A^{0.476}$	E	A between 1000 and 24000 sq. nu	U.S. Geological survey for Columbia	U_3	Coutagne,
		$Q = \left(\dfrac{44000}{A+700} + 20\right)A$	E	For frequent floods	Kuichling	U_4	K.K. Rao,
		$Q = \left(\dfrac{127000}{A+370} + 7.4\right)A$	E	For rare floods	Kuichling	U_5	K.K. Rao,
		$Q = 4600\,A^{0.894A^{-0.048}}$	E	Upper limit	Greager	U_6	Hunter & Wilnot,
		$Q = 1380\,A^{0.894A^{-0.048}}$	E	Upper limit	Greager	U_7	Hunter & Wilnot,
		$Q = 10000\,A^{0.5}$	E		Myer	U_8	Hunter & Wilnot,

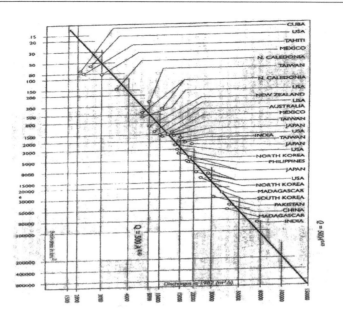

Figure 2.28 Maximum Flood Discharges in the World plotted against catchment area on log paper (after Rodier and Roche, 1984).

APPENDIX II

Table 2.20 Hydrologic soil groups (USSCS, 1964)

Soil Group	Description
A	**Lowest runoff potential.** Includes deep sand with very little clay and silt, and also deep, rapidly permeable loess.
B	**Moderately low runoff potential.** Mostly sandy soil less deeper than A, and loess less deeper or less aggregated than A, but the group as a whole has above average infiltration after thorough wetting.
C	**Moderately high runoff potential.** Comprises shallow soil and soil containing considerable clay and colloids, though less than those of group D. The group has below average infiltration after pre-saturation.
D	**Highest runoff potential.** Includes, mostly clay of high swelling per cent, but the group also includes some shallow soil with nearly impermeable sub-horizon near the surface.

Table 2.21 Seasonal rainfall limits for antecedent moisture condition classes (USSCS, 1964)

Antecedent moisture condition (AMC) class		5-day total antecedent rainfall (cm)	
		Dormant season	*Growing season*
I.	Optimum soil condition from about lower plastic limit to wilting point	Less than 1.25	Less than 3.5
II.	Average value for annual floods	1.25 to 2.75	3.5 to 5.25
III.	Heavy rainfall or light rainfall and low temperature during five days preceding the given storm	Over 2.75	Over 5.25

Table 2.22 Conversion of CN from AMC II to AMC I and AMC III

Curve Number at AMC II	Factor to convert the Curve Number at AMC II to	
	AMC I	AMC III
10	0.40	2.22
20	0.45	1.85
30	0.50	1.67
40	0.55	1.50
50	0.62	1.40
60	0.67	1.30
70	0.73	1.21
80	0.79	1.14
90	0.87	1.07
100	1.00	1.00

Table 2.23 Curve numbers for hydrologic cover complexes for watershed condition II, and $l_a = 0.2S$ (USSCS, 1964)

Land use cover	Treatment	Hydrologic condition	Hydrologic soil group			
			A	B	C	D
Fallow	Straight row	Poor	77	86	91	94
Row crops	Straight row	Poor	72	81	88	91
	Straight row	Good	67	78	85	89
	Contoured	Poor	70	79	81	86
	Contoured	Good	65	75	82	86
	Contoured and terraced	Poor	66	74	80	82
	Contoured and terraced	Good	62	71	78	81
Small grain	Straight row	Poor	65	76	84	88
	Straight row	Good	63	75	83	87
	Contoured	Poor	63	74	82	85
	Contoured	Good	61	73	81	84
	Contoured and terraced	Poor	61	72	79	82
	Contoured and terraced	Good	59	70	78	81
Close seeded legumes	Straight row	Poor	66	77	85	89
or rotational meadow	Straight row	Good	58	72	81	85
	Contoured	Poor	64	75	83	85
	Contoured	Good	55	69	78	83
	Contoured and terraced	Poor	63	73	80	83
	Contoured and terraced	Good	51	67	76	80
Pasture range		Poor	68	79	86	89
		Fair	48	69	79	84
		Good	39	61	74	80
	Contoured	Poor	47	67	81	88
	Contoured	Fair	25	59	75	83
	Contoured	Good	6	35	70	79
Meadow (permanent)		Good	30	58	71	78
Woodland		Poor	45	66	73	83
		Fair	36	60	73	79
		Good	25	55	70	77
Farmsteads	—		59	74	82	86
Roads, dirt	—		72	82	87	89
Roads, hard surfaces	—		74	84	90	92

APPENDIX III

Table 2.24 Values of runoff coefficient factor (C)

Vegetarion type	Slope range	Sandy loam soil	Loam Silt loam Clay soil	Stiff clay soil
Woodland and forests	0 – 5%	0.1	0.3	0.4
	5 – 10%	0.25	0.35	0.5
	10 – 30%	0.3	0.5	0.6
Grassland	0 – 5%	0.1	0.3	0.4
	5 – 10%	0.16	0.36	0.55
	10 – 30%	0.22	0.42	0.6
Agricultural land	0 – 5%	0.3	0.5	0.6
	5 – 10%	0.4	0.6	0.7
	10 – 30%	0.52	0.72	0.82

3

Flood Routing through Reservoirs and Channels

3.1 FLOOD ROUTING THROUGH RESERVOIRS—GENERAL

The process of computing the reservoir stage, storage volumes and outflow rates corresponding to a particular hydrograph of inflow is commonly referred to as flood routing. Flood routing studies are required in the design of spillways of reservoirs. The spillway capacity is fixed, taking into consideration the moderation caused by the flood absorption capacity of the reservoir, which is the capacity between the full reservoir and high flood level. The spillway is thus designed for discharging a flood peak at less than the peak of the incoming flood. Looking at it, from another angle, flood routing studies may be used to find out the high flood level, for a given spillway for a design flood.

Maximum reservoir elevations, obtained by routing a particular hydrograph through a reservoir reflects the integrated effect of—(i) initial reservoir level, (ii) rate and volume of inflow into the reservoir, (iii) rate of outflow, (iv) discharge of regulating outlets, power penstock, and (v) surcharge storage and rate of overflow.

In flood control reservoirs, flood routing studies are used to arrive at the most economical combination of storage and outflow capacity such that for the design flood, the outflow from the reservoir does not exceed the safe bankful capacity of the downstream channel. In a flood forecasting system flood routing studies are used to compute the probable flood levels with respect to time at downstream points, corresponding to a flood wave entering the valley at an upstream point.

3.2 BASIC PRINCIPLES OF ROUTING

The accumulation of storage in a reservoir or valley depends on the difference between the rates of inflow and outflow, if losses, by seepage and evaporations and direct accretion to storage by precipitation are ignored. If I is the rate of inflow and O, the corresponding rate of outflow at any instant and if 'dS' is the storage accumulation in a small-time interval 'dt' then $dS = (I\,dt - O\,dt)$. The various methods of flood routing essentially aim at a solution of the above one-dimensional mass continuity equation. The usual problem is one of finding the outflow hydrograph corresponding to a given inflow hydrograph. Some of the various methods of flood routing through reservoir are given as follows:

3.2.1 Pul's Method or Inflow-storage-discharge Method

The method developed by Puls [12] of the US Army Corps of Engineers is a semiempirical one. It makes use of graphs and tables for solving the basic equation as mentioned above. Let subscripts 1 and 2 denote the values corresponding to beginning and end of a small-time interval t and further, assuming, linear variation of the inflow and outflow during the time interval t, the basic storage equation can be expressed as

$$(S_2 - S_1) = \frac{1}{2}\,(I_1 + I_2)\,t - \frac{1}{2}\,(O_1 + O_2)\,t$$

or

$$S_2 = \left(S_1 - \frac{O_1 t}{2}\right) + \frac{(I_1 + I_2)}{2}\,t - \frac{O_2 t}{2} \tag{3.1}$$

The solution of the Eq. (3.1) is carried out as follows:

(i) From the given data prepare a table giving the storage S, rate of outflow O, $\left(S + \frac{1}{2}Ot\right)$ and $\left(S - \frac{1}{2}Ot\right)$ corresponding to different reservoir elevations. For this purpose a suitable time interval t should be chosen.

(ii) From the table prepare a diagram (Fig. 3.1) giving S, $\left(S - \frac{1}{2}Ot\right)$ and $\left(S + \frac{1}{2}Ot\right)$ curve against reservoir elevation.

(iii) When the reservoir level is rising starting with a reservoir storage S_1 at the beginning of a time interval t, the reservoir storage S_2 at the end of the time interval can be found by solving the above equation from S, $\left(S - \frac{1}{2}Ot\right)$ and $\left(S + \frac{1}{2}Ot\right)$ graphs (Fig. 3.2).

In Fig. 3.2, let the point B represents the storage and elevation at the beginning of the time interval t then $S_1 = AB$, $O_1 t/2 = BC$,

$$\left(S_1 - \frac{O_1 t}{2}\right) = AC$$

$$\left(\frac{I_1 + I_2}{2}\right) t = CD, \left(S_1 - \frac{O_1 t}{2}\right) + \left(\frac{I_1 + I_2}{2}\right) t = AC + CD$$

$$= AD = A'D';\ \frac{O_2 t}{2} = ED'$$

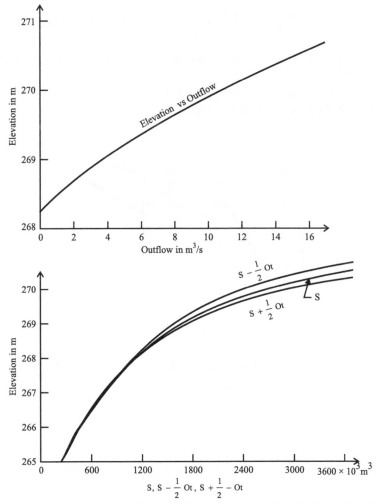

Figure 3.1 Reservoir elevation as a function of O, S, $S - Ot/2$ and $S + Ot/2$.

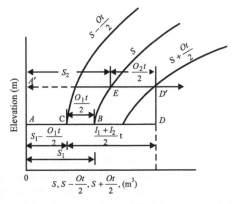

Figure 3.2 Flood routing with rising reservoir level.

$$\therefore \qquad \left(S_1 - \frac{O_1 t}{2}\right) + \left(\frac{I_1 + I_2}{2}\right)t - \frac{O_2 t}{2} = A'D' - ED' = A'E$$

hence, the point E represents the reservoir elevation and storage S_2 at the end of the time interval t.

(iv) When the reservoir level is falling, let P represents the storage and elevation at the beginning of the time interval t in Fig. 3.3.

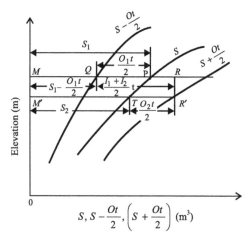

Figure 3.3 Flood routing with falling reservoir level.

Then
$$S_1 = MP, \quad \frac{O_1 t}{2} = QP$$

$$\therefore \qquad S_1 - \frac{O_1 t}{2} = MP - QP = MQ; \left(\frac{I_1 + I_2}{2}\right)t = QR$$

$$\therefore \qquad S_1 - \frac{O_1 t}{2} + \frac{I_1 + I_2}{2}t = MQ + QR = MR = M'R'$$

$$\frac{O_2 t}{2} = TR',$$

$$\therefore \qquad S_1 - \frac{O_1 t}{2} + \frac{I_1 + I_2}{2}t - \frac{O_2 t}{2}$$

$$= M'R' - TR' = M'T$$

Hence, point T represents the reservoir elevation and storage S_2 at the end of the time interval t. For gated reservoir the basic equation of continuity remains the same as before. However, the relation between storage and outflow will depend on the number of gates or valve openings. For a gated spillway with all similar gates the elevation discharge curves will be represented by a family of curves with the number of gates as parameter (Fig. 3.4). As

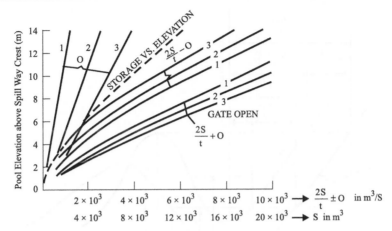

Figure 3.4 Elevation discharge, storage relationship for a gated spillway.

such, the curves $\left[\left(\dfrac{2S}{t}\right)\pm O\right]$ will be replaced by a family of curves with the number of gate openings. The operation of routing is, however, similar to the ungated problem except that the number of gate openings must be tabulated and the value of $\left[\left(\dfrac{2S}{t}\right)\pm O\right]$ interpolated depending on the number of gate openings.

3.2.2   Cheng's Graphical Method

The method is completely a graphical one and its working can be explained as outlined below—Herein the basic starting equation for solving the problem is taken as follows:

$$\left(\frac{I_1 + I_2}{2}\right)t + \left(S_1 - \frac{O_1 t}{2}\right) = \left(S_2 + \frac{O_2 t}{2}\right) \tag{3.2}$$

From the given data in the problem under investigation for the solution by the above method the various parameters, viz., $\left(S + \dfrac{1}{2}O\,t\right)$ (Ot), (It), etc. for a suitable time interval can be estimated. The result can then be shown graphically in Fig. 3.5. In the same figure the inflow hydrograph is also drawn. The procedure to be followed in routing may be outlined below:

   (i) Suppose P is a known point on the outflow hydrograph. Draw PV to meet $\left(S + \dfrac{1}{2}O\,t\right)$ curve at V.
  (ii) Draw VH parallel to Ot-line.
 (iii) Let A be the midpoint, of the time interval MN.
  (iv) Draw BC to meet VH at C.
   (v) Draw then CF to meet curve at F.
  (vi) Draw then FQ to meet NQ at Q, thereby establishing the requisite point.

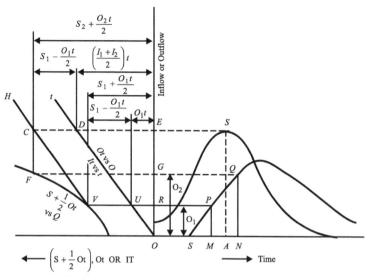

Figure 3.5 Flood routing by Cheng's method.

Steps (i) to (vi) complete one cycle. From the known point the procedure can be repeated to get the next unknown point on the hydrograph.

3.2.3 Working Value Method

The working values are functions of S, O and I. They are designated as under:

$$M_1 = S_1 + \frac{1}{2}O_1\,t,\; M_2 = S_2 + \frac{1}{2}O_2\,t$$

or

$$M_2 = M_1 + \frac{1}{2}(I_1 + I_2)\,t - O_1 t \tag{3.3}$$

$$N_1 = \frac{S_1}{t} + \frac{O_1}{2},\; N_2 = \frac{S_2}{t} + \frac{O_2}{2}$$

or

$$N_2 = N_1 + \frac{1}{2}(I_1 + I_2) - O_1 \tag{3.4}$$

Flood routing can be done using working values of either M or N. In the present case, routing, using the working value concept of N is explained. Using the data, the working values of N are prepared which are then plotted as a function of overflow (Fig. 3.6). This is called working value curve. Starting from a Known point P on the working value curve of known values of N_1 and O_1 how to get O_2 for the next point Q at the time interval t is explained in Fig. 3.6, P being on the N vs. O curve $AB = N_1$ and $PB = O_1$. From P, draw PC making $45°$ with AB. Then $CB = O_1$. Therefore, $AC = N_1 - O_1$. Set off

$$CD = \frac{I_1 + I_2}{2},$$

then

$$AD = (AC + CD) = (N_1 - O_1) + \frac{I_1 + I_2}{2} = N_2.$$

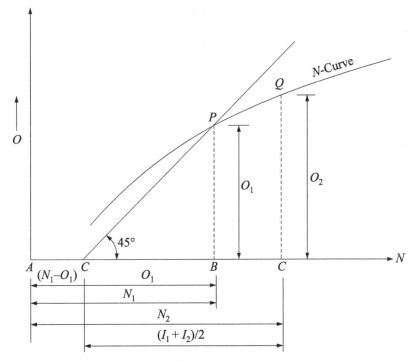

Figure 3.6 Flood routing by working value method.

Project point D to the curve. This projection point is the required point Q representing N_2 and O_2, AD being equal to N_2, DQ will be equal to O_2. This completes one cycle. Starting from the known point Q, the cycle is repeated to get the next point R and so on. The reservoir levels at various hours corresponding to the outflows can be had from the outflow elevation curve as has already been indicated.

The whole thing can also be done conveniently, using a strip of paper, on the edge of which, the common scale of O and N axis of those figures is marked. The use of the scale is explained below. From the known point B, using the paper scale, measure O_1 backwards and get point C. From C, using the paper scale, measure $\dfrac{I_1 + I_2}{2}$ forwards and get point D and its projection Q on the curve, which completes one cycle. Repeat the cycle with the new known point D and so on.

3.2.4 Other Methods

The other methods such as the Goodrich, Wisler-Brater; etc. solve the same fundamental inflow-storage-outflow equation but in different forms. In the Goodrich method, the fundamental equation is considered to be in the following form, viz.,

$$(I_1 + I_2) + \left(\frac{2S_1}{t} - O_1 \right) = \left(\frac{2S_2}{t} + O_2 \right) \tag{3.5}$$

and in the case of Wisler-Brater method it is taken in a form as given below.

$$\left(\frac{I_1 + I_2}{2}\right) + \left(\frac{S_1}{t} - \frac{O_1}{t}\right) = \left(\frac{S_2}{t} + \frac{O_2}{t}\right) \tag{3.6}$$

By constructing appropriate graphical relationships of the quantities as shown above, it is possible to get the outflow from a given inflow rate.

3.2.5 Electronic Analogue

The principle of the electronic analog is to utilize the analog between the equation for an electric circuit and the equation for routing the relationship governing rate of inflow I, rate of outflow O and the rate of change of storage $\dfrac{dS}{dt}$ is given by

$$I = O + \frac{dS}{dt} \tag{3.7}$$

With reference to electrical circuit shown in Fig. 3.7, it may be noted that the relationship between the current i, fed into the circuit, the current O through the resistor R and the rate of change $\dfrac{dQ}{dt}$ across the condenser C can be expressed as

$$i = O + \frac{dQ}{dt} \tag{3.8}$$

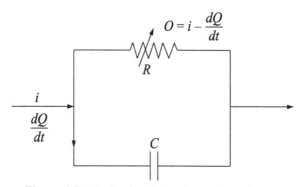

Figure 3.7 Basic circuit for electronic analogue.

It is thus obvious from the expressions above that the equation governing the outflow relationship through the reservoir is analogous to that of the flow of electric current in the circuit shown in Fig. 3.7. Thus, if the current i represents the flow into the reservoir and the charge across the condenser Q represents the storage and current O will represent the outflow. Therefore, it is possible to obtain a current O proportional to the spillway discharge from the electrical circuit, by pumping a current i proportional to the inflow hydrograph into the circuit and by making $\dfrac{dQ}{dt}$ vary in accordance as $\dfrac{dS}{dt}$. This can be accomplished by varying the time constant *R.C.* of the circuit, according to the time of emptying as obtained by dividing the surcharge storage by the corresponding spillway discharge. This can be noticed from the fact that the storage is given as *Ot*, where *O* is the spillway discharge and *t* is the time of emptying, whereas the electrical charge in the circuit is given by *ORC*. If the product of *RC* varies as *t*, *Q* varies as the storage and hence $\dfrac{dQ}{dt}$ varies as $\dfrac{dS}{dt}$. If the resistance *R* is made to vary, according to the time of emptying curve, the product *RC* too varies the same way if *C* is kept constant. Further details of the reservoir flood routing by electrical analogue can be had from the work of Murthy [13].

3.2.6 Mechanical Flood Router

Harkness [14] has designed a rolling type of flood router to solve equation of the form

$$\frac{dO}{dt} = \frac{I - O}{K} \tag{3.9}$$

where

K = coefficient.

It comprised two principal parts. Rigid *T*-frame and an undercarriage consists of a pair of freely rotating wheels mounted on an axle and a shaft extending perpendicular to the axle from the midpoint between the wheels (Fig. 3.8). The *T*-frame consists of a slotted vertical member,

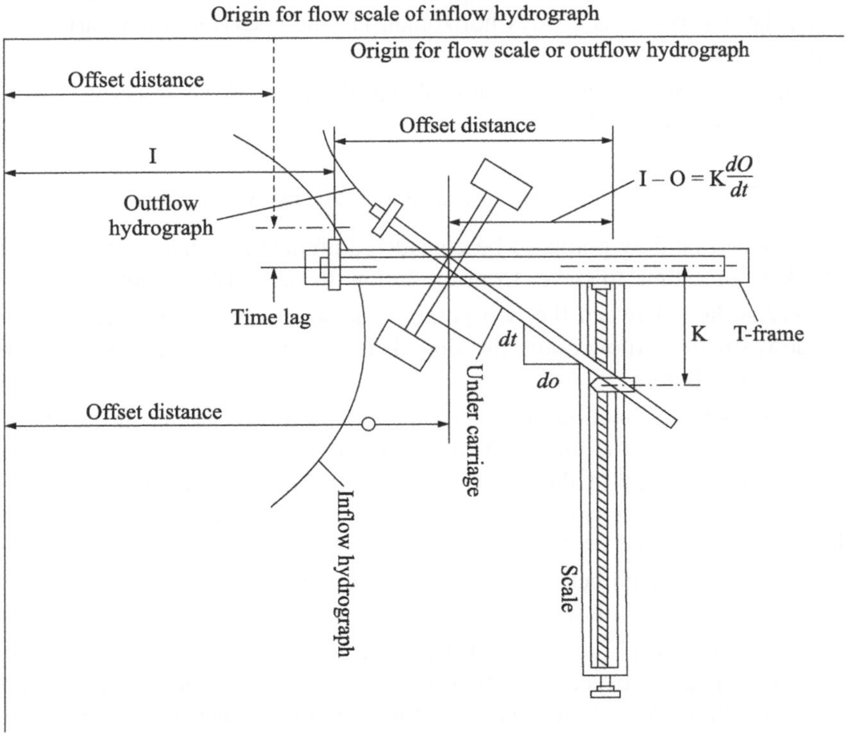

Figure 3.8 Harkness flood router.

a horizontal guide bar on which is mounted a *K*-scale, an indicator attached to a sliding cover on the guide bar, an adjusting screw for setting the indicator and a tracing pointer for following the inflow hydrograph. The torque of the undercarraige shaft is free to move along the guide bar under the sliding collar and the wheel assembly is allowed to move vertically in the slotted member of the *T*-frame. The tracing pointer is offset vertically to allow the free movement of the undercarriage. Provision is also made to offset the pointer horizontally to introduce a time lag. A scriber to trace the routed outflow is mounted on the underside of the carriage at the lower end of the shaft.

During use, the slotted member of the *T*-frame is always perpendicular to the base time of the inflow hydrograph. As the inflow hydrograph is traced the movement of the undercarriage is such that the angular relationship of the shaft and the horizontal member of the *T*-frame make the tangent of the angle = $(I_1 - O_1)/K$, or dO/dt. The scriber will then draw the outflow hydrograph.

3.3 ROUTING THROUGH RIVER CHANNELS

General: In reservoir routing dealt with earlier it may be remarked that:

   (i) It is as though the inflow and outflow points are very close and there is no time of travel required for the flood wave to pass from the inflow point to the outflow point.
  (ii) The inflow and outflow points being very close in the reservoir, the water surface between the two points is always horizontal, i.e. the reservoir water surface is always horizontal.
 (iii) The outflow and storage is a function of the elevation of the reservoir water surface which means that the outflow is a function of storage, i.e. for a particular storage there is a particular outflow.

In any reach of a river channel, the above assumptions are not valid, since.

   (i) The inflow and outflow points to be considered are some distance apart and the flood wave takes some time to travel from the inflow to the outflow point.
  (ii) The water surface between the two points is always sloping from the upstream point to the downstream outflow point; the slope being steeper when the flood level is rising, than when it is falling.
 (iii) The outflow is not a function of the elevation of the water surface at the outflow point alone but depends on the elevation of the water surface at the inflow point also, i.e. it depends on the slope of the water surface. This slope being more during the rising flood period, the outflow is more for that period for any water depth than for the falling period having the same depth.
  (iv) The storage in the reach between the inflow and the outflow point is a function of, (a) the water level at the inflow point, and (b) the water level at the outflow point.
   (v) From the above it would mean that for any particular outflow at the outflow point the storage in the reach between the inflow and outflow points is not one of fixed quantity. From (ii) it, therefore, follows that this storage is more when the flood level is rising than when it is falling.

The relationship between elevation of water surface and outflow for a reservoir and for the outflow section of a river reach are shown in Figs. 3.9 and 3.10 respectively. The relationship between the outflow and storage for a reservoir and river reach are shown in Figs. 3.11 and 3.12. The inflow and outflow hydrograph for a reservoir and river reach are shown in Fig. 3.13.

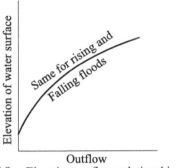

Figure 3.9   Elevation outflow relationship for a reservoir

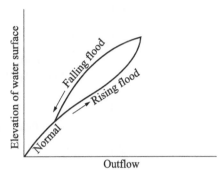

Figure 3.10   Elevation outflow relationship for a river reach.

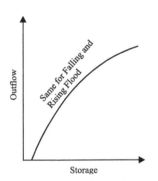

Figure 3.11   Outflow storage relationship for a reservoir.

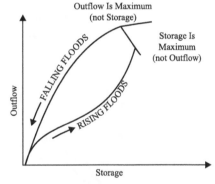

Figure 3.12   Outflow storage relationship for a river reach.

3.3.1   Muskingum Method

It will be evident from the Fig. 3.13 that the peak of outflow occurs when the outflow hydrograph crosses the inflow hydrograph in the case of a reservoir. This is so because the storage is maximum at that point of time and consequently the elevation of water surface is maximum and hence the outflow which is dependent on the upstream elevation is maximum. For a river cross-section the outflow does not depend on water's surface elevation and slope. The optimum combination occurs a little later than at the point of time when the inflow and outflow hydrograph crosses. In channel routing also the fundamental equation to be solved is the same, viz., Eq. (3.1).

$$(S_2 - S_1) = \left(\frac{I_1 + I_2}{2}\right) t - \left(\frac{O_1 + O_2}{2}\right) t.$$

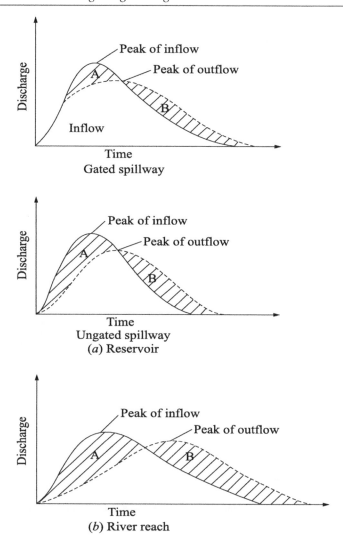

Figure 3.13 Inflow and outflow hydrograph for a reservoir and river reach.

Since in a channel reach, as the storage, inflow and outflow are interdependent, it is necessary to find an equation connecting these three parameters. The method adopted is usually based on variable storage discharge relationships. The method was developed by McCarthy of US. Army Corps of Engineers for the Muskingum Flood Control Project [15]. Here an equation connecting storage S, inflow I and outflow O is given as follows

$$S = KO + Kx \, (I - O) \tag{3.10}$$

where K and x are assumed to be constant by Muskingum, although in reality they do vary.

In a channel reach the total storage S = prism storage (KO) + wedge storage ($Kx \, (I - O)$), the physical meaning of which will be apparent from Fig. 3.14 when the inflow I is greater than O, flood wave is passing from section AA to section BB, the levels are rising in the reach,

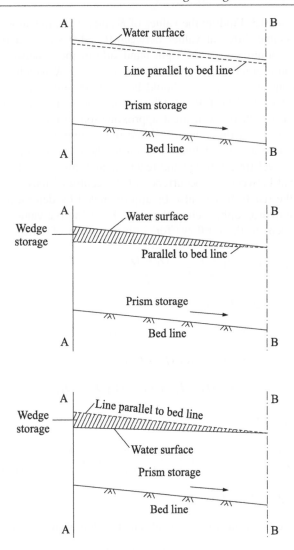

Figure 3.14 Channel storage components.

AB and the wedge storage is positive. When *I* is less than *O*, flood wave has passed from section *AA* to *BB*, the levels are falling in the reach *AB* and the wedge storage is negative. If the inflow and outflow figures of any flood for a particular reach of the river is known, then the value of *K* and *x* may be obtained as follows:

$$K = \frac{S}{x\,I + (1 - x)O}; \quad \text{where } S = \left[\frac{I_1 - O_1}{2} + \frac{I_2 - O_2}{2}\right]t \qquad (3.11)$$

in the above equation *K* and *x* are both unknown. However, the maximum value of *x* has been found to be equal to 0.5 based on a study of data of different rivers. For most streams it usually has a value between 0 and 0.3. Finding the values of K and x from the above equation is a matter of successive trials. Assume different values between 0 and 0.3 for x and compute the values of the numerator and denominator of the right hand side of the equation. Plot these values as abscissa and ordinate on a graph paper and join the points. As K which is the inverse slope of this graph is to be a constant, the points should lie on a straight line if the chosen value of x is correct, otherwise a loop will be formed. Choose the one which is nearest to a straight line and draw a straight line which is the nearest approximation to the plotted points. The inverse slope of this straight line gives the value of K and the assumed value of x which resulted in this graph being nearest to a straight line, is taken as the correct value. The constant K is a measure of the lag or travel time through the reach. It may also be determined by finding the lag, i.e. the time interval between the occurrence of the centre of mass of inflow and centre of mass of outflow over the reach. It may also be approximated by determining the time of travel of critical points on the hydrograph, such as the peak. Knowing the values of K and x for a river reach the method to calculate the outflows for a given pattern of inflows is described below.

Now
$$S_2 - S_1 = \frac{I_1 + I_2}{2}t - \frac{O_1 + O_2}{2}t$$

Again
$$S = KO + Kx\,(I - O)$$

or
$$S_1 = KO_1 + Kx\,(I_1 - O_1)$$

and
$$S_2 = KO_2 + Kx\,(I_2 - O_2)$$

∴
$$S_2 - S_1 = Kx\,(I_2 - I_1) + (1 - x)\,(O_2 - O_1) \tag{3.12}$$

or from Eqs. (3.12) and (3.1)

$$O_2 = \frac{Kx - 0.5t}{K - Kx + 0.5t}I_2 + \frac{Kx + 0.5t}{K - Kx + 0.5t}I_1 + \frac{K - Kx - 0.5t}{K - Kx + 0.5t}O_1$$

$$= C_0 I_2 + C_1 I_1 + C_2 O_1 \tag{3.13}$$

where C_0, C_1 and C_2 are the values of the coefficient in the above expressions and

$$C_0 + C_1 + C_2 = 1.$$

The routing constants have to be determined from observed hydrographs by empirical means and can be done by plotting I and O against storage S. Figure 3.15 indicates how storage S can be computed from the given hydrograph I and O. In the next step, S should be plotted against $[xI + (1 - x)O]$, which can be done by considering a number of values of x, viz., x_1, x_2, etc. and plotting S against $[x_1 I + (1 - x_1)O]$ and again S against $[x_2 I + (1 - x_2)\, O]$ in a second diagram and so on. The value of x will be deemed to be correct if the points for the various values of S, I and O are located in an almost straight line (Fig. 3.15). The slope of the line so plotted will provide the value of K. With the value of K and x thus known, the routing can now be carried with the Eq. (3.13), viz.,

$$O_2 = C_0 I_2 + C_1 I_1 + C_2 O_1.$$

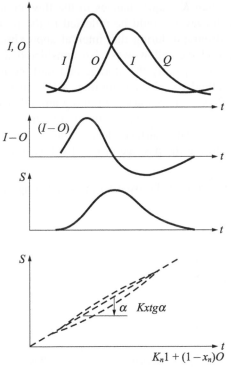

Figure 3.15 Determination of Muskingum routing coefficients.

As an example, say the values of C_0, C_1, and C_2 for a particular reach is $C_0 = 0.1$, $C_1 = 0.4$ and $C_2 = 0.5$ and the hydrograph of inflow I at the end of each period is given along with the initial value of O_2.

Period	1	2	3	4	5	6	7	8
inflow	2000	3000	5000	5500	5000	4000	3000	2000

Based on Eq. (3.13), the inflow hydrograph can be routed as follows (Table 3.1):

Table 3.1

Inflow	$C_0 I_2$	$C_1 I_1$	$C_2 O_1$	O_2
2000	—	—	—	2000
3000	300	800	1000	2100
5000	500	1200	1050	2750
5500	550	2000	1375	3925
5000	500	2200	1963	4663
4000	400	2000	2332	4732
3000	300	1600	2366	4266
2000	200	1200	2133	3533

In order to obtain the best results Chow [14] has stated that the best results are obtained when the routing period t is not less than $2Kx$ and not more than K, K being the travel time as stated earlier. If t is greater than K major changes in the flow could traverse the reach within the routing period and this, however, would be reflected in the routed outflow. Also t must be short enough so that the hydrograph during that interval approximates a straight line.

In the Muskingum method the inflow was also considered as a parameter affecting the slope of water surface and consequently the storage and outflow in the channel reach. Hence for adopting Muskingum method for channel routing, a known set of values of inflows with corresponding values of outflows is necessary for routing any other flood. Such data will not be available in many cases. In such cases the approximating assumptions that outflow and storage in the channel reach depend on water surface elevation as in a reservoir storage is made. The varying water surface slope during the passage of the flood wave is neglected. Then the routing can be done as follows [9]:

(i) By field survey take a number of cross-sections and a longitudinal section of a river reach (Fig. 3.16).

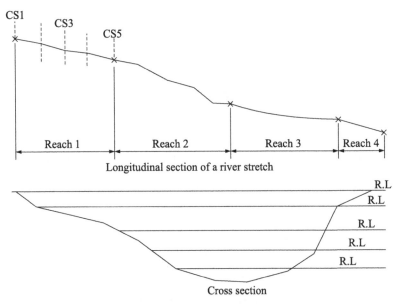

Figure 3.16 Longitudinal and cross-sections of a river reach.

(ii) Plot the longitudinal section and divide the stretch into a number of reaches such that in each reach, the river has one bed slope i_1, i_2, ..., etc. Plot the cross-sections and find the areas at different reduced levels. Assume a suitable value of rugosity coefficient n in the Manning's formula, i.e.,

$$v = \frac{1}{n} R^{2/3} i^{i/2} \tag{3.14}$$

the energy slope being considered equal to the bed slope. Using Manning's formula prepare the table for area and discharge for each cross-section giving values of outflow O for different levels from near the bed to well above the flood zone level. Using the values in table draw the curves of outflow vs. area of cross-section and outflow vs. water surface elevation (Fig. 3.17).

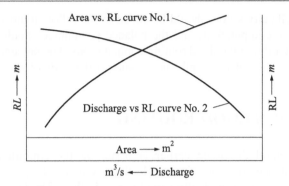

Figure 3.17 Discharge and water surface elevation as a function of flow area.

From the table and similar tables for the other cross-sections in the reach pick out values of velocities for approximately bankful discharges. Take the approximate average value of these velocities. Divide the length of the reach by this velocity and get the travel time t for the bankful discharge to pass to the reach. Round off the t to a lower convenient figure which will then be the routing interval. Using this value of t, compute the storage indication curve. Start with a small discharge and go up to a discharge slightly more than the value of the peak of the flood to be routed. The areas of cross-sections corresponding to different outflows for the table are obtained from the curve in Fig. 3.16. From the numerical values, then obtain the storage indication curve (Fig. 3.18). The flood routing table for the river reach is then obtained, which means the figures of outflow and the water surface elevation at the end of the reach. These figures then become the inflow figures for the next reach and the cycle is repeated for the next subsequent reaches.

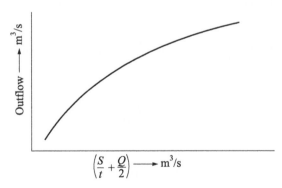

Figure 3.18 Storage indication curve.

Attempt should be made to select the reaches in such a way that any reach ends just before the confluence of a tributary to the main river. If the reaches are selected in this manner the flow from the tributary can be added to the outflow figure from the upper reach and the total volume taken as the inflow figures for the lower reach, before starting the flood routing computations for the lower reach. Compared to the huge inflow figures the flow from the small catchment on either side of the river banks can be neglected. If, however, the tributary joins the reach somewhere in between its two ends then the following procedure may be adopted.

A percentage of the tributary's flow should be added to the inflow figures into the reach before doing the flood routing computations and the balance percentage should be added to outflow figures of the reach after doing the flood routing computations. The percentages are discretionary and depend on the nearness of the confluence points to the ends of the reach.

3.4 HYDRAULIC FLOOD ROUTING

The flood routing method discussed earlier is based on the simplified approach basically involving the mass balance equation of unsteady flow. Actually the propagation of flood wave in a stream channel is a case of gradually varied unsteady flow. The laws governing such flows are the equations of continuity and momentum for unsteady flow.

3.4.1 Continuity Equation of Unsteady Flow

The law of continuity for unsteady flow may be established by considering the conservation of mass in an infinitesimal space between two channel sections. If u is the velocity and A the cross-sectional area at section x, then the velocity and cross-sectional area at another section $(x + dx)$ will be

$$\left(u + \frac{\partial u}{\partial x} \delta x\right), \left(A + \frac{\partial A}{\partial x} \delta x\right)$$

If B is the top width and y is the depth measured from channel bottom, then equating, (Inflow – Outflow) = Change in a storage during the time interval dt. One can write,

$$Au\, dt - \left(A + \frac{\partial A}{\partial x}\right)\left(u + \frac{\partial u}{\partial x} dx\right) dt = \left(\frac{\partial y}{\partial t} dt\right) Bdx$$

or

$$\frac{\partial(Au)}{\partial x} + B\frac{\partial y}{\partial t} = 0$$

or

$$\frac{\partial Q}{\partial x} + B\frac{\partial y}{\partial t} = 0 \tag{3.15}$$

where

Q = discharge = (Au).

3.4.2 Dynamic Equation of Unsteady Flow

Consider a fluid element which is subjected to the forces F_1, F_2 due to pressure and S as the shear due to friction (Fig. 3.19).

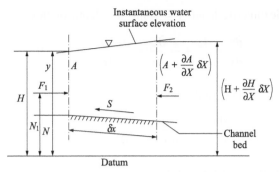

Figure 3.19 Definition sketch for dynamic equation.

Now $\quad F_1 = \gamma A (H - \bar{z})$ and $F_2 = \bar{z}\gamma \left[H + \dfrac{\partial H}{\partial x} \delta x - \left(\bar{z} + \dfrac{\partial \bar{z}}{\partial x} \delta x \right) \right] \left[A + \dfrac{\partial A}{\partial x} \delta x \right]$

where

$\quad$ H = piezometric head,

$\quad$ $\bar{z}$ = elevation of the centroid of water with respect to a datum line

and

$$S = \frac{\gamma f v^2}{8gR} A \, \delta x$$

where,

$\quad$ f = friction factor,

$\quad$ R = hydraulic radius and

$\quad$ g = acceleration due to gravity.

Now the resultant force = $(F_1 - F_2 - S)$

or,

$$F = -\gamma A \frac{\partial H}{\partial x} \delta x - \frac{\gamma A f v^2}{8gR} \delta x$$

$$\text{(neglecting higher order terms)}$$

Now $\qquad\qquad\qquad\qquad$ Force = Mass × Acceleration

where,

$\quad$ Mass of fluid element = $(\gamma/g \; A \; \delta x)$

Acceleration

$$\frac{dv}{dt} = \frac{\partial v}{\partial t} + v \frac{\partial v}{\partial x}$$

or

$$\frac{\gamma A}{g} \delta x \left[\frac{\partial v}{\partial t} + v \frac{\partial v}{\partial x} \right] = -\gamma A \frac{\partial H}{\partial x} \delta x - \frac{A \gamma \; f v^2}{8gR} \delta x$$

or

$$\frac{\partial v}{\partial t} + v \frac{\partial v}{\partial x} + g \frac{\partial H}{\partial x} + S_f = 0$$

where S_f = friction slope.

Now, $H = (y + z) =$ piezometric head, $y =$ being the depth of flow.

or

$$\frac{\partial y}{\partial x} + \frac{\partial z}{\partial x} + \frac{1}{g}\frac{\partial v}{\partial t} + \frac{v}{g}\frac{\partial v}{\partial x} + gS_f = 0$$

Since $\dfrac{\partial z}{\partial x} = S_0$, i.e. channel-bed slope, then

$$\frac{\partial y}{\partial x} + \frac{v}{g}\frac{\partial v}{\partial x} + \frac{1}{g}\frac{\partial v}{\partial t} = S_0 - S_f \tag{3.16}$$

3.4.3 Characteristics of Flood Waves

The propagation of flood wave in a channel of known topographic features is needed for design of engineering works. These are the height, form and speed of the wave and the subsidence and dissipation it undergoes as it progresses along the channel. Analytical studies of flood wave have been primarily related to waves in uniform prismatic channels and they are helpful in understanding the main characteristics of flood propagation even though flow in natural stream involve irregular section and frictional characteristics. The governing one-dimensional equation as has been derived earlier can be expressed as

$$\frac{\partial Q}{\partial x} + B\frac{\partial h}{\partial t} = 0 \tag{3.17}$$

$$\frac{\partial v}{\partial t} + v\frac{\partial v}{\partial x} + g\left(S_f + \frac{\partial h}{\partial x}\right) = 0 \tag{3.18}$$

where,

$Q =$ discharge;
$x =$ distance along river,
$B =$ surface width;
$h =$ height of water surface above datum,
$t =$ time,
$v =$ mean velocity over flow section in x-direction,
$g =$ acceleration due to gravity,
$S_f =$ friction slope.

The friction slope is generally assumed to be given by an equation of the Chezy type:

$$v = C\sqrt{RS_f} \tag{3.19}$$

The bed slope of the channel is often an important parameter in the propagation of the flood wave. Denoting the water depth by y, and bed slope by S_0

$$h = z + y; \; S_0 = -\frac{\partial z}{\partial x} \tag{3.20}$$

Equations (3.18) and (3.19) may be combined and rewritten as

$$Q = CA\left[R\left(S_0 - \frac{\partial y}{\partial x} - \frac{v}{g}\frac{\partial v}{\partial x} - \frac{1}{g}\frac{\partial v}{\partial t}\right)\right]^{1/2} \tag{3.21}$$

where $A = Q/v$. The four terms in the bracket on the right-hand side of Eq. (3.21) have the form of slopes. For a wide rectangular channel, $R \simeq y$ and hence

$$Q = CBy \left[y \left(S_0 - \frac{\partial y}{\partial x} - \frac{v}{g} \frac{\partial v}{\partial x} - \frac{1}{g} \frac{\partial v}{\partial t} \right) \right]^{1/2} \tag{3.22}$$

The last two slope terms are of the order of $F^2 \left(\dfrac{\partial y}{\partial x} \right)$ [see Appendix II] where $F = \dfrac{v}{(g y)^{1/2}}$. Records show that except in mountain torrents the value of $F < 1$. In steep rivers the bed slope S_0, predominates, while in rivers of gentle slopes ($F^2 < 1$) the last two slope terms are of the order of $F^2 \dfrac{\partial y}{\partial x}$ and only the first two S_0 and $\dfrac{\partial y}{\partial x}$ may be retained. For average slopes it may be necessary to retain all.

3.4.3.1 Flood wave on steep slopes

3.4.3.1.1 *Kinematic wave*
In this case $Q = f(y)$ alone and only the term S_0 needs to be retained. As such the continuity equation can be written as:

$$\frac{dQ}{dy} \frac{\partial y}{\partial x} + B \frac{\partial y}{\partial t} = 0$$

or

$$\frac{\partial y}{\partial t} + \frac{1}{B} \frac{dQ}{dy} \frac{dy}{dx} = 0 \tag{3.23}$$

After comparing with the expression:

$$\frac{dy}{dt} = \frac{\partial y}{\partial t} + \frac{\partial y}{\partial x} \frac{dx}{dt} = 0 \tag{3.24}$$

one finds that to an observer moving at a velocity $c = \dfrac{1}{B} \dfrac{dQ}{dy}$, the depth y and hence Q will appear to be constant. This type of wave which is obtained from the equation of continuity alone has been termed as 'Kinematic wave'. For a constant value of c and in a wide rectangular channel, the velocity of the kinematic wave comes to $1.5v$ [see Appendix II].

3.4.3.1.2 *Dynamic wave*
Lighthill et al. [16] studied the effect of other slope terms and they found that the main flood wave moves almost like a kinematic wave with the leading and trailing wave fronts moving with speeds $v \pm (gy)^{1/2}$. In this case, the wave is called dynamic.

The dynamic forerunner moves downstream with a speed of $v \pm (gy)^{1/2}$ which is greater than $1.5v$ (i.e. kinematic speed) of the main wave provided $F < 2$, which holds good for natural floods. It may further be shown that if $F < 2$, a progressive wave will be attenuated rapidly unless

$$\frac{dy}{dt} \geq \frac{g y_0 S_0 (2 - F_0)(1 + F_0)}{3 v_0} \tag{3.25}$$

where the suffix '0' denotes initial waves. For normal rivers the right-hand side yields a value for the rate of rise of water level of the order of about 10 cm per minute which is quite high and is not attained in normal floods. Thus in such cases the dynamic forerunner will disappear rapidly. If $F > 2$, the main kinematic wave will tend to overtake the forerunner with consequent steepening and formation of a progressive surge.

3.4.3.1.3   *Monoclinal wave*

With the steepening of the wave, the slope terms which were neglected in the derivation of kinematic wave become increasingly more important. The kinematic wave may then assume a stable form of what is known as a monoclinal wave, which is nothing but a single elongated step in the water surface. This has been described by Lighthill and Whitlam as a kinematic shock. The possible wave forms are shown in Fig. 3.20. For case (a) the wave will maintain a stable wave form as it moves downstream. Such a wave will occur in big long rivers as the effective spread of the wave form is of several tens of kilometres for normal river parameters. The waves on steep slopes as discussed above do not lengthen or disperse, so that they do not subside in their progress along the river.

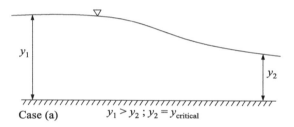

Case (a) $y_1 > y_2$; $y_2 = y_{\text{critical}}$

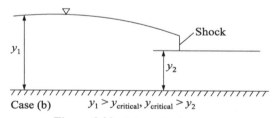

Case (b) $y_1 > y_{\text{critical}}$, $y_{\text{critical}} > y_2$

Figure 3.20   Monoclinal wave.

3.4.3.2   Flood waves on gentle slopes

When the bed slope is relatively small the water surface slope is also relatively important and needs to be considered. In that case the dynamic equation becomes

$$Q = CBy \sqrt{y \left(S_0 - \frac{\partial y}{\partial x} \right)} \qquad (3.26)$$

Based on the constancy of Chezy's C value, the profile of the wave, its speed and subsidence characteristics can be found out. The instantaneous wave profile is shown in Fig. 3.21 with the crest at A and the water surface sloping down on both sides. The discharge Q is not a function of y alone. As such Q is not maximum at A but is at B, downstream of A, where

$\frac{\partial y}{\partial x} = 0$. As a result, the depth at B is the maximum that may be attained at the place. Thus the local crest is at B and the level at B is $<$ than A which shows that the crest subsides as it moves downstream. The wave is mild, and it rises slowly as well as falls slowly. The wave velocity of a subsiding wave may be expected to differ from the kinematic velocity since Q is no longer a function of y alone. However, it may be shown that the wave velocity at the river crest A and local crest B differ from each other and from the kinematic wave velocity $\frac{3}{2}v$ in second order terms only.

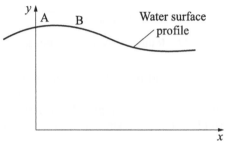

Figure 3.21   Instantaneous wave profile.

3.4.3.2.1   *Rating curve for unsteady flow*
Let Q_0 denotes the uniform flow discharge given by $Q_0 = CBy\sqrt{yS_0}$. Then during the passage of a flood wave the discharge at the same depth will be as given in Eq. (3.26) which on replacing CBy by $Q_0/(yS_0)^{1/2}$ becomes $Q = Q_0\left(1 - \frac{\partial y}{\partial x}\frac{1}{S_0}\right)^{1/2}$. For quantitative estimation of Q the water surface slope is needed which, however, is extremely difficult to determine in practice. Assuming kinematic wave approximation, we have

$$\frac{\partial y}{\partial x} = \frac{dy}{dQ}\frac{\partial Q}{\partial x}.$$

Again $\frac{\partial Q}{\partial x} = -B\frac{\partial y}{\partial t}$ vide continuity equation for unsteady flow, therefore, $\frac{\partial y}{\partial x} = -B\frac{dy}{dQ}\frac{\partial y}{\partial t}$.

Further, as has already been shown that the kinematic wave celerity $c = \frac{1}{B}\frac{dQ}{dy}$, therefore, from above mentioned relationship $\frac{\partial y}{\partial x} = -\frac{1}{c}\frac{\partial y}{\partial t}$. Accordingly, the discharge equation becomes

$$Q = Q_0\left[1 + \frac{1}{cS_0}\frac{\partial y}{\partial t}\right]^{1/2} \tag{3.27}$$

The above is known as Jones' formula, and it exemplifies the well-known loop rating curve (Fig. 3.22). An observer stationed on the river bank observes first the point C of the maximum discharge, then the point B of the maximum stage and then as the water level is falling the point A where the flow appears to be momentarily uniform. In practice it is only possible to use

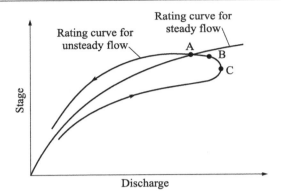

Figure 3.22   Rating curve due to unsteady flow.

the rating curve for steady discharge and then by using the above relationship the rating curve for unsteady flow can be constructed from a known hydrograph. Conversely, if the discharge changes rapidly during the measurement the adjusted discharge or the normal discharge can be obtained using Jones' formula for the preparation of a normal stage discharge relationship.

3.4.4   Characterisation of Flood Waves by Rating Curve

The flood waves have also been characterized within the framework of loop or hysterisis rating curves. Mishra et al. (38) defined the flood wave characteristics in natural channel using quantified hysterisis η and criteria were developed for the identification of kinematic (KW), diffusion (DW) and dynamic (DYW) wave. Physical significance of hysterisis has been explained by relating it with the channel and flow characteristics and energy loss.

The hysterisis η is an index that measures the size or area of the loop in a non-dimensional rating curve and can be considered as a rough index of energy expended by a flood wave at a site under consideration during its time period. The non-dimensional form of shape and discharge values are as follows:

$$h = \frac{H - H_{\min}}{H_{\max} - H_{\min}}, q = \frac{Q - Q_{\min}}{Q_{\max} - Q_{\min}} \qquad (3.28)$$

where $H_{\max}$, $H_{\min}$ are the maximum and minimum computed depth at a site of interest, h = dimensionless stage, similarly Q stands for the discharge and q = dimensionless discharge. The speed of travel of the wave is defined as

$$C = \Delta x/\Delta t, \qquad (3.29)$$

where C = speed of travel in m/s, Δx = reach length in m and Δt = time of travel of the flood peak in second. The phase difference ϕ is computed from the relationship

$$\phi = 2\pi/T \, (t_{ph} - t_{PQ}) \qquad (3.30)$$

where ϕ = phase difference in radian, T = time period in hours, t_{ph} = time of rise of stage wave (hours) and t_{PQ} is the time of rise of discharge wave (hours). The attenuation is described by logarithmic decrement as per the following relationship

$$\delta = \frac{CT}{\Delta x} (I_n Q_J - I_n Q_{J+1})$$ (3.31)

where δ = logarithmic decrease in one period of wave travel (dimensionless), T is in seconds, Q_J and Q_{JH} are peak discharges cumec at successive locations J and $J + 1$ respectively, Δx is in meters. Greater the value of δ, greater is the attenuation.

According to Fread (39) the dam break flood encompasses the whole dynamic range of the waves and its nature changes from DYW to DW or KW as it travels downstream. The dam break analysis results of the Bargi dam located in Central India has, therefore, been considered for generating flood waves in the downstream valley of the dam to characterize the flood waves. The analysis has been carried out by the software DAMBRK developed by NWS of USA (by Scientist of N.H. Roorkee). The results are shown in Fig. 3.23 which show the plan view of the river reach, the slope variation along the reach and η variation respectively. The slope of the river reach generally varies from 3×10^{-5} and 5×10^{-5} and where the slope is flatter, i.e. 115 to 307 km of the reach the η values are higher than those where the slope is steeper reach (307 km to 465 km) and adjacent downstream reach. Further, where the slope changes from steeper to mild at 115 km the η assumes a higher value than where the slope changes from milder to steeper, i.e. at 307 km. The channel is generally of uniform width. On the hysterisis curve characterization of various flood wave zone has been indicated in the Fig. 3.23.

3.4.5 Flood Waves in Natural Streams

The analytical methods described earlier provide description of flood wave characteristics in uniform channels which serve as qualitative descriptions of flows in natural irregular channels. In the study of actual floods in such channels, however, special analytic-cum-numerical methos are employed to take into account the existence of wide flood plains which will have a favourable influence on the attenuation of flood waves because of the large storage capacity for flood water.

In the computation of storage of such practical problems it is often assumed that the same water level exists in the river and flood plain and the velocity of rise and fall of the water level is the same in the river and flood plain. In that case the storage capacity is proportional to both the width of the flood plain and the velocity of rise of water level, i.e. $\partial h/\partial t$. This type of storage is termed as uniform storage.

Usually the assumption of uniform storage is not quite correct. When the river is in spate and overtops the banks, water will start flowing on the flood plains. Because of small flow depths and in view of the presence of bushes, small dykes and other obstructions which offer additional resistance, the flow is retarded. This results in a difference of water level between low water bed and flood plain. As the water rises further, the inflow resistance decreases and at the crest of the flood wave the levels often become equal (Fig. 3.24). Flood plain storage in the first period of a flood wave has a negligible influence on crest height. The maximum effect being obtained during storage just before the top of the wave. Accordingly, the retardation of inflow and outflow is of advantage sometimes in reducing flood level.

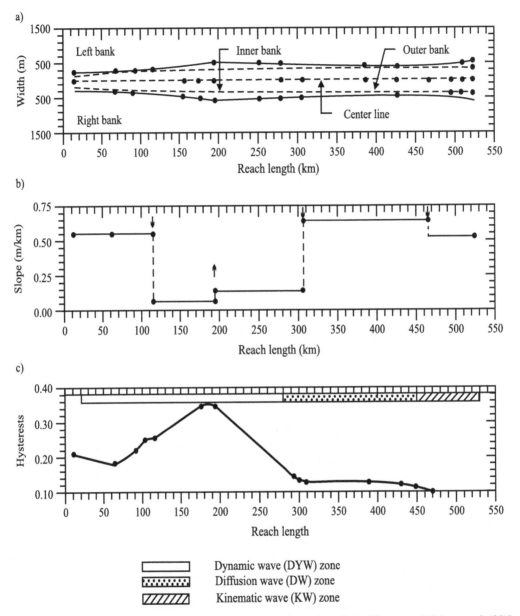

Figure 3.23 River reach characteristics (0 km corresponds to dam site): [Courtsey Mishra et al (38)]
  (a) Plan view of river valley. The available cross-sections are presented by dots.
  (b) Bed slope variation along the river valley. Up and down arrows marks indicate respectively steepening and flattening of adjoining downstream reaches.
  (c) Hysterisis variation along the river valley.

In many practical problems such an exact analysis is not required and the solution based on a simplified analysis of the momentum equation yields satisfactory results.

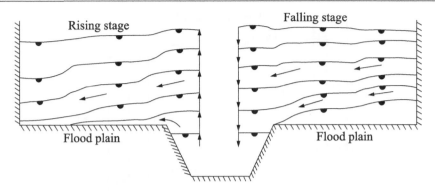

Figure 3.24 Effect of resistance on inflow and outflow from plain. (Ref. Bibliography 2)

3.4.5.1 Simplified analysis

By suitable algebraic manipulation, the momentum Eq. (3.16) can be written in terms of Manning's resistance equation as follows:

$$Q = \frac{1}{n} R^{2/3} A \left[S_0 - \frac{\partial y}{\partial x} \frac{1}{g} \frac{\partial v}{\partial t} - \frac{v}{g} \frac{\partial v}{\partial x} \right]^{1/2} \tag{3.32}$$

where

n = Manning's rugosity coefficient
R = hydraulic radius.

In Eq. (3.32) when the bed slope S_0 is large compared to other terms representing the water surface slope, acceleration slope, etc. the momentum equation simply takes the form of Manning's equation, viz.,

$$Q = \frac{1}{n} R^{2/3} A \sqrt{S_0} \tag{3.33}$$

When the bed slope is relatively small, the water surface slope is also relatively important and needs to be considered. In that case the Eq. (3.33) becomes

$$Q = \frac{1}{n} R^{2/3} A \left(S_0 - \frac{dy}{dx} \right)^{1/2} \tag{3.34}$$

Accordingly, the inflow-outflow response of a particular stretch of a stream may be obtained as per Eq. (3.33) or Eq. (3.34). This approach is usually designated as kinematic method according to Henderson [17].

Based on the above concept, the highly variable flow characteristics of the channel system can be represented by dividing it into a number of reaches and using a time step which ensures stability in the simulation. Inflow can take place both laterally and from the upper end of the reach and calculations normally proceed from the top reach in a downstream direction. The procedure for the computation is as follows. For a given reach sketch the profile at the initial

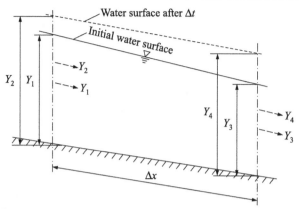

Figure 3.25 Definition sketch of computational scheme.

and final instant by extending the known profile from the upstream reach (Fig. 3.25). Measure the depths Y_1, Y_2, Y_3 and Y_4, and bed slope S_0, at the sections and then find out the water areas A_1, A_2, A_3 and A_4, the hydraulic radii R_1, R_2, R_3 and R_4 and the water surface slope. Therefore, obtain Q_1, Q_2, Q_3, and Q_4 in accordance with Eqs. (3.33) or (3.34), as the case may be. The above quantity should then satisfy the continuity relations which means

$$(Q_1 + Q_2 - Q_3 - Q_4) \Delta t = (A_1 + A_4 - A_2 - A_3)\Delta x \qquad (3.35)$$

where

Δt = time interval,

Δx = length of the channel reach.

In case the continuity relation vide Eq. (3.35) is not satisfied by the given flow, revise the profiles and repeat the process until the check becomes satisfactory.

In a compound section when the discharge increases the bankful condition, to estimate the discharge Q for the channel as a whole it is the usual practice, and quite satisfactory (as a simplified approach), to divide the cross-section into channel flow and flood-plain flow assuming that the water surface profile across the width of the channel and flood plain to be horizontal. In practice, however, the water surface profile would be a curved one.

3.4.5.2 Methods of characteristics

For practical purposes, apart from the simplified approach outlined above a graphical solution based on the general method of characteristics can also be made. Any disturbance which occurs at some point at time $t = 0$, in open channel flow propagates along the channel in time and in two directions upstream and downstream. Thus the perturbation which takes place at point Q influences the shaded region (Fig. 3.26) limited by curves C_+ and C_- which represents the paths of disturbance propagation. If the perturbations form shallow water waves of small amplitude, the lines forming the boundaries of these regions are called characteristics. They can be defined as lines in the (x, t) plane along which the disturbances propagate. Coming back to the mathematical definitions we shall define disturbances as discontinuities in the first and higher derivatives of the dependent variable and the physical parameters which appear in the flow equations. Thus the discontinuity in the water surface slope $(\partial y/\partial x)$ or in the velocity

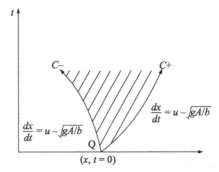

Figure 3.26 Zone of influence of characteristics.

gradient ($\partial u/\partial x$) propagate along the characteristics with a celerity equal to that of shallow water waves:

$$\frac{dx}{dt} = u \pm \left(\frac{gA}{b}\right)^{1/2} \tag{3.36}$$

For prismatic channel of constant cross-sectional shape and bottom slope S_0, the continuity and dynamic equation can be expressed as

$$\frac{\partial h}{\partial t} + \frac{A}{b}\frac{\partial u}{\partial x} + u\frac{\partial h}{\partial x} = 0 \tag{3.37}$$

$$\frac{\partial u}{\partial t} + u\frac{\partial u}{\partial x} + g\frac{\partial h}{\partial x} + g(S_f - S_0) = 0 \tag{3.38}$$

The above equations will be transferred into their characteristics form, by introducing a new variable c to replace h.

$$c = \left(\frac{gA}{b}\right)^{1/2}$$

By differentiating c^2, with respect to x and t and noting that $\partial A/\partial h = b$, we have,

$$2c\frac{\partial c}{\partial x} = g\frac{\partial h}{\partial x};\ 2c\frac{\partial c}{\partial t} = g\frac{\partial h}{\partial t}$$

Substituting into Eqs. (3.33) and (3.34) lead to the following system of two equations

$$2\frac{\partial c}{\partial t} + 2u\frac{\partial c}{\partial x} + c\frac{\partial u}{\partial x} = 0 \tag{3.39}$$

$$\frac{\partial u}{\partial t} + 2u\frac{\partial c}{\partial x} + u\frac{\partial u}{\partial x} + E = 0 \tag{3.40}$$

where
$E = g\ (S_f - S_0).$

The above equations are added and then subtracted to obtain the following pair of equations the so-called characteristics form of the flow equation:

$$\left\{ \frac{\partial}{\partial t} + (u + c) \frac{\partial}{\partial x} \right\} (u + 2c) + E = 0 \tag{3.41}$$

$$\left\{ \frac{\partial}{\partial t} + (u - c) \frac{\partial}{\partial x} \right\} (u - 2c) + E = 0 \tag{3.42}$$

Now $(u + c) = W_+$ is a velocity and so is $(u - c) = W_-$. Indeed according to Eq. (3.42) functions c and u are differentiated along curves in the (x, t) plane which satisfy the differential equations $[dx/dt] = (u \pm c)$. The differential operators are nothing more than total derivatives along these curves:

$$\frac{D_+}{Dt} = \left\{ \frac{\partial}{\partial t} + W_+ \frac{\partial}{\partial x} \right\}, \frac{D_-}{Dt} = \left\{ \frac{\partial}{\partial t} + W_- \frac{\partial}{\partial x} \right\}$$

Hence, one may write

$$\frac{D_+}{Dt} (u + 2c) = -E \tag{3.43}$$

$$\frac{D_-}{Dt} (u - 2c) = -E \tag{3.44}$$

Thus for any point moving through the fluid with the velocity $(u \pm c)$, the relationship (3.39) is true along the characteristics curves C_+ defined by:

$$\frac{dx}{dt} = u + c \tag{3.45}$$

While the relationship (3.44) is valid along the characteristics curve C_- defined by

$$\frac{dx}{dt} = u - c \tag{3.46}$$

With the system of 4, differential Eqs. (3.43) to (3.46) one can find 4, partial derivatives of dependent variables $\left(\frac{\partial h}{\partial t}, \frac{\partial h}{\partial x}, \frac{\partial u}{\partial t}, \frac{\partial u}{\partial x} \right)$ around a line element in (x, t) plane and then continue the solution in the vicinity of such an element. For frictionless horizontal channel $E = 0$, hence as per Eqs. (3.43) and (3.44) $u + 2c = \text{const.} = J_+$ along C_+ characteristics, while $u - 2c = J_-$ along C_- characteristics.

These quantities J_+ and J_- are called as Reimann invariants of the fluid motion.

Therefore, the characteristic directions $\left(\frac{dx}{dt} \right)_\pm = C_\pm$ and the Reimann invariants J_+ and J_-

provide a method for solving problems of nearly horizontal flow [18]. This method is called the method of characteristics. The quantities J_+ and J_- form the elementary solution on which pseudoviscosity terms are superimposed as carrection terms. It is, therefore, necessary to know what relationship corresponding to the Reimann invariants is obtained for a channel with bed slope and frictional resistance represented by shear drag τ.

For a gentle bed slope the quasi-invariants may be written as

$$\left[u \pm 2\sqrt{gy} \right]_t^{t+\Delta t} = \pm \int_t^{t+\Delta t} gs\, dt \tag{3.47}$$

For a nearly horizontal bed with bed shear stress τ, the corresponding form may be obtained as follows. When a shear stress τ acts in the negative direction (i.e. opposite to that of positive u), the elementary surge dy not only has to provide an increase in momentum du but also has to cover the loss in momentum due to shear stress. This relationship can be expressed as $(\tau/\rho y)\, dy$, ρ being the density of fluid. Hence, we have

$$du + \left(\frac{\tau}{\rho y} \right) dt = \pm \left(\frac{g}{y} \right)^{1/2} dy \tag{3.48}$$

and by integration across the C_+ and C_- characteristics we have,

$$\left[u \pm 2\sqrt{gy} \right]_t^{t+\Delta t} = \int_t^{t+\Delta t} \left(\frac{\tau}{\rho y} \right) dt \tag{3.49}$$

where

$$-\frac{\tau}{\rho y} = \frac{gu^2}{C^2 y} + \left(\frac{y_2 u_2^2 - y_1 u_1^2}{ly} \right) \quad \text{vide (Fig. 3.27)}$$

The bracketed quantities shown in Eqs. (3.47) and (3.49) are quasi-invariants and their application as correction terms is done in u, $2\,(gy)^{1/2}$ plane, known as the hodograph plane [18].

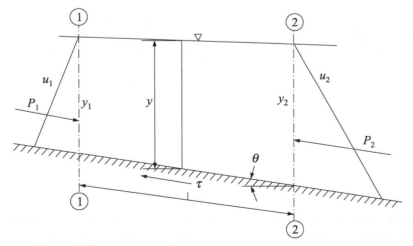

Figure 3.27   Derivation of correction terms due to bed shear stress.

3.4.5.2.1   *Method of finite difference*

Simulation of long flood waves in open channels can be described mathematically by one dimensional St. Venant equations which for a gravity oriented co-ordinate system with x-measured horizontally along the channel and y-measured in the vertical can be expressed as:

$$\frac{\partial A}{\partial t} + \frac{\partial Q}{\partial x} = 0; \tag{3.50}$$

$$\frac{1}{gA}\frac{\partial A}{\partial t} + \frac{1}{gA}\frac{\partial Q}{\partial x}(Q^2/A) + \frac{\partial h}{\partial x} + S_f = 0; \tag{3.51}$$

where

    A = flow cross-sectional area measured normal to x,

    Q = the discharge through A,

    h = the water surface elevation above a reference horizontal datum

    S_f = friction slope

    t = time

    g = acceleration due to gravity.

Friction slope S_f for turbulent flow can be expressed in terms of Manning's formula as

$$S_f = \left(\frac{nQ}{AR^{2/3}}\right)^2 \tag{3.52}$$

where

    n = Manning's roughness factor,

    R = hydraulics radius and

    n = 1.486 for fps unit and n = 1 for *SI* units.

Solutions of the St. Venant equations have to be obtained by numerical techniques using various degrees of simplification. Diffusion wave models approximate the friction slope by the slope of the water surface $\left(\dfrac{\partial h}{\partial x}\right)$ whereas kinematic wave models let the friction slope equal to the channel bottom. For flood routing in open channels dynamic and diffusion wave models can truly account for the downstream flow effects in sub-critical flows. Herein the application of a diffusion wave model to simulate flood propagation for flow in a rectangular channel by finite difference method of numerical solution is discussed.

In the model first two terms of Eq. (3.51) are neglected i.e.

$$S_f = \frac{-\partial h}{\partial x} \tag{3.53}$$

Accordingly, Eq. (3.52) becomes

$$Q = \left(\frac{1}{nAR^{2/3}}\right)\left(-\frac{\partial h}{\partial x}\right)^{1/2} \tag{3.54}$$

The equation accounts for the reversal of flow in the channel as well as a positive discharge towards downstream. The model is based on simultaneous solution of Eqs (3.50) and (3.54) by use of implicit finite difference scheme.

Finite difference equations: The computational grid system in the $(x - t)$ plane in the numerical scheme is shown in Fig. 3.28. The vertical flow lines represent flow sections along a channel numbered in increasing order from upstream to downstream extremity. Channel properties at these sections are assumed to remain the same within the channel segments bounded by vertical dashed lines at equal distance on both sides of each interior flow section. The horizontal lines in the figure represent different time levels which is assumed as constant in the present case.

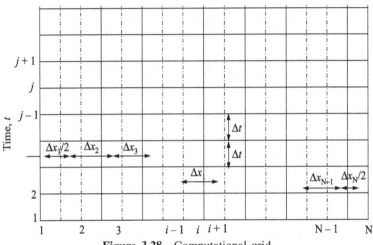

Figure 3.28 Computational grid.

Accordingly Eq. (3.50) is written in finite difference form for the grid between the generic flow sections i and $i + 1$ as

$$\frac{(A_{i+1}^{j+1} + A_i^{J+1} - A_{i+1}^j - A_i^j)}{2\Delta t} + \frac{2(Q_{i+1}^{j+1} - Q_i^{j+1})}{\Delta x_{i+1} + \Delta x_i} = 0, \ i = 1, 2, ... N-1 \qquad (3.55)$$

In which the superscripts j and $j + 1$ indicate the stages. Likewise the finite difference form of Eq. (3.54) is

$$Q_i^{j+1} = \frac{K_n}{n_i} A_i^{j+1} (R_i^{j+1})^{2/3} \frac{-\dfrac{2(h_i^{j+1} - h_{i-1}^{j+1})}{(\Delta x_i + \Delta x_{i-1})}}{\left| \dfrac{2(h_i^{j+1} - h_{i-1}^{j+1})}{\Delta x_i + \Delta x_{i-1}} \right|^{1/2}}, \ i = 1, 2, ... N-1 \qquad (3.56)$$

Or simplifying

$$Q_i^{j+1} = \frac{1.41 \, K_n \, A_i^{j+1} (R_i^{j+1})^{2/3}}{n_i (\Delta x_1 + \Delta x_{i-1})^{1/2}} \frac{(h_{i-1}^{j+1} - h_i^{j+1})}{\left| h_{i-1}^{j+1} - h_i^{j+1} \right|^{1/2}}, \ i = 1, 2, ... N-1 \qquad (3.57)$$

It shows that the flow rate at the upstream estremity of a channel cannot be evaluated from Eq. (3.57). However Q_1^{J+1} is prescribed function of time and or h_1^{j+1} is specified as the upstream boundary condition. Combining Eq. (3.57) and the upstream boundary condition with Eq. (3.56) and writing for the grid between the flow sections $i = 1$ and $i = 2$. One obtains

$$\frac{A_2^{j+1} + A_1^{j+1} - A_2^j - A_1^j}{2\Delta t} + \frac{2^{3/2}}{\Delta x_2 + \Delta x_1} \left\{ \frac{K_n A_2^{j+1} (R_2^{j+1})^{2/3}}{n_2 (\Delta x_2 + \Delta x_1)^{1/2}} \frac{(h_1^{j+1} - h_2^{j+1})}{\left| h_1^{j+1} - h_2^{j+1} \right|^{1/2}} - Q_1^{j+1} \right\} = 0 \quad (3.58)$$

For the remaining $(N-2)$ grids, combination of Eqs. (3.56) and (3.57) yields

$$\frac{A_{i+1}^{j+1} + A_i^{j+1} - A_{i+1}^j - A_i^j}{2\Delta t} + \frac{2^{3/2} K_n}{\Delta x_{i+1} + \Delta x_i} \left\{ \frac{A_{i+1}^{j-1} - (R_{i+1}^{j+1})^{2/3}}{n_{i+1} (\Delta x_{i+1} + \Delta x_i)^{1/2}} \frac{(h_i^{j+1} - h_{i+1}^{j+1})}{\left| h_i^{j+1} - h_{i+1}^{j+1} \right|^{1/2}} \right.$$

$$\left. - \frac{A_i^{j+1} (R_i^{j+1})^{2/3}}{n_i (\Delta x_1 + \Delta x_{i-1})} \frac{(h_{i-1}^{j+1} - h_i^{j+1})}{\left| h_{i-1}^{j+1} - h_i^{j+1} \right|^{1/2}} \right\} = 0 \; i = 2, 3, \ldots N - 1 \quad (3.59)$$

In Eqs. (3.58) and (3.59) the quantities with the superscript j are known either from initial conditions or from the previous one time step computations. The quantities R_i^{j+1} and A_i^{j+1} are prescribed functions of h_i^{j+1} determined from the channel geometry. The upstream discharge Q_i^{j+1} is specified by the upstream boundary condition. Hence the only unknown parameters are h_i^{j+1} for $i = 1, 2, \ldots, N$. Therefore, Eq. (3.59) written for the last $(N-2)$ grids and Eq. (3.58) constitute system of $(N-1)$ non-linear algebraic equations of N unknowns. An additional equation is provided by the downstream boundary condition which may be a stage discharge relationship or a stage hydrograph.

Solution of the finite difference equations: Let G_1 denote Eq. (3.58) G_i denote Eq. (3.59) for $i = 2, 3, 4. \ldots \ldots \ldots \ldots \ldots$, N-1 and G_N. denote the downstream boundary condition. Then the system of N non-linear algebraic equations can be written in the form

$$\left. \begin{aligned} &G_1 \; (h_1^{j+1}, h_2^{j+1}) = 0 \\ &G_2 \; (h_1^{j+1}, h_2^{j+1}, h_3^{j+1}) = 0 \\ &\ldots\ldots\ldots\ldots\ldots\ldots\ldots \\ &\ldots\ldots\ldots\ldots\ldots\ldots\ldots \\ &G_i \; (h_{i-1}^{j+1}, h_i^{j+1}, h_{i+1}^{j+1}) = 0 \\ &\ldots\ldots\ldots\ldots\ldots\ldots\ldots \\ &\ldots\ldots\ldots\ldots\ldots\ldots\ldots \\ &G_N \; (h_{N-1}^{j+1}, h_N^{j+1}) = 0 \end{aligned} \right\} \quad (3.60)$$

The Newton iteration method is adopted to solve this system of non-linear Eq. (3.60). Computation for the iterative procedure begins by assigning a set of trial values to the N-unknowns h_i^{J+1} for $i = 1, 2, \dots N$. Substituting the trial values into Eqs. (3.60) yields the residuals $r(G_i)$. New values of h_i^{J+1} are estimated for the next iteration to make the rey siduals approach zero. This is accomplished by computing the corrections dh_i^{J+1} such that the total differentials of the functions G_i are equal to the negative of the calculated residuals, i.e.

$$\frac{\partial G_i}{\partial h_{i-1}^{j+1}} \Delta h_{i-1}^{j+1} + \frac{\partial G_i}{\partial h_i^{j+1}} \Delta h_i^{j+1} + \frac{\partial G_i}{\partial h_{i+1}^{J+1}} \Delta h_{i+1}^{J+1} = - r(G_i) \tag{3.61}$$

Equation (3.6l) written for $i = 1, 2, 3 \dots N$ forms a set of N linear algebraic equations in N unknowns, dh_i^{J+1}. In matrix notation this linear system of equations

$$\begin{bmatrix} \dfrac{\partial G_1}{\partial h_1^{j+1}} & \dfrac{\partial G_1}{\partial h_2^{j+1}} & 0 & 0 & \cdots & 0 & 0 & 0 & 0 \\[2ex] \dfrac{\partial G_2}{\partial h_1^{j+1}} & \dfrac{\partial G_2}{\partial h_2^{j+1}} & \dfrac{\partial G_2}{\partial h_3^{j+1}} & 0 & \cdots & 0 & 0 & 0 & 0 \\[2ex] 0 & \dfrac{\partial G_3}{\partial h_2^{j+1}} & \dfrac{\partial G_3}{\partial h_3^{j+1}} & \dfrac{\partial G_3}{\partial h_4^{j+1}} & \cdots & 0 & 0 & 0 & 0 \\[2ex] \vdots & \vdots & \vdots & \vdots & \vdots\ \vdots & \vdots & \vdots & \vdots \\[2ex] 0 & 0 & 0 & 0 & \cdots & 0 & \dfrac{\partial G_{N-1}}{\partial h_{N-2}^{j+1}} & \dfrac{\partial G_{N-1}}{\partial h_{N-1}^{j+1}} & \dfrac{\partial G_{N-1}}{\partial h_N^{j+1}} \\[2ex] 0 & 0 & 0 & 0 & \cdots & 0 & 0 & \dfrac{\partial G_N}{\partial h_{N-1}^{j+1}} & \dfrac{\partial G_N}{\partial h_N^{j+1}} \end{bmatrix} \begin{bmatrix} \Delta h_1^{j+1} \\ \Delta h_2^{j+1} \\ \\ \\ \\ \Delta h_{N-1}^{j+1} \\ \Delta h_N^{j+1} \end{bmatrix} \begin{bmatrix} -\gamma(G_1) \\ \\ \\ \\ \\ \\ -\gamma(G_N) \end{bmatrix} \tag{3.62}$$

And this system of N linear Eqs. (3.62) is obtained by the following recurrence formula

$$\Delta h_i^{j+1} = H_i - B_i\ \Delta h_{i+1}^{j+1} \tag{3.63}$$

$$B_i = \frac{\left(\dfrac{\partial G_1}{\partial h_{i+1}^{j+1}} \right)}{\left(\dfrac{\partial G_i}{\partial h_i^{j+1}} - \dfrac{\partial G_1}{\partial h_{i-1}^{j+1}} B_{i-1} \right)} \tag{3.64}$$

$$H_1 = \frac{-\gamma(G_1) - \dfrac{\partial G_i}{\partial h_{i-1}^{j+1}} H_{i-1}}{\dfrac{\partial G_i}{\partial h_i^{j+1}} - \dfrac{\partial G_i}{\partial h_{i-1}^{j+1}} B_{i-1}} \qquad (3.65)$$

The quantities B_i and H_i are first evaluated from Eqs. (3.64) and (3.65) for $i = 1, 2, ..., N$ noting that $\left(\dfrac{\partial G_i}{\partial h_{i-1}^{j+1}}\right) = 0$ for $i = 1$. Then the corrections Δh_i^{j+1} are evaluated from Eq. (3.63) starting with $i = N$ and noting that $\Delta h_{i+1}^{J+1} = 0$ for $i = N$.

This procedure is repeated until the corrections are reduced to within a tolerable magnitude. The number of iterations required for the desired accuracy depends on the closeness of the first trial values to the actual solution. Normally previous time step solution values are chosen to be the first trial values and an error of 0.1 per cent of flow depth at each flow section is tolerated.

A computer software programme developed is placed in Appendix IV. It has been run for a simple case of routing a triangular shaped discharge hydrograph in a channel of rectangular shape having a weir type downstream boundary condition. The input data for the problem has been given at the end of the programme.

Problem 3.1

Tabulated below in Tables 3.2 and 3.3 are the elevation, storage and elevation discharge data for a small reservoir.

Table 3.2

Elevation (m)	Storage (m³) 10^3	Discharge (m³/s)
262.7	0	0
263.6	49.339	0
265.2	246.696	0
266.7	616.740	0
268.2	1233.480	0
268.8	1504.846	2.832
269.4	2010.572	6.513
270.1	2800.000	11.157
270.7	3885.462	16.990

If the inflow to the reservoir during a flood as given below, determine the maximum pool elevation and peak outflow rate, assuming the pool elevation at 266.7 m at 24.00 hrs on the 1st day. Also draw the outflow hydrograph.

Table 3.3

Date	Hours	Inflow (m³/s)
1	24.00	1.13
	12.00	0.99
2	24.00	1.05
	12.00	3.54
3	24.00	9.63
	12.00	16.28
4	24.00	20.44
	12.00	20.95
5	24.00	19.05
	12.00	12.91
6	24.00	9.06
	12.00	6.94
7	24.00	5.44
	12.00	4.08
8	24.00	3.34
	12.00	2.69
9	24.00	2.26
	12.00	1.90
10	24.00	1.59
	12.00	1.42
11	24.00	1.19

Solution

The solution of this problem is given in different methods as shown below.

Puls method or I.S.D. method—*i.e.* **Inflow discharge method:** The solution of the problem is made as per the following steps:

1. From the given data prepare the elevation vs. outflow curve (Fig. 3.29), and then Table 3.1 is prepared giving the storage S, rate of outflow O, $S + \dfrac{1}{2}Ot$ and $S - \dfrac{1}{2}Ot$ corresponding to different reservoir elevations for the time interval $t = 12$ hr.

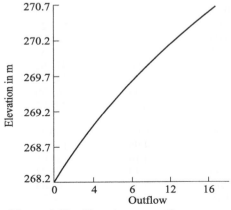

Figure 3.29 Elevation vs. outflow curve.

2. From the Table 3.4, Fig. 3.30 is prepared.

Table 3.4 Computations for Drawings $(S + \frac{1}{2}Ot)$ and $(S - \frac{1}{2}Ot)$ Curves. Choose Time Interval $t = 12$ hrs.

Reservoir elevation, (m)	Storage, $(10^3 \times m^3)$	Rate of outflow, O $(m^3 \times s)$	$\frac{1}{2}Ot$ $(m^3 \times 10^3)$	$(S + \frac{1}{2}Ot)$ $(m^3 \times 10^3)$	$(S - \frac{1}{2}Ot)$ $(m^3 \times 10^3)$
1	2	3	4	5	6
262.7	0	0	0	0	0
263.6	49.339	0	0	49.339	49.339
265.2	246.696	0	0	246.696	246.696
266.7	616.740	0	0	616.740	616.740
268.2	1233.480	0	0	1233.480	1233.480
268.8	1504.846	2.832	61.13	1565.973	1443.72
269.4	2010.572	6.513	141.05	2151.62	1869.52
270.1	2800.00	11.157	240.99	3040.99	2559.01
270.7	3885.462	16.99	366.98	4252.44	3518.48

3. From the given data, the inflow hydrograph is drawn (Fig. 3.31).
4. From the given inflow data, columns 1 to 5 are prepared for Table 3.5.

Table 3.5 Flood Routing by Puls Method or I.S.D. Method Routing Period $t = 12$ hr.

Date	Time (hr.)	Inflow I_1 (m^3/s)	$(I_1 + I_2)$ (m^3/s)	$\left(\frac{I_1 + I_2}{2}\right)$ $(m^3 \times 10^3)$	Reservoir elevation (m)	Outflow, O_1 (m^3/s)
1	2	3	4	5	6	7
1	24	1.13	2.12	45.89	266.70	0
2	36	0.99	2.04	44.04	266.79	0
2	48	1.05	4.59	99.10	266.91	0
3	60	3.54	13.17	284.47	267.19	0
3	72	9.63				
4	84	16.28	25.91	559.62	267.89	0
4	96	20.44	36.72	793.28	268.93	3.40
5	108	20.95	41.39	894.20	269.53	7.08
5	120	19.05	40.00	864.24	269.95	10.19
6	132	12.91	31.96	690.53	270.21	12.45
6	144	9.06	21.96	474.62	270.31	13.30
7	156	6.94	16.00	345.58	270.28	13.02
7	168	5.44	12.38	267.30	270.17	12.17
8	180	4.08	10.52	205.52	270.02	10.90
8	192	3.34	7.42	160.25	269.87	9.48
9	204	2.69	6.03	130.27	269.69	8.07
9	216	2.26	4.95	107.03	269.50	7.08
10	228	1.90	4.16	89.91	269.34	5.80
10	240	1.59	3.49	75.23	269.15	4.67
11	252	1.42	3.01	64.84	269.00	3.82
11	264	1.19	2.61	56.27	268.89	3.26
					268.77	2.55

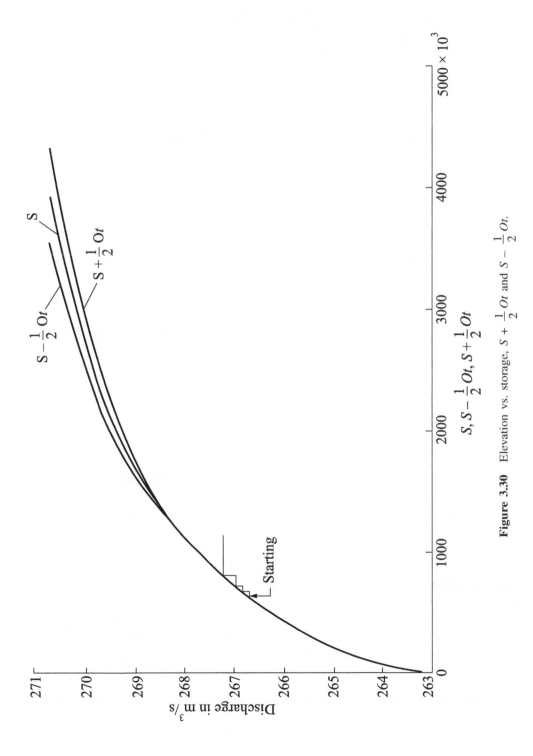

Figure 3.30 Elevation vs. storage, $S + \frac{1}{2}Ot$ and $S - \frac{1}{2}Ot$.

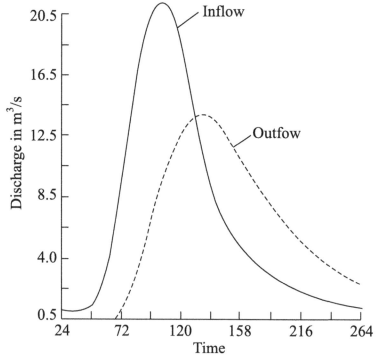

Figure 3.31 Inflow and routed outflow hydrograph.

5. From the given data enter the first figure in column 6 which is reservoir elevation when the flood impinges at 24 hrs, i.e. *RL* 266.70.

6. The outflow starts only after *RL* 268.8 m has reached, till then the inflow $\left(\dfrac{I_1 + I_2}{2}\right)$ *t* goes on adding to the storage and the reservoir elevation goes on rising.

7. Starting with a reservoir storage, S_1 at the beginning of a time interval *t*, the reservoir storage S_2 at the end of the time interval can be found as explained in the theory.

8. Knowing the storage one can find out reservoir elevation and corresponding outflow with the help of Figs. 3.29 and 3.30.

9. The routed hydrograph is then plotted in Fig. 3.31.

Cheng's graphical method: This is completely a graphical method. The steps involved in its solution are as follows:

1. From the given data, fill in columns 1 to 3 of Table 3.6. Choose a suitable time interval *t* = 12 hrs and then calculate the figures in columns 4 and 5. The outflow starts only after the reservoir level reaches *RL* 267.89 till then inflow only builds up the storage and at 72 hrs *RL* becomes 267.89 and the storage at that *RL* is 1080 m³.

2. From Table 3.6, draw $S + \dfrac{1}{2} Ot$ vs. outflow curve in Fig. 3.32. Also draw the line of *O t* vs. *O*, which has to pass through the origin as *O t* is zero when *O* is zero.

Table 3.6 Flood Routing by Cheng's Method: Routing Period $t = 12$ hr.

Routing Equation $\left(\dfrac{I_1 + I_2}{2}\right)t + \left(S_1 - \dfrac{O_1 t}{2}\right) = \left(S_2 + \dfrac{O_2 t}{2}\right)$

Reservoir elevation, (m)	Storage (10^3 m^3)	Outflow (m^3/s)	$\frac{1}{2}Ot$ (10^3 m^3)	$S + \frac{1}{2}Ot$ (10^3 m^3)
1	2	3	4	5
262.7	0	0	0	0
263.6	49.34	0	0	49.339
265.2	246.696	0	0	246.696
266.7	616.740	0	0	616.740
267.89	1080.00	0	0	1080.00
268.2	1233.480	0	0	1233.480
268.8	1504.846	2.832	61.13	1565.973
269.4	2010.572	6.513	141.05	2151.62
270.1	2800.00	11.157	240.99	3040.99
270.7	3885.462	16.99	366.98	4252.44

Note: The outflow starts only after the reservoir level reaches RL 268.2 till then the inflow only builds up the storage. At 72 hrs the RL becomes 267.89 as already calculated in Table 3.6. The storage at RL 267.89 equals $1080.0 \times 10^3 \text{ m}^3$.

3. From the given data draw the inflow hydrograph (Fig. 3.32).

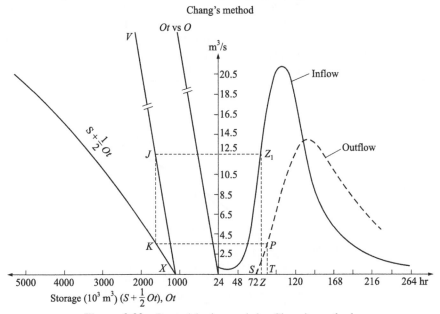

Figure 3.32 Routed hydrograph by Cheng's method.

4. In the problem starting point 'S' is the time 72 hrs when outflow is O. From S draw SX to meet the $S + \dfrac{1}{2}Ot$ curve at X. Draw XY parallel to $O\,t$ line. Z_1 is the midpoint of the time interval $\overline{ST}$. Draw ZZ_1 to meet the inflow hydrograph at Z_1. Draw Z_1J to

meet *XY* at *J*. Draw *JK* to meet $S + \frac{1}{2}O\,t$ curve at *K*. Draw *KP* to meet *TP* at *P* which is a point on the outflow hydrograph. Repeat the steps till the outflow hydograph is obtained.

Working value method: The solution of the problem is made as per the following steps based on the computational procedure with working values of *N*.

(i) With the given data prepare Table 3.7 of working values of *N*.

(ii) Using Table 3.7, draw Fig. 3.33 giving *N* vs. *O* curve which is called the working value curve.

Table 3.7 Working Values of *N*.

RL (m)	Storage ($10^3 m^3$)	S/t (m^3/s)	O (m^3/s)	O/2 (m^3/s)	Working value N or (S/t + O/2) (m^3/s)
262.7	0	0	0	0	0
263.6	49.34	1.132	0	0	1.132
265.2	246.696	5.717	0	0	5.717
266.7	616.740	14.263	0	0	14.263
267.89	1080.0	25.102	0	0	25.102
268.2	1233.480	28.026	0	0	28.526
268.8	1504.846	34.089	2.832	1.416	36.22
269.4	2010.572	46.525	6.513	3.257	49.782
270.1	2800.00	64.779	11.157	5.579	70.358
270.7	3885.462	89.881	16.991	8.495	98.376

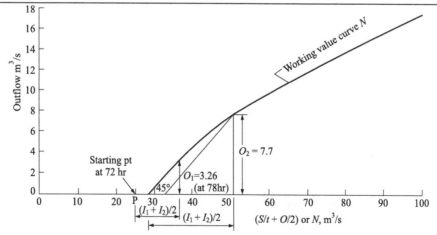

Figure 3.33 Routed flood hydrograph by working value method.

(iii) Starting from a known point *P*, i.e. at 72 hr, the value of N_1 is 25.1 m^3/s and *O* is zero. This value, subsequent values and construction lines are marked in Fig. 3.33 for easy understanding.

(iv) The values of outflows are shown in Table 3.8.

Table 3.8 Reservoir Flood Routing by Working Value Method $t = 12$ hr

Time (hr)	Instantaneous inflow (m³/s)	$(I_1 + I_2)^2$ (m³/s)	Outflow (m³/s)	Remarks
72	9.62	12.96	3.265	At 72 hrs $N_1 = 25.10$
84	16.27	18.37	7.22	and $O_1 = O$ as given
96	20.43	20.69	10.33	
108	20.94	20.01	12.58	
120	19.05	15.99	13.16	
132	12.91	10.98	12.74	
144	9.06	8.01	11.60	
156	6.93	6.20	10.47	
168	5.43	4.75	9.20	
180	4.08	3.71	8.07	
192	3.34	3.03	6.79	
204	2.69	2.49	5.66	
216	2.26	2.09	4.81	
228	1.90	1.76	4.10	
240	1.59	1.50	3.40	
252	1.42	1.30	2.83	
264	1.19			

Problem 3.2

The inflow for a river reach and the outflow for the first time interval are given in Table 3.9. The values of the routing constants after Muskingum for the reach have been found to be $K = 23.6$ hrs. and $x = 0.20$. Obtain the outflows for the successive time intervals.

Table 3.9

Time (hr)	Inflow (m³/s)	Outflow (m³/s)
6	28.3	
12	113.2	
18	270.0	
24	388.0	
30	352.0	
36	288.0	
42	232.0	
48	165.0	
54	140.0	
60	119.0	
66	104.0	
72	91.5	
78	78.8	
84	68.5	
90	61.0	
96	53.4	

Solution

Muskingum method: Considering routing period t as 6 hr, $K = 23.6$ hr and $x = 0.20$.

$$C_0 = -\frac{Kx - 0.5t}{K - Kx + 0.5t}$$

$$= -\frac{23.6 \times 0.20 - 0.5 \times 6}{23.6 - 23.6 \times 0.2 + 0.5 \times 6} = -\frac{1.72}{21.88}$$

$$= -0.079$$

$$C_1 = \frac{Kx + 0.5t}{K - Kx + 0.5t} = \frac{4.72 + 3.00}{21.88} = \frac{7.72}{21.88} = 0.313$$

$$C_2 = \frac{K - Kx - 0.5t}{K - Kx + 0.5t} = \frac{23.6 - 4.72 - 3}{21.88} = \frac{15.88}{21.88} = 0.726$$

Check $C_0 + C_1 + C_2 = -0.079 + .353 + .726 = 1.00$

With the above values known,

$$O_2 = C_0 I_2 + C_1 I_1 + C_2 O.$$

Accordingly, the successive values of O_2 are calculated and is shown in Table 3.10 and is self-explanatory.

Table 3.10   Flood Routing for a River Reach by Muskingum Method with Routing Interval $t = 6$ hr

Time (hr)	Inflow (m³/s)	C_0I_2 (m³/s)	C_1I_1 (m³/s)	C_2O_1 (m³/s)	O (m³/s)
1	2	3	4	5	(3) + (4) + (5)
6	28.3	—	—	—	28.3
12	113.2	− 8.94	9.99	20.55	21.6
18	270.0	− 21.33	39.96	15.68	34.31
24	388.0	− 30.65	95.31	24.90	89.56
30	352.0	− 27.81	136.93	65.02	174.17
36	288.0	− 22.75	124.26	125.45	227.96
42	232.0	− 18.33	101.66	165.50	248.85
48	165.0	− 13.04	81.90	180.67	249.53
54	140.0	− 11.06	58.25	181.16	221.18
60	119.0	− 9.40	49.42	160.57	194.36
66	104.0	− 8.22	42.01	141.11	170.59
72	91.5	− 7.23	36.71	123.85	149.92
78	78.8	− 6.23	32.30	180.84	131.25
84	68.5	− 5.41	27.82	95.29	114.66
90	61.0	− 4.82	24.19	83.24	100.55
96	53.4	− 4.22	21.53		

CASE STUDY 1

Application of the Method of Characteristics

Based on the graphical approach the routing of a flood wave in a river reach with no downstream control has been carried out by the author [19], under the following simplifying assumptions:

  (a) The river is uniformly wide and the flow is one-dimensional.

  (b) The bed slope is constant at 1 in (30×10^3).

  (c) The bed roughness height is constant and equal to 0.025 m.

  (d) The input hydrograph has a rise of 6 m in 24 hours and is over 6 m initial depth of water with a total base period of 46 hours (Fig. 3.34).

  (e) For calculation of the resistance coefficient, the Chezy resistance formula $C = \dfrac{u}{(yS_f)^{1/2}}$

     and the relation $C = 18 \log(12 \, y/d)$ are employed, d being equal to the height of the bed roughness.

  (f) Pseudoviscosity terms such as bed slope and shear stress have been incorporated. The Chezy coefficient has been taken as an index of the resistance variation. The choice is purely a matter of convenience and any other resistance law could have been used.

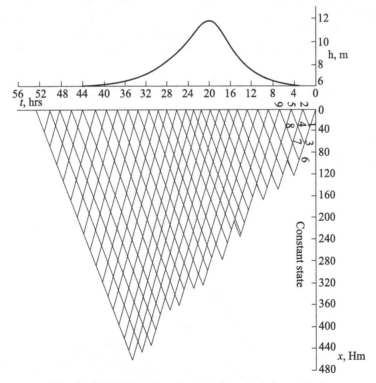

Figure 3.34 Flood routing by characteristic method.

The starting condition of the problem has been taken as the point at which normal upland flow prevails throughout the reach of the river. On this starting condition a flood hydrograph in terms of stage is superimposed at location $x = 0$ (Fig. 3.34). The problem is to route the flood wave or hydrograph along the reach. To solve the problem in the manner outlined in the theory, two plots are necessary: one showing the characteristic grid network, which is known as the physical plane or (x, t) plane; and the other incorporating the correction due to the pseudoviscosity terms, which is known as the hydograph plane or u, $2(gh)^{1/2}$ plane.

As illustration at the grid point O (Fig. 3.34) we have $u = 3.16$ km/hr and $h = 6$ m. Therefore, the value of C_+ and C_- characteristics can be computed as equal to 30.77 km/hr and –24.45 km/hr, respectively. From O, draw line with slope 30.77, which gives the direction along with which the disturbance propagates in the positive x-direction. Select point 1 lying on this characteristic. From point 1, proceed to point 2 after incorporating the correction terms due to slope and friction in the hydograph plane. The value of u, h at the grid point 2 being known computation is carried on for the successive grid points. The grid size is varied depending on the steepness of the input hydrograph. The solution is carried out to 286 grid points and covers the whole duration of the input hydrograph and the region of interest. However, to investigate the duration after which the original normal state will return, further grid points need to be worked out. With regard to the propagation of the disturbance, Fig. 3.34 shows that the disturbance extends to 460 km from the starting station at $x = 0$. The zone of no disturbance is indicated by the constant state region.

Starting from the physical plane solution, one can prepare Figs. 3.35 and 3.36. These show the routed profiles in the hx, ux ht, and ut planes, respectively. The figures show that the peak rapidly decreases within 180 km from the start, the reduction at locations 60, 120 and 180 km being 30.8, 50.1 and 65.0 per cent in stage and 26.4, 52.8 and 66.9 per cent in velocity.

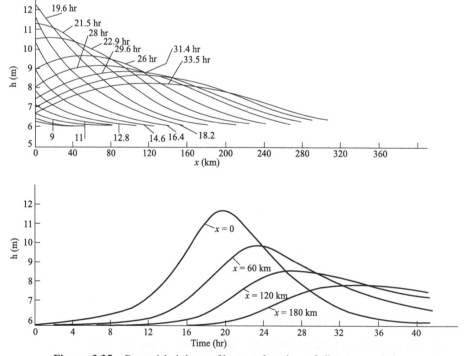

Figure 3.35 Routed height profile as a function of distance and time.

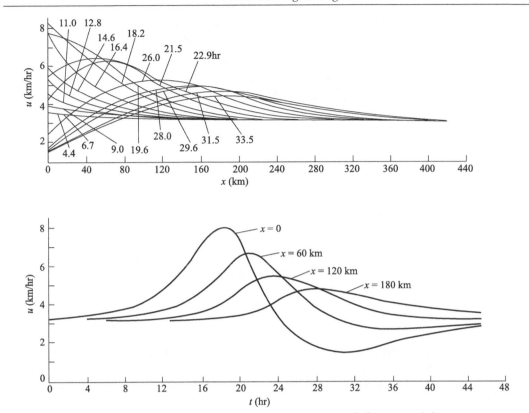

Figure 3.36 Routed velocity profile as a function of distance and time.

People concerned with river gauging know that the discharge is greater for a given stage when the stream is rising then when the stream is falling. The difference depends on the channel geometry downstream and the rate of rise and fall. In this case, the stage discharge relationships at locations 0, 60, 120 and 180 km are shown in Fig. 3.37. The discharge curves show the existence of the usual loop patterns, which progressively decrease, as the degree of steepness of the hydrograph decreases.

The loop may be plotted from stage and discharge measurements of a flood. This curve may be used as an approximation for other floods of about the same magnitudes and duration. For floods with multiple peaks the loop may be complicated. A better procedure is to relate for the same stage the normal discharge Q_N under steady uniform flow to the measured unsteady discharge Q_M by

$$\frac{Q_M}{Q_N} = \left[1 + \frac{1}{V_c S_0}\frac{dh}{dt}\right] \qquad (3.66)$$

where V_c = velocity of flood wave, S_0 = channel slope = water surface flow for uniform flow, dh/dt = rate of change of stage, $V_c = 1.4\,V$ where V = average velocity as per Manning's equation for the given stage and S_0 is to be replaced by S_f. From measurements dh/dt and $1/\,V_c S_0$ are computed. The term $1/V_c S_0$ is plotted against stage. The actual discharge for an observed stage is first determined by eliminating dh/dt for that stage and reading $1/\,V_c S_0$, from the plot.

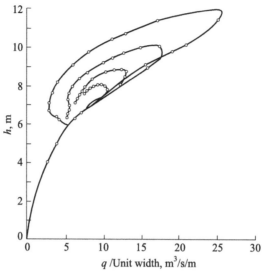

Figure 3.37 Stage discharge relationship.

COMMENTS

In real engineering problems flood propagation takes place over large basins having interlaced river system and wide plains intersected by natural and artificial barriers. For solution of such problems, a suitable mathematical model has to be developed followed by appropriate computer programming.

Problem 3.3

The steady state discharge in a wide rectangular river having a bed slope equal to 0.15×10^{-4} is given as 1500 m³/s. The average velocity of flow has been estimated to be 1.5 m/s. Compute the discharge at the same stage due to a flood wave in its rising stage when there is a change of 15 cm water level in 3 hr.

Solution
The unsteady discharge

Now

$$Q = Q_0 \left(1 + \frac{1}{S_0 c} \frac{\partial h}{\partial t} \right)^{\frac{1}{2}}$$

$$S_0 = 0.15 \times 10^{-4}$$

$$c = \text{celerity of the wave} = 1.5 \ V \text{ m/s}$$

where

V = average velocity assuming the channel shape to be wide rectangular

$$\frac{\partial h}{\partial t} = \left[\frac{0.15}{3 \times 60 \times 60} \right]$$

$$Q = 1500 \left[1 + \frac{0.15}{0.15 \times 10^{4} \times 1.5 \times 1.5 \times 3 \times 3600} \right]^{\frac{1}{2}}$$

$$= 1000 \left[1 + \frac{1}{2.25 \times 1.08} \right]^{\frac{1}{2}}$$

$$= 1782 \text{ m}^3/\text{s}$$

APPENDIX I

Rating Curve Characteristics

In the case of uniform flow a direct relationship exists between the discharge and the water depth at a given cross-section. This is called the stage-discharge relationship or rating curve (Fig. 3.38). The equation can be expressed in terms of Chezy's law and in the special case of a wide rectangular cross-section it can be written as:

$$Q = bC\sqrt{S}h^{3/2}$$

where

b = width,
h = depth of flow and
S = bed slope.

The above equation indicates that if C is assumed to be constant, the discharge varies as 3/2th power of depth. However, because of several reasons the above simple relationship is not followed and in practice this occurs simultaneously in a complex manner. As such, the rating curve should be determined by measurement of discharge and corresponding stage. For making a rough estimate and to provide guidelines the following cases may prove useful.

(i) The Chezy coefficient generally will not be independent of water depth.

(ii) The conveying cross-section usually depends on depth as for example consider the case of a triangular cross-section (Fig. 3.39). For this the cross-sectional area and hydraulic radius are $h^2 \tan^2 \theta$ and $\frac{1}{2} h \tan \theta \sin \theta$ respectively.

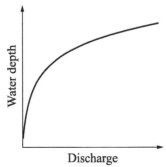

Figure 3.38 Rating curve for uniform flow.

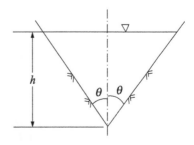

Figure 3.39 Triangular cross-section.

Accordingly,
$$Q = \tan^{5/2} \theta \, C \sqrt{\left(\frac{1}{2} \sin \theta \, S \right) . h^{5/2}} \tag{3.67A}$$

Again assuming C to be constant, the discharge varies as, 5/2th power of h instead 3/2th power. Consider the second example for a river with flood plains. When the flood plains carry flow over the full cross-section the discharge consists of two parts:

$$Q_r = h \, b_r \, C \sqrt{h S} \tag{3.68A}$$

and

$$Q_f = (h - h_r) \, (b - b_r) \, C \left[\sqrt{(h - h_r) \, S} \, \right] \tag{3.69A}$$

Assuming that the flood plains have the same slope and the same bottom roughness as the river, the total discharge becomes

$$Q = b_r \, C \sqrt{S} \left[h^{3/2} + (b/b_r - 1) \, (h - h_r)^{3/2} \right] \tag{3.70A}$$

which is shown in Fig. 3.40 in graphical form and it shows dramatic changes in the rating curve as soon as the flood plains are flooded. (iii) The third reason for deviation from 3/2th power relationship is nonuniformity of the flow. From the examination of backwater curve it can be noticed that the local depth does not depend on the discharge alone but also on the downstream situation. Accordingly, the depth can be either larger or smaller than the normal depth based on which the discharge relationship is worked out. (iv) The fourth reason for deviation is due to unsteadiness of flow and its effect has been considered in the text.

Formulation of Rating Curve or Stage–discharge Relationship [2]

Principles of measurement
The two variates to be measured are the stage and discharge in the channel. Stage is determined by the height of the water level above a fixed datum line measured by a reference gauge. The reference gauge shall be located as closely as possible to the measuring station in the case of measurement using a current meter or near the mid point of measuring reach in case any

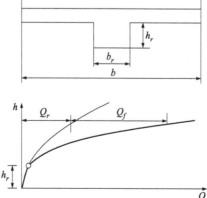

Figure 3.40   Cross-section and rating curve for river with flood plain.

other method for measurement is adopted. The reference gauge shall be a vertical staff gauge or an inclined gauge. The markings on the gauge shall be clear and accurate to within 2 mm and the lowest marking should be at least 0.3 m above the highest anticipated flood level. The gauge shall be securely fixed to an immovable grid support and the gauge datum has to be related by precise leveling to the G.T.S. datum. It shall have stalling arrangements whenever necessary so that the water level can be read accurately. Indian Bureau of Standard (IS 2914-1964) have given forms for recording gauge statements of river during normal period as well as during flood season and these are placed in the Appendix V.

Discharge is expressed as the volume of water per unit of time, passing a particular cross section normal to the direction of flow. It is given by the product of the area of that cross section upto the water surface and the mean velocity of flow in that section. The area of the cross section is computed by finite integration method from measurement of the width at the water surface along the section. The velocity is measured by a float, current meter or any other suitable method. IBS has furnished appropriate forms for computation of discharge by velocity area method and these are placed in Appendix V.

The estimation of discharge by direct measurement is laborious and time consuming. It, therefore, cannot be adopted for providing daily information of the flow in the river. Hence, it is the usual practice to establish a relationship commonly known as rating curve between the stage at a particular gauging station and the discharge. Thereafter, from the observation of the stage, the flowrate can be estimated.

The relationship between the stage and discharge depends on the cross-sectional shape and nature of the bed roughness. Accordingly, such a relationship is established empirically. Afterwards, conti-nued measurements have to be taken in order to verify and make necessary modifications to suit the actual conditions.

Normal shape of the gauge discharge curve drawn on an ordinary graph paper is parabolic, the slope of the curve decreasing with increasing discharge. On a log-log paper the curve will follow a straight line. Any abnormality in the shape of the curve is an indication of the variation in the characteristics of the river. A sudden flattening of the parabolic curve at bankful stage is caused on account of overbank spills which drastically reduce the rate of rise of water level with increased discharge. Apart from that, during the passage of a flood wave, the rate of rise of a gauge on a rising flood is generally higher than the rate of fall on a falling flood. This results in the release of a bigger discharge at a particular gauge on the rising stage than on the falling stage. Thus the rising stage lies some what to the right of falling stages producing what is known as loop curve which in turn results in a single curve at a medium or low stage. The above loop in the rating curve at the peak is irrespective of the stability of the channel characteristics and variation in suspended concentration. It derives its origin from the characters of flood wave. Instability of the river bed also results in similar phenomenon.

In alluvial rivers there usually occurs scour and deposition. In some rivers such as the Ganges the beds scour during the flood season and deposition occurs to almost to the same amount at the end of the flood season. However, in rivers like Brahmaputra the scour and deposition do not compensate and there occurs rise in the river bed. Normally a rating curve equation will be of the form

$$Q = a \, (Z - Z_0)^b \tag{3.71A}$$

where Q and Z are discharge and corresponding stage, Z_0, stage at zero discharge, a and b are parameters of the particular station to be determined. The parameters can be determined by a computer method using a non-linear least squares method. Alternately, it can be ascertained first by the determination of a and b by logarithmic transformation. The evaluation of Z_0 can be found out by successive trials. With an assumed value of Z_0 the curve on the log-log paper may become concave upwards or concave downwards. In the former it indicates the assumed value to be lower and in the latter it is higher than the correct values. A few trials will give the value of Z_0 which will make the curve a straight line. Another approach for determination of Z_0 is as follows. Select three values of discharge Q_1, Q_2, Q_3 from the smooth curve drawn by visual method, such that $Q_2^2 = Q_1$, Q_3 the corresponding values of stages being Z_1, Z_2, Z_3. Then following Eq. (3.71A), Z_0 can be expressed as

$$Z_0 = \frac{Z_1 Z_3 - Z_2^2}{Z_1 + Z_3 - 2Z_2} \tag{3.72A}$$

After determination of Z_0, the values of Q plotted against $(Z - Z_0)$ on a log-log scale, will be approximately on a straight line between shifts in control. This will enable us to locate the possible discontinuities in the portions of the curve for separate treatment. The station parameters a and b are found out from the logarithmic transformation of Eq. (3.71A), which means,

$$\log Q = \log a + b \log (Z - Z_0)$$

or
$$y = bx + C$$

where $y = \log Q$, $x = \log (Z - Z_0)$ and $C = \log a$.

From above, the parameters a and b are found out for each continuous portion of the curve, following method of least squares. It is thus possible to obtain the mathematical equation of the rating curve. Apart from above, every rating curve is accompanied by an statistical analysis giving the confidence limit, for which the standard deviation $\sigma_{\bar{Q}}$ of the rating curve is found out from the following relationships

$$\sigma_{\bar{Q}} = \frac{\sigma_{\bar{Q}}}{n} \tag{3.73A}$$

$$\sigma_{\bar{Q}} = \sqrt{\frac{1}{n-1 \sum\limits_{1}^{n} (Q_m - Q_r)^2}} \tag{3.74A}$$

where
$\quad Q_m$ = measured discharge,
$\quad Q_r$ = discharges read from the rating curve for the corresponding stage,
$\quad n$ = number of observations within the interval.

A pair of curves can then be drawn passing through points at a distance of twice the standard deviation $\sigma_{\bar{Q}}$ of the mean from the rating curve and these are called the 95 per cent confidence limits of the rating curve.

Rating curves for each year or for each flood season in a year could be obtained by fitting a smooth curve through the plotted gauge-discharge data if the scatter is small. If there is considerable scatter it may be examined if there is a loop function with the rising and falling flood. In such a case two curves may be used. In case scatter is indicated without a loop formation it will be due to considerable variation from flood to flood. These are observed in alluvial channels where bed level variation occurs very frequently with heavy floods changing stage discharge relationships. In such cases separate stage discharge curves for individual floods should be plotted to provide more weightage to the observed values.

COMMENTS

The steps to be followed in the preparation of a rating curve may be summarised briefly as follows:

1. Selection of a suitable gauge site. For this purpose it is necessary to have stability of the channel and this is ensured by suitable channel control. Controls can be a particular reach of a river, if the bed and banks are rigid, and stable and variation in cross-sectional area due to bed scour and deposition is absent. Alternatively, artificial control is formed by constructing low sills and notches and they provide effective control at low stages.
2. Equipment, in the form of inclined gauges or vertical shaft gauges is usually provided for recording stage. In case where rapid fluctuations in streams occur an automatic recording gauge may be installed. Further, when the gauge site is expected to be affected for variable discharge as a result of say backwater effect from downstream control, it will be necessary to install two gauges to determine the slope of the water surface in the reach.
3. Discharge measurement site should preferably be located at the gauge site. Measurement of discharge should be undertaken using area velocity method with a current meter both for the initial preparation of the rating curve and its subsequent checking.
4. In case the gauge site is affected by backwater, or due to passage of floodwater, necessary corrections should be applied to the rating curve (see worked-out problem).

Problem 3.4

The gauging site at a particular station is affected by backwater. The measured discharge at the main gauge and the corresponding stages recorded by the main and auxiliary gauges are given in Table 3.11.

Table 3.11

Measured discharge (m³/s)	Main gauge reading (m)	Auxiliary gauge reading (m)
2535	3.51	3.24
2160	2.92	2.62
6534	4.83	4.53
729	1.80	1.50
14931	7.49	7.29
11259	6.31	5.97
15120	6.53	6.06
11367	7.05	6.83
6507	4.34	3.91
3024	3.23	2.86
9855	5.49	5.17
18954	8.17	7.82
15390	7.94	7.67
8035	6.48	6.29
3483	4.12	3.91

Develop a suitable relationship for applying necessary corrections to the rating curve at the main gauging station. Hence, find the discharge corresponding to a stage of 8 m at the main gauge and difference of 0.434 m between the main and the auxiliary gauge.

Solution

Theoretical background: If Q_1 and Q_2 are two different discharges passed by the channel at the same gauge but at different slopes S_1 and S_2 then

$$\frac{Q_1}{Q_2} = \left(\frac{S_1}{S_2}\right)^2 \tag{3.75A}$$

Considering the energy slope S to be a function of fall, i.e. the difference between the water levels at the two gauging stations in the reach, i.e. $S = f_{nc}(F)$ then

$$\frac{Q_1}{Q_2} = f_{nc}\left(\frac{F_1}{F_2}\right) \tag{3.76A}$$

Examination of field data indicates that the above relationship can be expressed in the form

$$\frac{Q_1}{Q_2} = \left(\frac{F_1}{F_2}\right)^n \tag{3.77A}$$

where n is exponent.

The procedure usually adopted in evaluating the relationship is as follows. Draw the rating curve corresponding to a constant fall of 0.3 m. Estimate the discharge corresponding to 0.3 m fall. Then find the ratio of mesured discharge to that computed as well as the ratio of measured fall to the constant fall. Plot the ratios in a log-log graph paper and find out n.

Following the above procedure the example has been worked out and is shown in graphical and tabular form below (Figs. 3.41 and 3.42, Table 3.12)

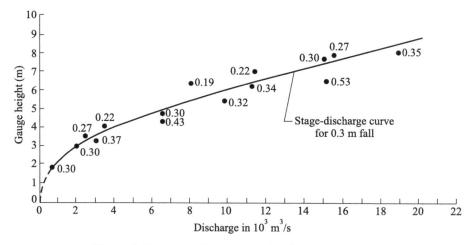

Figure 3.41   Stage discharge relationship for 0.3 m fall.

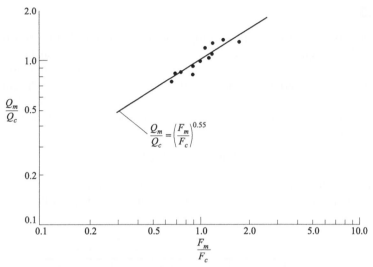

Figure 3.42 Formulation of discharge relationship.

Finally, the equation:

$$\left[\frac{Q_m}{Q_c}\right] = \left[\frac{F_m}{F_c}\right]^{0.55}$$

Accordingly, for 8 m stage and 0.3 m fall, the $Q_c = 16800$ m³/s as read from the curve:

Therefore,
$$Q_m = Q_c \left[\frac{0.434}{0.3}\right]^{0.55} = 16800 \times 1.2252$$

$$= 20583.2 \text{ m}^3/\text{s}.$$

Table 3.12 Computation for Stage-discharge Relationship

No. of observa-tions	Primary gauge reading (m)	Discharge for 0.3 m fall, i.e., Q_c (m³/s)	Dischage that has been obtained Q_m(m³/s)	$\left[\dfrac{Q_m}{Q_c}\right]$	Auxiliary gauge reading (m)	$\left[\dfrac{F_m}{F_c}\right]$	Adjusted Q_m (m³/s)
1	3.51	3000	2535	0.845	3.24	0.90	2831
2	2.92	2160	2160	1.00	2.62	1.00	2160
3	4.83	6534	6534	1.00	4.53	1.00	6534
4	1.80	729	729	1.00	1.50	1.00	729
5	7.49	14931	14931	1.00	7.19	1.00	14931
6	6.31	10700	11259	1.05	5.97	1.13	11444
7	6.53	11500	15120	1.31	6.06	1.77	15472
8	7.05	13200	11367	0.86	6.83	0.73	11102
9	4.34	4800	6507	1.36	3.91	1.43	5843
10	3.23	2300	3024	1.31	2.86	1.23	2577
11	5.49	8000	9855	1.23	5.17	1.07	850
12	8.17	17500	18954	1.08	7.82	1.17	19078
13	7.94	16600	15390	0.93	7.67	0.90	15465
14	6.48	11200	8035	0.72	6.29	0.63	8686
15	4.12	4200	3483	0.83	3.91	0.70	3457

Problem 3.5

The storm runoff from a business park development is to be controlled by constructing a storm water storage culvert adjacent to a river vide Fig 3.43. The culvert shape is rectangular in cross-section 1.5 m wide and 2.0 m height and is 100 m long. The design storm inflow 'I' to the culvert is assumed to start when the culvert is empty. The inflow hydrograph is assumed to be triangular in shape having base length 3 hours and peak inflow of 0.10 m³/s occurring 1 hour after the start of the storm. The discharge from the culvert is though an orifice, with head discharge characteristics

$$Q(\text{m}^3/\text{s}) = 0.06\sqrt{H} \ .$$

where H, is the differential head across the orifice. Find the peak outflow from the culvert to the river as well as its time of occurrence relative to the start of the inflow.

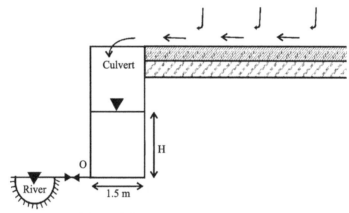

Figure 3.43 Storm water storage culvert.

Solution

Let the time interval of routing the hydrograph be 600 sec or 10 mins. Using working value method of flood routing. One can wirte

$$N = \left(\frac{S}{\Delta t} + \frac{O}{2} \right)$$

here

$$S = 100 \times H \times 1.5 \text{ and } O = 0.06\, H^{1/2}$$

so

$$N = \left(\frac{150H}{600} + \frac{O}{2} \right)$$

or, by eliminating H, in terms of 'O'
we have

$$H = 400(O^2)$$

or

$$N = \left(\frac{120 \times 400}{600} O^2 + \frac{O}{2} \right)$$

$$N = 80 \times O^2 + \frac{O}{2}$$

or
$$2N = 160\,O^2 + O$$

or
$$O = \frac{-1 + \sqrt{1 + 1280N}}{320}$$

The computational sequence for estimating the outflow O at any given time following the commencement of storm in flow I, to the storage culvert is shown in Table 3.13.

Table 3.13 Routing of storm runoff through storage culvert.

T (minutes)	I (m³/s) *minus*	I_m(m³/s) *plus*	O (m³/s)	N
0	0.0000	0.0083	0.0000	0.0000
10	0.0167	0.0250	0.0075	0.0083
20	0.0333	0.0417	0.0151	0.0258
30	0.0500	0.0583	0.0226	0.0524
40	0.0667	0.0750	0.0302	0.0880
50	0.0833	0.0917	0.0377	0.1328
60	0.1000	0.0958	0.0453	0.1868
70	0.0917	0.0875	0.0514	0.2373
70	0.0833	0.0792	0.0554	0.2734
90	0.0750	0.0708	0.0579	0.2971
100	0.0667	0.0625	0.0592	0.3101
110	0.0583	0.0542	0.0595	0.3134
120	0.0500	0.0458	0.0590	0.3080
130	0.0417	0.0375	0.0577	0.2948
140	0.0333	0.0292	0.0556	0.2747
150	0.0250	0.0208	0.0527	0.2483
160	0.0167	0.0125	0.0490	0.2164
170	0.0083	0.0042	0.0444	0.1800
180	0.0000	0.0000	0.0388	0.1397

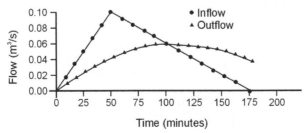

Figure 3.44 Routing of flow through storage culvert.

From the Fig. 3.44 it can be seen that the peak outflow from the culvert storage is 0.06 m³/s and it occurs after 110 minutes from the start of storm.

Problem 3.6

The flow depth in a 50 m wide rectangular channel laid at a slope of $S_0 = 3 \times 10^{-4}$ having Manning's $n = 0.024$ is measured by an observer. Initially the flow depth is 1.20 m and the water level rise at a rate of 1.1 m/h. Find the initial discharge at a distance of 1 km downstream find the relative magnitude of the acceleration terms in St. Venant equation and also determine whether flood wave attenuates as it propagates downstream.

Solution

The initial upstream discharge is

$$Q_u = Av = Bh\frac{1}{n}R^{2/3}S_0^{1/2}$$

$$= 50 \times 1.20 \frac{1}{0.024}(1.145)^{2/3}(3 \times 10^{-4})^{1/2}$$

$$= 2500(1.145)^{2/3} \times 1.73 \times 10^{-2}$$

$$= 43.25(1.145)^{2/3} - 43.25 \times 1.094$$

$$= 47.32 \text{ m}^3/\text{s}$$

$$R = \frac{A}{P} = \frac{50 \times 1.20}{52.40} = 1.45$$

From continuity equation

$$\Delta Q = \frac{-B\Delta h\Delta x}{\Delta t}$$

$$= -50\frac{1}{3600}1.1 \times 1000 \text{ m}^3/\text{s}$$

$$= -15.28 \text{ m}^3/\text{s}$$

The downstream discharge

$$Q_d = Q_u - \Delta Q$$

$$Q_d = (47.32 - 15.28) \text{ m}^3/\text{s}$$

$$= 32.04 \text{ m}^3/\text{s}$$

The downstream flow depth is

$$h_d = \left(\frac{nQ}{BS_0^{1/2}}\right)^{3/5} = \left(\frac{0.024 \times 32.04}{50\sqrt{3 \times 10^{-4}}}\right)^{0.6}$$

$$= \left(\frac{0.77 \times 10^2}{86.5}\right) \text{n}$$

$$= (0.89)^{0.6} \text{ m}$$

$$= 0.92 \text{ m}$$

The *u/s* and *d/s* velocities are respectively

$$V_u = \frac{Q_u}{Bh_u} = \frac{47.32}{50 \times 1.20} = 0.79 \text{ m/s}$$

$$V_d = \frac{Q_d}{Bh_d} = \frac{32.04}{50 \times 0.92} = 0.69 \text{ m/s}$$

Over a distance of 1 km the flow depth changes by 0.28 m and $\Delta v = 0.1$ m/s. The terms in St. Venent equation are

(2) $$S_0 = 3 \times 10^{-4}$$

(3) $$\frac{\partial h}{\partial x} = \frac{-0.28}{1000} = -2.8 \times 10^{-4}$$

(4) $$\frac{V}{g}\frac{\partial V}{\partial x} = \frac{-0.74}{9.81} \times \frac{(0.1)}{1000} = -0.75 \times 10^{-5}$$

Given the flood wave celerity $C = \dfrac{5}{3}v = \dfrac{5}{3} \times 0.79$ m/s $= 1.23$ m/s

The 1 km distance travel time is

$$\frac{1000}{C} = \frac{1000}{1.23} = 813 \text{ s}$$

(5) $$\frac{1}{g}\frac{\partial V}{\partial t} = \frac{1}{9.81} \times \frac{0.1}{813} = 1.25 \times 10^{-5}$$

The friction slope

$$S_f = S_0 - \frac{\partial h}{\partial x} - \frac{V}{g}\frac{\partial V}{\partial x} - \frac{1}{g}\frac{\partial V}{\partial t}$$

$$S_f = 3 \times 10^{-4} + 2.8 \times 10^{-4} + 7.5 \times 10^{-4} - 0.125 \times 10^{-4} = 13.17 \times 10^{-4}$$

The Froude number

$$F_r = \frac{V_u}{\sqrt{gh_u}} = \frac{0.79}{\sqrt{9.81 \times 1.2}} = 0.23$$

which is less than unity and term (4) is smaller than term (3).

The flood wave diffusivity term is

$$1 - (\beta - 1)^2 (F_r)^2 = 1 - \left(\frac{5}{3} - 1\right)^2 (F_r)^2$$

$$= 1 - \frac{4}{9}(0.23) = 1 - 0.02\iota$$

$$= 0.976$$

So the flood wave is diffusive and attentuates as it propagates down stream.

APPENDIX II

(A) *To prove that the acceleration slope terms, viz., Eq. (3.22), are of the order of* $F^2 \left(\dfrac{\partial x}{\partial y}\right)$, *where* F = Froude number, $v/\sqrt{g\,y}$.

Proof: (i) $\dfrac{v}{g} \dfrac{\partial v}{\partial x} = \dfrac{v}{g} \dfrac{\partial}{\partial x}\left(\dfrac{q}{y}\right)$; $v = (q/y)$, q being the discharge/unit width.

Or

$$\frac{v}{g} q \frac{\partial}{\partial x}(y^{-1}) = \frac{v}{g} q\, y^{-2}\left(\frac{\partial y}{\partial x}\right)$$

$$\frac{v}{g} \frac{v y}{y^2} \frac{\partial y}{\partial x} = \frac{v^2}{g y}\cdot\frac{\partial y}{\partial x}$$

$$\simeq F^2 \left(\frac{\partial y}{\partial x}\right)$$

(ii)

$$\frac{1}{g}\frac{\partial v}{\partial t} = \frac{1}{g}\frac{\partial}{\partial t}(q/y) = \frac{q}{g}\cdot\frac{\partial}{\partial t}(y^{-1})$$

$$= \frac{v y}{g} y^{-2}\cdot\frac{\partial y}{\partial x}\cdot\frac{\partial x}{\partial t} = \frac{v}{g y}\cdot v\frac{\partial y}{\partial x} = \frac{v^2}{g y}\frac{\partial y}{\partial x}$$

$$\simeq F^2\left(\frac{\partial y}{\partial x}\right)$$

(B) *To prove that wave celerity 'c' in channel of rectangular cross-section equals* 1.5 V, *where V equals mean channel flow velocity:*

Proof : The celerity

$$c = \left[\frac{1}{B}\frac{\partial Q}{\partial y}\right] = \frac{\partial Q}{\partial A}$$

Again $Q = AV = AC \sqrt{(A/P)S}$ since $V = C \sqrt{RS}$

by Chezy's formula

or
$$\frac{\partial Q}{\partial A} = \frac{3}{3} A^{1/2} C \sqrt{S/P} = \frac{3}{3} C \sqrt{RS}$$

$$= \frac{3}{2} V$$

Hence celerity $c = \frac{3}{2} V$.

APPENDIX III

The derivation of the pseudoviscosity correction terms due to bed shear stress is as follows.

With reference to Fig. 3.27 consider the flow between sections 1-1 and 2-2. From the momentum equation we have

$$(\rho q u_2 - \rho q u_1) = P_1 - P_2 + W \sin \theta - \tau l$$

where q is the discharge per unit width, P_1 and P_2 are the pressures at sections 1-1 and 2-2, W is weight of the fluid enclosed between the sections, and θ is the inclination of the bed to the horizontal.

Now
$$q = h_1 u_1 = h_2 u_2$$

$$\rho(h_2 u_2^2 - h_1 u_1^2) = (\gamma/2)(h_1^2 - h_2^2) - \tau l$$

or
$$\frac{(h_2 u_2^2 - h_1 u_1^2)}{l} = gh \frac{(h_1 - h_2)}{l} - \frac{\tau}{r}$$

or
$$-\frac{\tau}{\rho h} = \frac{(h_2 u_2^2 - h_1 u_1^2)}{lh} - \frac{(h_2 - h_1) g}{l}$$

or
$$-\frac{\tau}{\rho h} = \frac{(h_2 u_2^2 - h_1 u_1^2)}{lh} + \frac{(h_2 - h_1) g}{l}$$

$$-\frac{\tau}{\rho h} = \frac{(h_2 u_2^2 - h_1 u_1^2)}{lh} + S_f g$$

since $(h_2 - h_1)/l$ can be taken as an energy gradient

$$-\frac{\tau}{\rho h} = \frac{(h_2 u_2^2 - h_1 u_1^2)}{lh} + \frac{g u^2}{C^2 h}$$

APPENDIX IV

```
C
C+++   Program for one dimensional rivers   (natural or tidal)
C      real * 8  h(100),  q (100),  width (100), aman (100), hydr (100), area (100),
C      1         ubh (100), ubt (100), ubq (100), ubu (100), dbh (100), dbt (100),
       2         qo (100), ho (100), z0 (100), u (100), ap (100), aq (100)
       real * 8  alpha (100),  beta (100), gamma (100), delta (100), areao (100)
       integer   ubound, dbound
C
C+++ Input and Output data files
C
       open (5, file = 'impinp. dat', status = 'UNKNOWN')
       open (6, file = 'impout. dat', status = 'UNKNOWN')
C
C+++ Input data of Channel Geometry
C
       read (5, *) nx, clength
       read (5, *) (width (i) , i =1, nx)
       read (5, *) aman(i), i=1, nx)
       read (5, *) z 0 (i), i =1, nx)
       read (5, *) akn, delt, tidperiod, epsv, maxncyc


C
C+++ Grid Computation
C

       delx = clength/float (nx)


C
C+++ Boundary Conditions
C

C      (a) Upstream Boundary Conditions
C              ubound =     1 Water level
C                           2 Discharge
C                           3 Velocity

       read (5, *) ubound
       go to (10, 20, 30, 40) ubound
10     continue
       read (5, *) uncord
       read (5, *) (ubt (i), i=1, uncord)
       read (5, *) (ubh (i), i=1, uncord)

       go to  40
```

```
20      continue
        read (5, *) uncord
        read (5, *) (ubt (i), i = 1, uncord)
        read (5, *) ubq (i), i = 1, uncord)

        go to 40
30      continue
        read (5, *) uncord
        read (5, *) ubt (i), i = 1, uncord)
        read (5, *) (ubu (i), i = 1, uncord)
40      continue

C
C       (a) Dpstream Boundary Conditions
C           dbound =      1 Water level
C                    2 Q = 19.25* (y – 0. 60)** 1.5
C                    3 Q = 5 * y** 3.1
C                    4 sf = s 0

        read (5, *) dbound
        go to (11, 21, 21, 21) dbound

11      countinue
        read (5, *) dncord
        read (5, *) (dbt (i), i = 1,  dncord)
        read (5, *) (dbh (i), i = 1, dncord)
21      continue

C
C+++ Initial conditions
C

        read (5, *) hinit, qinit
        do  i=1, nx
            q (i) = qinit
            h (i) = hinit
        enddo

C
C+++ Computational Time started from here
C

        ncycle = 1
550     continue

        time-sec = 0.0
```

```
100    continue

C
C+++ Iteration started from here
C
       iter = 1
200    continue

C
C+++ Equations of sections
C
C+++ Calculating Boundary Conditions
C+++ Downstream Boundary conditions

       go to  (12, 22, 22, 22) dbound

12     continue
       do i = 1, dncord −1
         if (time_sec.ge.dbt (i).and.time_sec.le.dbt (i + 1)) then
           dh = dbh (i + 1) − dbh (i)
           if (dh.ge.0.0) then
             h (nx) = dbh (i)  + dh/ (dbt (i+1) − dbt (i))  * (time_sec − dbt (i))
           else
                     h (nx)  = dbh (i)   −  dh/  (dbt (i+1)  −  dbt (i))  * (dbt (i+1)  −  time_
       sec)
           endif
         endif
       enddo
22     continue

C+++ Upstream Boundary Conditions

       go to (13, 23, 33, 43) ubound

13     continue
       do i = 1, uncord − 1
         if (time_Sec.ge.ubt (i).and.time_Sec.le.ubt (i + 1)) then
           dh = ubh (i + 1) − ubh (i)
           if (dq.ge.0.0) then
             h (1) = ubh (i) + dh/ (ubt (i + 1) − ubt (i)) * (time_sec − ubt (i))
           else
             h (1) = ubh (i) − dh/ (ubt (i+1) − ubt (i)) * (ubt (i+1) − time_sec)
           endif
         endif
       enddo
```

```
23      continue
        do i = 1, uncrord − 1
          if (time_Sec.ge.ubt (i).and.time_Sec.le.ubt (i + 1)) then
              dq = ubq (i + 1) − ubq (i)
              if  (dq.ge.0.0) then
                q (1) = ubq (i) + dq/ (ubt (i + 1) − ubt (i)) * (time_Sec − ubt (i))
              else
                q (1) = ubq (i) − dq/(ubt (i+1) − ubt (i)) * (ubt (i + 1) − time_Sec)
              endif
          endif
        enddo
33      continue

        do i = 1, uncord − 1
          if (time_Sec.ge.ubt (i).and.time_sec.le.ubt (i + 1)) then
              du = ubu (i + 1) − ubu (i)
              if (dq.ge.0.0 ) then
                u (1) = ubu (i) + du/ (ubt (i+1) − ubt (i)) * (time_sec − ubt (i))
              else
                u (1) = ubu (i) − du/ (ubt (i+1) −ubt (i)) * (ubt (i + 1) − time_sec)
              endif
          endif
        enddo
43      continue

C
C+++ Calculating Hydraulic parametrs
C
        do i = 1, nx
          area (i) = width (i) * h (i)
          hydr (i) = area (i) / (2 * h (i) + width (i))
        enddo

C
C+++ Storing old values of h and q
C

        do i = 1, nx
          ho (i) = h (i)
          qo (i) = q (i)
          areao (i) = width (i) *ho (i)
        enddo

C
C+++ Calculations Boundary conditions
C
```

```
C+++ (a) u/s  boundary conditions
      if (ubound.eq.1) then
         dhdx = (h (2)  –  h (1)) / delx
         adhdx = dhdx * dhdx
         adhdx = sqrt ( adhdx)
         if  (adhdx. eq. 0.0 ) adhdx = 1.e – 10

         q (1) = akn / aman (1) * area (1)  * hydr (1) ** ( 2./3.) *
      1     ( – dhdx / sqrt (adhdx))
      else if (ubound.eq.2) then
         area (1) = area (1) – (q (2) – q (1)) * delt/delx
         h (1) = area (1) / width (1)
      else if (ubound.eq.3) then
         q (1) = u (1) * area (1)
         area (1) = area (1) – (q (2) – q (1)) * delt/delx
         h (1) = area (1) / width (1)
      endif

C+++ (b) d/s boundary conditions

C           dbound =1    Water level
C                  2    Q = 19.25 * (y – 0.60) ** 1.5
C                  3    Q = 5 * y ** 3.1
C                  4    sf = s0
      if (dbound.eq.1) then
         dhdx = (h (nx) – h (nx – 1)) / delx
         adhdx = dhdx*dhdx
         adhdx = sqrt (adhdx)
         if (adhdx. eq. 0.0) adhdx = 1.e – 010

         q (nx) = akn/aman (nx)* area (nx)* hydr (nx)** (2./3.)*
      1     (– dhdx/sqrt (adhdx))
      else if (dbound.eq.2) then
         qq = h (nx – 1) – 0.6
         if (qq.le.0.0) qq = 0.0001
         q (nx) = 19.25 * qq ** 1.5
         h (nx) = h (nx – 1)

      else if (dbound.eq.3) then
         qq = h (nx – 1)
         if (qq.le.0.0) qq = 0.0001
         q (nx) = 5 * qq ** 3.1
         h (nx) = h (nx – 1)
      else if (dbound.eq.4) then
         dhdx = (z0 (nx) – z0 (nx – 1))/ delx
         adhdx = dhdx*dhax
```

```
            adhdx = sqrt (adhdx)
            if (adhdx. eq.0.0) adhdx = 1.e – 010

            q (nx) = akn / aman (nx) * area (nx) * hydr (nx) ** (2./3.) *
     1      (– dhdx / sqrt (adhdx))
            h (nx) = area (nx) / width(nx)
          endif

C
C+++ Continuity and Momentum Equation coefficients
C

            do i = 2, nx – 1
              dh = (h (i  –  1) – h (i))
              adh = dh*dh
              adh1 = sqrt (adh)
              if (adh.eq.0.0) adh1 = 1.e – 010

              alpha (i) =  – 2 ** 1.5 *akn / (2 * delx) * (area (i) * hydr (i) ** (2./3.)) /
     1                  (aman (i) * (2 * delx) ** 0.5) * 1. /adh1 ** 0.5

              dh = (h (i) – h ( i+ 1))
              adh = dh * dh
              adh =  sqrt (adh)
              if (adh.eq.0.0) adh = 1.e – 010

              beta (i)   = 2 ** 1.5 *akn / (2 * delx) * (area (i + 1) * hydr (i + 1) ** (2./3.)) /
     1                  (aman (i + 1) * (2 * delx) ** 0.5) * 1. / adh ** 0.5 +
     2                  2 ** 1.5 * akn / (2 * delx) * (area (i) * hydr (i) ** 2./3.)) /
     3                  (aman (i) * (2 * delx) ** 0.5) * 1. /adh1 ** 0.5

              gamma (i)= – 2 **1.5*akn / (2 * delx) * (area (i + 1) *hydr (i+1) ** (2./3.)) /
     1                  (aman (i + 1) * (2 * delx) **  0.5) * 1. / adh ** 0.5

              delta (i)   = (area (i +1) + area (i) – areao (i+1) – areao (i) ) / (2 * delt)
            enddo

C+++ Boundaray Condition u/s

            i = 1
            dh = (h (i) – h (i + 1))
            adh = dh * dh
            adh = sqrt (adh)
            if (adh.eq.0.0) adh = 1.e – 010

            beta (i)      = 2 ** 1.5 * akn / (2 * delx) * (area (i + 1) * hybr (i + 1) ** (2./3.))
```

```
/
1                        (aman (i + 1) * (2 * delx)  ** 0.5) * 1. / adh ** 0.5 +
2                        2 ** 1.5 * akn / (2 *delx) * (area (i) * hydr (i)  ** (2. /3. )) /
3                        (aman (i) * (2 * delx) ** 0.5) * 1. /adh1 ** 0.5

     gamma (i)  = − 2 ** 1.5 * akn / (2 * delx) * (area (i + 1) hydr (i+1)  ** (2./3.))
/
1                        (aman (i +1) * (2 *delx) ** 0.5) * 1. /adh ** 0.5

     delta (i)    = 2 ** 1.5 * akn / (2 * delx) *  q (i) − (area (i + 1) + area (i) −
1                        areao (i +1)  − areao (i) ) / (2 *delt)

     do i = 1, nx − 1
       if (i.eq.1) then
         ap (i) = gamma (i) / beta (i)
         aq (i) = delta (i) / beta (i)
       else
         ap (i) = gamma (i) / (beta (i)  − ap (i − 1) * alpha (i))
         aq (i) = (delta (i) − aq (i − 1) * alpha (i)) / (beta (i) − ap (i − 1) * alpha (i))
       endif
     enddo

     do i = 2, nx − 1
       h (i) = aq (i) − ap (i) * h (i + 1)
     enddo

     do i = 2, nx − 1
       dhdx = (h (i) − h (i −1)) / delx
       adhdx = dhdx * dhdx

       adhdx = sqrt (adhdx)
       if (adhdx. eq.0.0) adhdx = 1.e − 010

       q (i) = akn / aman (i) * area (i) * hydr (i) ** (2./3.) *
1               (− dhdx / sqrt (adhdx))
     enddo

C
C+++ Checking for convergency
C

     amax = 1.e + 10
     do i = 1, nx
     errh = abs (ho (i) − h (i))
     if (amax.gt.errh) amax = errh
     enddo
```

```
if (amax.gt.epsv.or.iter.eq.1) then
    iter = iter + 1
    go to 200
endif

write (*, *) 'time_sec =', time_sec
write (*, *) 'Water levels'
write (*, *) (h (i), i = 1, nx)
write (*, *) 'Discharge'
write (*, *) (q (i), i = 1, nx)

if (ncycle.eq.maxncyc) then
    write (6, *) 'time_sec = ', time_sec
    write (6, *) 'Water Levels'
    write (6, *) (h (i), i = 1, nx)
    write (6, *) 'Discharge'
    write (6, *) (q (i), i = 1, nx)
endif

if (time_sec.lt.lidperiod) then
    time_sec =  time_sec + delt
    go to 100
endif

if (ncycle.lt.maxncyc) then
    ncycle = ncycle + 1
    go to 550
endif

close (5)
close (6)
stop
end

7 210
2.0 2.0 2.0 2.0 2.0 2.0 2.0
0.03 0.03 0.03 0.03 0.03 0.03 0.03
0.1 0.1 0.1 0.1 0.1 0.1 0.1
1.1.3600 0.001 3
2
7
0.0 600. 1200. 1800. 2400.  3000. 3600.
0. 3 0.9 2.4 2.8 2.4 0.9 0.3
2
5.0.3
```

APPENDIX V

IS : 2914 - 1964

FORM 2
INFORMATION ON GAUGE SITE AND OPERATIONS

(1) Give plan of the gauging site showing longitudes and latitudes along with the river section to the following scales:

    (a) Plan of gauging site showing 5 km (3 miles) above and below the site .... Scale 1 : 30,000 (approximate)

    (b) River Cross-section:
    Vertical scale 1 : 100 (approximate).
    Horizontal scale to be suitably selected so that the length of the plotted section is between 0.3 and 0.6 m (or 1 ft or 2 ft) (approximate).

(2)   (a) Is any control existent helping the rating curve to stabilize; if so, what type?

    (b) For how many years has the site been under observation without feeling the necessity for a change?

    (c) If the river spills over the bank at the discharge site during high floods, indicate the stage at which the spilling takes place on one bank or both banks. What is the approximate percentage of spill discharge as against the contained discharge?

(3) Whether the river bed is mobile or inerodible?

(4) Is the river reach stable in the vicinity of the gauging site?

(5) Does the river cross-section change during floods?

(6) Are the water level and discharge observations made at the same site?

(7) What type of gauge installed?

(8) What is the frequency of gauge observations?

(9) What method is used for discharge measurement?

(10) What is the frequency of discharge observation?

(11) Any correction to rating curve deemed necessary on account of —

  (a) gauge and discharge sites being separated,

  (b) discharge being unsteady,

  (c) because of over bank spill, and

  (d) variable slopes.
    Are these applied?

Name of the Observer.. Signature.............................

Designation .... Date......................................

FORM 3

GAUGE STATEMENT OF RIVER .... **AT** ....

For $\overline{\underset{\text{third}}{\overset{\text{first}}{\text{second}}}}$ ten days during the month of .... 20 ....

(Gauge No. ....)

DATE	GAUGE READING			ZERO R.L. OF GAUGE	WATER LEVEL			WATER TEMPERATURE* °C			RE-MARKS
	Morning	Noon	Evening		Morning	Noon	Evening	Morning	Noon	Evening	
(1)	(2)	(3)	(4)	(5)	(6)	(7)	(8)	(9)	(10)	(11)	(12)

* The water temperature is taken 30 cm (or 1.0 ft) below the surface. When the depth is less, temperature is taken at the bed level.

Name of the Observer ....   Signature ....

Designation ....   Date ....

IS : 2914-1964

FORM 4
STATEMENT SHOWING HOURLY FLOOD GAUGES

On River .... at .... for the Month of .... 20.......

Zero R.L. of Gauge ....
Gauge No. ....

DATE	HOURS																								MAXIMUM WATER LEVEL			RE-MARKS
	1st	2nd	3rd	4th	5th	6th	7th	8th	9th	10th	11th	12th	13th	14th	15th	16th	17th	18th	19th	20th	21st	22nd	23rd	24th	During the Days	During the Month	During the Year	
(1)	(2)	(3)	(4)	(5)	(6)	(7)	(8)	(9)	(10)	(11)	(12)	(13)	(14)	(15)	(16)	(17)	(18)	(19)	(20)	(21)	(22)	(23)	(24)	(25)	(26)	(27)	(28)	(29)

Observer ....

Note: Col 26 and 27 should be filled by the Supervisor before submitting the statement.

FORM 5

COMPUTATION OF DISCHARGE BY VELOCITY AREA METHOD

(METRIC UNITS)

IS : 2914-1964

River..

To.................

Site.................. Date.................. Time From..................

Meter No. and Make..................

Spin before Measurement..................

Equation.................. After.................. Date of Last Rating..................

Description of Floats..................

Float Run Marked with.................. Length of Float Run.................. Metres.

Soundings Taken with.................. Weight Used..................

Method of Suspending Meter..................

Section line Marked with..................

Time Piece Used..................

Weight Used..................

Condition of Water { Fairly Clear.................. Ordinarily Silty.................. Intensely Silty..................

River Water Temperature (C°)..................

Atmospheric Temperature (C°) Max.................. Min..................

Weather..................

Direction of Wind..................

Current

GAUGE	PERMANENT			TEMPORARY		
	L.B.	R.B.	Mean	L.B.	R.B.	Mean
Zero R.L						
Beginning						
End						
Mean						

Very Slight
Slight
Strong
Very Strong

(*Continued*)

IS : 2914-1964

FORM 5 (Continued)
COMPUTATION OF DISCHARGE BY VELOCITY AREA METHOD

R.D. (Reduced Distance on Section)	Water Depth	Difference of Depth	Increase in Bed	Time Seconds	Meter Revolutions	Mean Velocity	Angle of Current with Section	Correction for Angle of Current	Corrected Mean Velocity	Drift (Metres)	Correction in Velocity for Drift	Final Corrected Mean Velocity	Water Depth X Corrected Velocity (Col 2 × Col 13)	Correction + or – for Unequal Segments	Remarks
(1)	(2)	(3)	(4)	(5)	(6)	(7)	(8)	(9)	(10)	(11)	(12)	(13)	(14)	(15)	(16)
Total		Total											Total		
Multiply by common width of segments		Add Surface width											Multiply by common width of segments		
Product		P = wetted perimeter											Product		
Deduct													Deduct the total of col. 15.		
A = Area													Q = Discharge		

IS : 2914 - 1964

Rugosity Coefficient

(1) $\bar{v}$ = Mean velocity = $\dfrac{Q}{A}$ = =

(2) R_A = H.M.D. = $\dfrac{P}{P}$ = =

(3) $C = \dfrac{\bar{v}}{\sqrt{R_h S}}$ = =

(4) $X = 23 + \dfrac{0.00155}{S}$ = =

(5) $Y = \dfrac{CX}{\sqrt{R_h}}$ = =

(6) $Z = C - X$ = =

(7) $n = \dfrac{\sqrt{(Z^2 + 4Y)} - Z}{2Y}$ = =

(Kutter)

(8) $n_a = \dfrac{1}{C} R_h^{1/2}$ =

(Lacey)

(9) $n = \dfrac{1}{C} R_h^{1/2}$ =

(Manning)

(10) $f = \dfrac{8g}{C^2}$ =

(Darcy-Weisbach)

Note: The value of 'C' in equations (5), (6), (8), (9) and (10) shall be obtained from (3) and not assumed.

Surface Slope Observed

	Right Bank			Left Bank		
	Back	Fore	Difference	Back	Fore	Difference
500 m D/S						
0						
500 m U/S						

Fall in 1,000 Metres

Mean =

$S = 0.00$

	Datum Area	Scour or Fill
Of date		
Previous		

OTHER INFORMATION

(1)   Character of river bed ....

(2)   Class of Roughness under which it falls ....

(3)   Monthly or every change, a free hand sketch should be made of the configuration of the river 500 m upstream and downstream of the discharge site, showing direction of general flow of the river and position of permanent and temporary gauges and other permanent marks and their distances from the cross line.

Observed by ....   Entered by ....

Checked by ....   Compared by ....

Note 1: Mean velocity (col 7) will generally be velocity at 0.6 depth. Where mean velocity is deduced from surface velocity, the coefficient employed should be noted in remarks column. Unless specially warranted, coefficient should be taken as 0.89.

Note 2: If no drift occurs, it has to be shown as 'NIL' in column 11, the column is never to be left blank.

Note 3: When floats are used or more than one meter observation is taken at the same section, each observation of time and revolutions shall be recorded in a separate line in columns 5 and 6, respectively.

Note 4: In columns 1 and 2 all the lines relating to one station will be bracketed and R.D. on section and water depth will be recorded once.

Note 5: Col 15 = (Common width of segments − ½ the sum of segments on either side of the R.D.) × col 14.

REMARKS BY INSPECTING OFFICER ....

IS : 2914-1964

FORM 6

STATEMENT SHOWING DETAIL OF DISCHARGE OBSERVATIONS MADE ON THE RIVER .... AT ....

For $\dfrac{\text{first}}{\text{second}}$ ten days during the month of .... 20...........
$\overline{\text{third}}$

Position of Discharge Site ....

Section Line Marked with..................................

Soundings Taken with/Weight Used.................

R.L. of Zero of Permanent Gauge ....

When was the Latter Last Checked ....

R.L. of Zero of Temporary Gauge ....

Date	Name of observer	Weather	Central Line Water Levels		Total Discharge in m³/s	Discharge Elements												Remarks
			Left Bank Mean	Right Bank Mean		Q = Discharges of Main Channel in m³/s	S = Slope	A = Area of Section m²	$\bar{\upsilon}$ = Velocity col 7 col 9 m/s	P = Wetted Perimeter m	R_r = Hydraulic Mean Depth m	$\sqrt{RrS}$	C	Manning n	Mean or Surface Velocity m/s	Wind Direction	
(1)	(2)	(3)	(4)	(5)	(6)	(7)	(8)	(9)	(10)	(11)	(12)	(13)	(14)	(15)	(16)	(17)	(18)
Total																	
Average																	

Wind Direction key:
⟶⟶ Very Slight
⟶ Slight
↠ Strong
⟶⟶ Very Strong

Observer ....

Note 1: Actually observed discharges to be entered in red.
Note 2: Tabular discharge to be entered in black.

IS : 2914 - 1964

FORM 7

DAILY STAGE-DISCHARGE DATA

River... Site... Year, 20.....................

Zero R.L. ....

Date	Month.........................		Month..................................	
	Water Level	Discharge	Water Level	Discharge
	(m) or (feet)	(m^3/s) or (cusecs)	(m) or (feet)	(m^3/s) or (cusecs)
1				
2				
3				
4				
5				
6				
7				
8				
9				
10				
Total				
Average				
11				
12				
13				
14				
15				
16				
17				
18				
19				
20				
Total				
Average				
21				
22				
23				
24				
25				
26				
27				
28				
29				
30				
31				
Total				
Average				
Monthly	Total			
	Average			

Note 1: Actually observed discharges to be entered in red. Entered by........................

Note 2: Tabular discharges to be entered in black. Checked by........................

IS : 2914 - 1964

FORM 8

DAILY STAGE-DISCHARGE DATA

(To be used for flow under non-uniform conditions)

River.. Site.. Year, 20.....................

Zero R.L.'s U/S Gauge... D/S Gauge....................

Distance Between Gauges...

Date	Month.........................			Month.................................		
	Water Level		Discharge	Water Level		Discharge
	U/S Gauge m (or feet)	D/S Gauge m (or feet)	m³/s (or cusecs)	U/S Gauge m (or feet)	D/S Gauge m (or feet)	m³/s (or cusecs)
1						
2						
3						
4						
5						
6						
7						
8						
9						
10						
Total						
Average						
11						
12						
13						
14						
15						
16						
17						
18						
19						
20						
Total						
Average						
21						
22						
23						
24						
25						
26						
27						
28						
29						
30						
31						
Total						
Average						
Monthly	Total					
	Average					

Note 1: Actually observed discharges to be entered in red.

Note 2: Tabular discharges to be entered in black.

Entered by.......................

Checked by......................

4

Design of Spillways

4.1 FUNCTIONS OF SPILLWAYS

Spillways are invariably provided in storage dams for safe disposal of the excess water flowing into the reservoir after it has filled up. Spillways may have simple overflow crests or may have controlled crests with gates. The controlled crests permit controlled release of surplus water in excess of the reservoir capacity and convey the same into the river channel below the dam or to any other drainage area in such a manner that the foundations are protected from erosion and scour. Overflow on the dam at floods cannot be allowed unless the downstream face is designed for passing the floods with requisite depth and the bed well protected to take the impact of falling waters. Many failures of dams have been attributed to inadequate spillways capacity with consequent overtopping of bund. In case of earthen dams if the water overtops there is no possibility of saving the dam from failure. Adequate spillway is a very important factor in the design of earthen dams.

Thus the object of the spillway design should be to provide a safe and adequate spillway structure at most economical cost, without impairing the safety of the dam. Dam failure due to inadequate spillway capacity may result in heavy loss of life and property.

In the hydraulic design of a spillway the following procedure is generally followed:

(i) First, hydrological analysis to determine the magnitude of design flood (already dealt with in Chapter 2).
(ii) And thereafter detailed hydraulic analysis.

4.2 SPILLWAY CLASSIFICATION

Spillways may be classified as controlled crest spillways and uncontrolled crest spillways. In the former case the operations of spillways are controlled by the operation of gates. The uncontrolled or automatic spillway come into action immediately after the water level reaches the crest of the spillway. Typical examples of this type are the ordinary simple weirs provided on small dams or the overflow dams, the siphon spillways and the shaft spillways. By having a controlled crest for the spillway the flood lift can be reduced. The gates can be lifted at the onset of floods so that the reservoir is kept low and the flood passed on. As the flood subsides the gates can be lowered and the storage effected.

4.3 TYPES OF SPILLWAYS

Nearly all spillways fall into one of six types or are made of combinations of these types. These are:

   (i) overfall (free or gated),
   (ii) side channel,
   (iii) chute or trough,
   (iv) shaft or morning glory,
   (v) siphon, and
   (vi) emergency.

The overfall type is the most common and is adopted in masonry dams having sufficient crest length to provide the requisite capacity. Overfall spillways with crest gates will act as orifices under partial control of gates and as open crest weirs when full gate openings are made. Chute spillways are commonly adopted for earth dams, whereas the side channel and shaft spillways are most suitable in narrow canyons. The siphon spillways are usually used to provide approximately constant headwater flow under varying flow. Emergency spillway is an auxiliary spillway provided to take care of the possibility of having a flood greater than the spillway design flood.

4.3.1 Hydraulic Design—Overfall or Ogee Spillways (Bib. 10)

The crest of an overflow dam is generally given to fit the shape so that the overflowing water at the selected design head takes the shape of the lower nappe of freely falling jet over a sharp-crested weir (Fig. 4.1). When the space between sharp-crested weir and the lower nappe is filled with concrete or masonry the weir so formed will have a profile similar to an Ogee (a 'S'-shaped curve in section). It is, therefore, called an Ogee spillway or weir. The upper and lower nappes for such a spillway due to 1 ft (0.305 m) head of water on the crest for a dam with vertical face and sufficient height is shown in Fig. 4.2. The profile for any other head, say design head can be obtained by multiplying all values taken from the curve by that head. The lower nappe of the overflow crest is also designated as a standard crest and it can be closely represented by the following relationship

$$x^{1.85} = 2.0h_d^{0.85}y \qquad (4.1)$$

The plotted form of the crest and the meaning of the symbols can be had from Fig. 4.1.

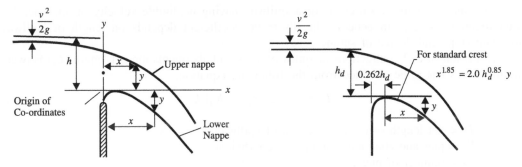

Figure 4.1 Overfall and the standard crest spillway.

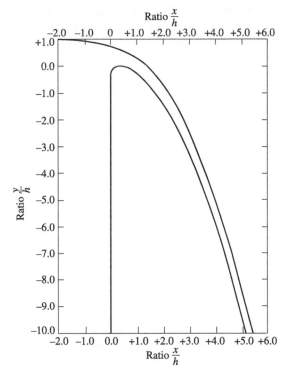

Figure 4.2 Co-ordinates of the overfall spillway.

4.3.1.1 Discharge

The discharge over the spillway can be expressed as :

$$Q = C L h^{3/2} \tag{4.2}$$

where

Q = discharge,

L = length of spillway,

h = head on the crest including head due to approach velocity,

C = coefficient of discharge.

The usual discharge coefficient for a spillway having negligible velocity of approach has been found to be 2.2. In general, the discharge coefficient depends on length of spillway, upstream slope and degree of submergence.

As a result of end contractions at spillway piers and abutments the net length (L_e) between the pier faces has to be obtained from the following equation:

$$L_e = L - 2 \left[K_p n - K_a \right] h \tag{4.3}$$

where

L_e = net length or effective length of spillway,
K_p, K_a = pier and abutment contraction coefficient,
n = number of piers.

The values of K_p and K_a depend mainly on the shape of the piers and that of abutment. The greater is the divergence from streamline flow, the greater the contraction coefficient and lesser the effective length of the crest. The values of the K_p and K_a are given in tabular form below and the various shapes of piers are shown in Fig. 4.3.

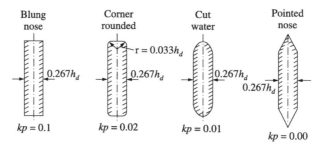

Figure 4.3 Various shapes of piers.

Pier shape	K_p
1. Square-nosed piers without any rounding.	0.1
2. Square-nosed piers with corners rounded on radius equal to 0.1 of pier thickness.	0.02
3. Rounded nose piers and 90° cut water-nosed piers.	0.01
4. Pointed nose piers.	0.0

Abutment shape	K_a
1. Square abutment with headwall at 90° to the direction of flow.	0.2
2. Rounded abutment with headwall at 90° to the direction of flow.	0.1

4.3.1.2 Discharge from gate controlled spillway

The discharge for a gated Ogee crest at partial large openings is computed with the help of following equation

$$Q = 2/3 \sqrt{2g} \; CL \left[h_1^{3/2} - h_2^{3/2} \right] \tag{4.4}$$

where

h_1 = head on upper edge,
h_2 = head on lower edge,
C = orifice discharge coefficient.

The value of C will vary with different gates and crest arrangement and is also affected by upstream and downstream flow conditions. Figure 4.4 shows an approximate variation of C for various ratios of gate openings d to total head h_1. Usually the rating curves from controlled gates with partial openings are determined from model tests for various operating conditions such as *(i) adjacent gates completely opened and (ii) adjacent gates completely closed.*

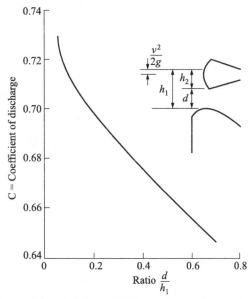

Figure 4.4 Variation of C for various gate openings.

4.3.2 Side Channel Spillways

The side channel spillway is used in sites where the sides are steep and rise to a length greater than the dam. Here the flow after passing over the crest is carried away in a channel running parallel to the crest, to the stream below the dam (Fig. 4.5). Analysis of flow in the side channel is made based on the law of conservation of linear momentum [27].

Consider a short reach of length dx (Fig. 4.6). At the upstream end let the velocity and discharge be V and Q and the corresponding values at the downstream end being $(V + dV)$ and $(Q + dQ)$, where dQ is the additional discharge between the two sections.

If the channel is laid in a slope equal to S_0, then the weight component ($W \sin \theta$) of the water between the two sections in the direction of flow can be expressed as

$$\gamma S_0 \left(A + \frac{dA}{2} \right) dx \simeq S_0 \, A \, dx$$

where

$$\left(A + \frac{dA}{2} \right) = \text{average area and}$$

$$A = \text{area at the upstream section}$$

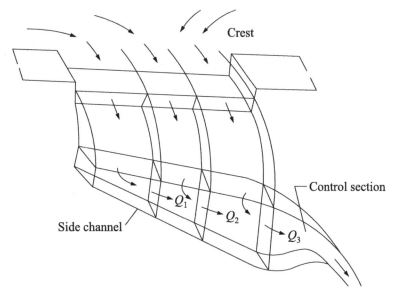

Figure 4.5 Side channel spillway.

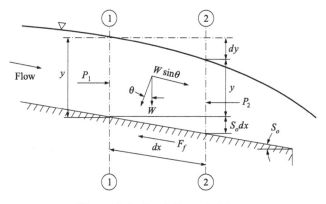

Figure 4.6 Spatially varied flow.

Similarly, the friction, loss between the two sections can be expressed as

$$h_f = S_f \, dx$$

Accordingly, the frictional force F_f along the channel wall can be expressed as equal to

$$\gamma(A + 1/2 \, dA) \, S_f \, dx \simeq \gamma A S_f \, dx$$

Again the total pressure force P_1 acting on the upstream section equal $\gamma \overline{Z} A$
where $\overline{Z}$ = depth of the centroid of A below the surface of flow. The total pressure P_2 on
section (2–2) equal $\gamma(\overline{Z} + dy) \, A + \dfrac{\gamma}{2} d \, A \, dy = \gamma(\overline{Z} + dy) A$.

Therefore, the resultant pressure acting on the body of water between the two sections is
$(P_1 - P_2) = -\gamma A \, dy$.

The momentum change of the water body

$$= [\rho(Q + dQ)(V + dV) - \rho QV] = \rho[QdV + (V + dV)dQ]$$

and this will equal all the external forces acting on the body, i.e. pressure, weight component and frictional, which means:

$$\rho[QdV + (V + dV)dQ] = -\gamma A\,dy + \gamma S_0\,Adx - \gamma AS_f\,dx \qquad (4.5)$$

Considering the differential as finite increments the above Eq. (4.5) can be written as

$$\frac{\gamma}{g}[Q\,\Delta V + (V + \Delta V)\,\Delta Q] = -\gamma \int_0^{\Delta y} Ady + \gamma S_0 \int_0^{\Delta x} Adx - \gamma S_f \int_0^{\Delta x} Adx$$

$$= -\gamma\overline{A}\,\Delta y + \gamma S_0\,\overline{A}\,\Delta x + \gamma S_f\overline{A}\,\Delta x \qquad (4.6)$$

where
$\overline{A}$ = average area.

As the discharge varies with finite increment of the channel length, the average area comes to $\overline{A} = (Q_1 + Q_2)/(V_1 + V_2)$. Further, taking $Q = Q_1$ and $V + \Delta V = V_2$ and on simplification the above equation can be expressed as

$$\Delta y = -\frac{Q_1(V_1 + V_2)}{g(Q_1 + Q_2)}\left(\Delta V + \frac{V_2}{Q_1}\Delta Q\right) + S_0\,\Delta x - S_f\,\Delta x \qquad (4.7)$$

The drops in the channel surface elevation between 1 and 2 can be found out from the following relationship (Fig. 4.6).

$$y + S_0\,dx = y + dy + dy_1$$

or

$$dy_1 = -dy + S_0\,dx \qquad (4.8)$$

Converting the differential to finite increments

$$\Delta y_1 = -\Delta y + S_0\,\Delta x \qquad (4.9)$$

Hence, from the above equation

$$\Delta y_1 = \frac{Q_1(V_1 + V_2)}{g(Q_1 + Q_2)}\left(\Delta V + \frac{V_2}{Q_1}\Delta Q\right) + S_f\,\Delta x \qquad (4.10)$$

The first term on the right-hand side of the equation represents the loss due to impact and the second the frictional effect. By the use of above equation, the water surface profile can be found out for any particular side channel spillway by considering successive short reaches from a starting section by trial. The starting section is commonly a control section where critical flow occurs.

4.3.3   Chute Spillway

The chute or trough spillway is the commonest type of water conductor used for the conveyance of flow between the control section and the downstream. Here the water flows over the crest of the spillway into a steep-sloped open channel called a chute or trough (Fig. 4.7). The channel is usually cut into rock abutments and either left unlined or lined with concrete slabs. The flow in a chute spillway is super-critical, which means the depth is below critical or the Froude number is greater than one and as a result particular attention is needed for providing a correct shape to guide walls in the plane. This is because when high velocity flowing water collides with an improperly aligned guide wall standing waves will occur which will result in piling up of water against walls, overtopping, cavitation, etc.

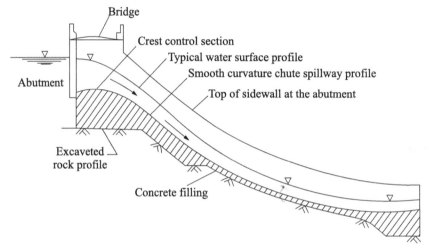

Figure 4.7   Chute spillway.

A chute spillway has geneally flared type of inlet and the hydraulic design of the inlet involves the design of the weir length which is determined from the formula

$$Q = CLH^{3/2} \tag{4.11}$$

Where Q = peak discharge rate in m³/s. C = discharge coefficient = 1.77, L = weir length in m, h = head of flow in, m. For determining the value of L, h is assumed or for a given h, L is determined depending on site condition. The width or length of the spillway is established by the required discharge to be passed where the controlling section meet the crest and generally it will be larger than the remaining length of the chute. This requires a transition or a flared type spillway. When the slope of the chute conforms to the topography, minimum excavation will be needed. Apart from this if it is necessary to provide vertical curves this should be gradual so that there occurs no separation of flow from the channel bottom. Figure 4.8 shows a chute spillway constructed in an earth fill dam with energy dissipation arrangement at the exit end.

4.3.4   Shaft Spillway

In a shaft or morning glory spillways the water drops through a vertical shaft to a horizontal pipe which conveys the water down a horizontal shaft or tunnel. It is often used where there is not enough space for other types of spillways. Three possible types of flow conditions are possible in a shaft spillway. At low heads the outlet conduit flows partly full and with increase

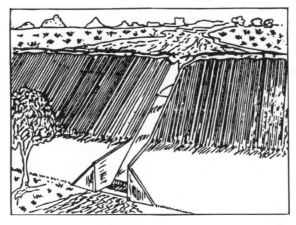

Figure 4.8   Chute spillway constructed on an earth fill dam (Based on USDA, Farmer's Bul., 2171).

in the head the outlet flow is partly full or full. In partly full condition the discharge varies as 3/2th power of head over the weir and under full condition the discharge varies as 1/2th power of total head acting on the outlet. Further, it has been observed that under submerged conditions, an increase in total depth results in only a very slight increase in discharge.

The form of the spillway is largely controlled by the discharge to be accommodated and the permissible depth of overflow, for the length of crest must be sufficient for the required discharge at the maximum head. Accordingly, large discharges and small depths of overflow give rise to a large intake diameter section. The outlet tunnel size is determined by the discharge and fall and is commonly constructed so that the tunnel will flow full throughout its length and will not cause a backwater effect in the spillway crest under conditions of maximum discharge. The basic equation for the discharge over a nappe-shaped profile for a circular weir can be expressed as (Fig. 4.9):

$$Q = C\ 2\pi R H^{3/2} \tag{4.12}$$

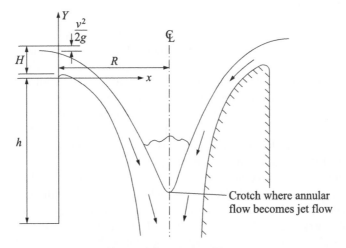

Figure 4.9   Shaft spillway.

where

C = coefficient of discharge,

R = radius of the sharp-crested weir,

H = head acting on the weir including velocity head.

The coefficient C is related to H and R since it is affected by submergence and back pressure in terms H/R and h/R, where h height of weir as shown in Fig. 4.10.

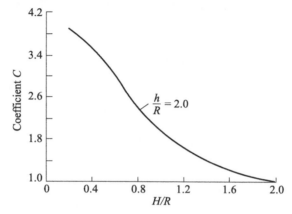

Figure 4.10 Variation of C vs. (H/R) and (h/R).

An abrupt transition between the shaft and outlet conduit is avoided by providing an elbow. The flow condition at elbow in large structures should be investigated by suitable model tests. However, model results should be used with caution for the air pressure in the model is not reduced to model scale. Another undesirable feature of shaft spillway is the danger to clogging with debris and for that purpose suitable preventive measures are adopted.

4.3.5 Siphon Spillway

In case there exists limitation in space and where the discharge is not extremely high, siphon spillway is generally very suitable. It is also adopted in providing automatic regulation of water surface within a certain narrow limit. A siphon spillway essentially consists of a siphon pipe, one end of which is kept on the upstream side in contact with the reservoir and the other end discharges water at the downstream end. Figure 4.11 shows the cross-section of a siphon, where it is installed in the body of the dam. There is an air vent which connects with the siphon pipe, the level of which may be kept at normal pool level. The entry point of the siphon pipe may be kept further lower in order to prevent the entry of logs, debris, etc. in the siphon. The outlet of the siphon may be submerged to prevent entry of air from its downstream end. This is not always necessary as an ejector action can be introduced by placing a bend or lip in the downstream end which deflects water when flowing over the crest at a slight depth to the opposite side of the barrel. As a result, the lower end of the barrel will be sealed. The flowing water will carry the air from the inside and gradually will reduce the pressure inside, sufficient for the siphonic action to start. After the siphon has started unless it is vented, the upstream water level will be drawn up to the level of the entrance before the flow stops. But

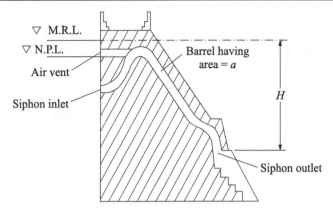

Figure 4.11   Siphon spillway.

usually the magnitude of the drawdown is controlled with the help of air vent. The discharge through the siphon is given by the equation

$$Q = C_d \, a \, \sqrt{2gH}$$

(4.13)

where

C_d = coefficient of discharge usually about 0.90,

a = area of the siphon barrel,

H = effective head, i.e. difference of water level in the reservoir and downstream end or the outlet level of the siphon.

4.3.6   Emergency Spillway

An emergency spillway is sometimes provided to take into consideration the probability of a higher flood occurring than the design flood within the life span of the project. For this purpose a suitable spillway capacity up to, say 25 per cent of the normal spillway is provided. In the event extraordinary floods occur, the damage caused by the operation of the emergency spillway may vary from the loss of the control apparatus to the complete washing out of the spillway structure and its foundation.

The location of the emergency spillway can be a low divide in the reservoir periphery. If the elevation of the divide is very low then a dyke is put at such an elevation so that it will be overtopped before the main dam overtops.

Problem 4.1

An overfall Ogee-shaped spillway has to be designed to pass the routed peak flood discharge of 5000 m³/s. The height of the spillway crest is at *RL* 185 m, and the average stream bed at the site equals *RL* 100 m. The spillway length consists of 10 spans having a clear width of 10 m. The piers provided are of 90° cut water nose type with K_p = 0.01 and the abutments are rounded having K_a = 0.1. The pier thickness can be taken as equal to 2.5 m. Assume the upstream face of the dam to be vertical and the downstream face to have a slope of IV : 0.7 *H*.

Solution

In the present case height of spillway from the river bed equals $185 - 100 = 85$ m. As such coefficient of discharge C may be taken as equal to 2.2. Again effective length of spillway comes to

$$L_e = L - 2 (n K_p + K_a) h$$

Now $Q = CL_e h^{3/2}$ first consider $Le \simeq L$, that is 100 m. Hence, $5000 = 2.2 \times 100 \times h^{3/2}$

or $h = \left[\dfrac{50}{2.2} \right]^{2/3} = 8.03$ m. Consider the actual value of h to be slightly more than the approximate value calculated above, i.e. $h = 8.20$ m

$$L_e = 100 - 2(9 \times 0.01 + .1) \times 8.2$$

$$= (100 - .38 \times 8.2) = (100 - 3.12)$$

$$= 96.88 \text{ m.}$$

Hence $5000 = 2.2 \times 96.88 \, h^{3/2}$

or $$h = \left[\dfrac{5000}{2.2 \times 96.88} \right]^{2/3} \simeq 8.20 \text{ m}$$

i.e. assumed value.

The downstream profile or a vertical upstream face is given by the equation (Fig. 4.1)

$$x^{1.85} = 2h_d^{0.85} \, y$$

or $$y = \dfrac{x^{1.85}}{2(8.2)^{0.85}} = \dfrac{x^{1.85}}{11.96}$$

The various coordinates of the downstream profile will be found out up to the tangent point corresponding to downstream dam slope of IV : 0.7 H.

i.e. $$\dfrac{dy}{dx} = \dfrac{1}{0.7}$$

or, differentiating the functional form of the *d/s* profile with respect to x we get:

$$\dfrac{dy}{dx} = \dfrac{1.85 \, x^{0.85}}{11.96} = \dfrac{1}{0.7}$$

or $$x^{0.85} = \dfrac{11.96}{1.85 \times 0.7}$$

or $$x = \left[\dfrac{11.96}{1.85 \times 0.7} \right]^{1/0.85} = [9.236]^{1.176}$$

$$= 13.66 \text{ m}$$

The coordinates from $x = 0$ to $x = 13.66$ m are given below:

x	1	3	5	7	9	11	13	13.66
$y = \dfrac{(\text{m})}{(\text{m})}\left(\dfrac{x^{1.85}}{11.96}\right)$	0.084	0.638	1.64	3.06	4.87	7.06	9.618	10.54

Thereafter, the *d/s* face is extended up to the river bed level with slope IV : 0.7 H. At the toe of the dam usually a reverse curve having a radius equal to about 0.25 times the height of the dam is provided. The upstream profile up to the distance equal to $0.262\, h_d$ from the crest may be obtained from the following relationship:

$$y = \frac{0.724\,(x + 0.262\, h_d)^{1.85}}{h_d^{0.58}} + 0.126\, h_d$$

$$- 0.432\, h_d^{0.375}\,(x + 0.226\, h_d)^{0.625}$$

or after substituting $h_d = 8.20$, the above relationship becomes

$$y = \frac{0.724\,(x + 2.15)^{1.85}}{5.98} + 1.033 - 0.95\,(x + 2.15)^{0.6}$$

$$= 0.12\,(x + 2.15)^{1.85} + 1.033 - 0.95\,(x + 2.15)^{0.6}$$

The curve should go up to $x = -2.15$ m and thereafter the face will be drawn vertical. The various coordinates corresponding to the above are:

x (m)	0.5	-1.0	-2.0	-2.15
y (m)	.037	.152	.746	1.033

Problem 4.2

A rectangular side channel spillway 100 m long is to be designed to carry a varying discharge of 4.0 m³/S per metre length. The longitudinal slope of the channel is 0.150, starting at an upstream elevation $Z_0 = 24$ m. Assuming Manning's roughness coefficient $n = 0.015$ and $\alpha = 1.0$, compute the flow profile for the design discharge.

Solution

The first step is to determine the control section. For its determination, the computation of the drop in water surface necessary to maintain a flow at the critical depth throughout the full length of the channel is made and is shown in Table 4.1. The cumulative drop in water surface is plotted as the heavy dashed line in Fig. 4.12 starting from an arbitrary elevation. The critical depths are then plotted from the dashed line and is shown by the chain, dotted line.

Table 4.1 Computation for the Determination of the Control Section.

x (m)	Δx (m)	Q (m³/s)	Q₁+Q₂ (m³/s)	Y_c (m)	V_c (m/s)	V₁+V₂ (m/s)	ΔQ (m³/s)	ΔV (m/s)	Δy_m (m)	R = $\frac{V^2 n^2}{R^{4/3}}$ Δx (m)	S_f Δx (m)	Δy₁ (10+11) (m)	Σ Δy₁
1	2	3	4	5	6	7	8	9	10	11	12	13	14
0													
5	5	20	20	1.37	3.66	3.66	20	3.66					
10	5	40	60	2.20	4.65	8.31	20	0.99	1.62	1.06	.023	1.643	1.643
20	10	80	120	3.44	5.81	10.42	40	1.16	2.48	1.26	.056	2.536	4.179
30	10	120	200	4.51	6.65	12.46	40	0.84	2.12	1.39	.064	2.184	6.363
40	10	160	280	5.46	7.32	13.97	40	0.67	1.90	1.46	.073	1.973	8.336
50	10	200	360	6.34	7.89	15.21	40	0.57	1.75	1.52	.080	1.830	10.166
60	10	240	440	7.16	8.38	16.27	40	0.49	1.63	1.56	.087	1.717	11.883
70	10	280	520	7.93	8.82	17.20	40	0.44	1.55	1.60	0.94	1.644	13.527
80	10	320	600	8.67	9.22	18.04	40	0.40	1.48	1.63	0.100	1.580	15.107
90	10	360	680	9.38	9.59	18.81	40	0.37	1.41	1.65	.106	1.516	16.623
100	10	400	760	10.06	9.93	19.52	40	0.34	1.31	1.69	0.111	1.421	18.044

Note: $\Delta y_m = \dfrac{Q_1(V_1+V_2)}{g(Q_1+Q_2)}\left[\Delta V + \dfrac{V_2}{Q_1}\Delta Q\right]$

$\Delta y_1 = \Delta y_m + S_f \Delta x$

$y_c = (q^2/g)^{1/3}$, where $q = Q/B$, B being the width of the channel

$V_c = (gy_c)^{1/2}$

$= $ hydraulic radius

Table 4.2 Computation of the Flow Profile of the Example.

Z_0	Δy_1	Z	y	A	Q	V	Q_1+Q_2	V_1+V_2	Q	ΔV	Δy_m	R	$S_f\Delta x = \dfrac{v^2 n^2}{R^{4/3}}\Delta x$	Δy_1 (14+16)
(m)	(m)	(m)	(m)	(m²)	(m³/s)	(m/s)	(m³/s)	(m/s)	(m³/s)	(m/s)	(m)	(m)	(m)	(m)
3	4	5	6	7	8	9	10	11	12	13	14	15	16	17
9.00	—	19.06	10.06	40.24	400.00	9.93	720	17.01	80	2.85	4.11	1.7	0.11	4.22
12.00	4.24	23.30	11.30	45.20	320.00	7.08	560	12.85	80	1.31	2.03	1.68	0.08	2.11
15.00	2.11	25.40	10.40	41.60	240.00	5.77	400	10.13	80	1.41	1.77	1.64	0.04	1.81
18.00	1.77	27.17	9.17	36.68	160.00	4.36	280	7.86	40	0.83	0.78	1.62	0.02	0.80
19.50	0.83	28.00	8.50	34.00	120.00	3.53	200	6.13	40	0.93	0.67	1.59	0.01	0.68
21.00	0.70	28.70	7.70	30.80	80.00	2.60	120	4.09	40	1.11	0.51	1.54	0.00	0.51
22.50	0.50	29.20	6.70	26.80	40.00	1.49	60	2.30	20	0.68	0.19	1.51	0.00	0.19
23.25	0.20	29.40	6.15	24.60	20.00	0.81								
24.00	0.067	29.47	5.47											

Note: The value of Δy_1 between $x = 0$ and $x = 5$ m is taken as $\dfrac{2V^2}{2g}$, i.e. $\dfrac{2\times 0.81^2}{2\times 9.81} = 0.067\ m$

x (1) (m)	Δx (2) (m)
100	0
80	20
60	20
40	20
30	10
20	10
10	10
5	5
0	5

The chain dotted line, therefore, represents the bottom of a fictitious channel where the flow is critical throughout. A tangent parallel to the bottom of the actual channel is then drawn to the bottom of the fictitious channel. The point of tangency at which the two bottoms have the same slope gives the location of the critical section which in the present case is at location 100 m. The flow profile computation is then carried out from the control section and is shown in Table 4.2. The final flow profile is also shown in Fig. 4.12.

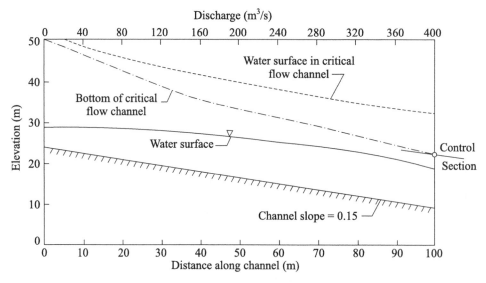

Figure 4.12   Computation of flow profile.

5

Flood Mitigation through Planning of Reservoir Capacities and Operation of Reservoirs

5.1 INTRODUCTION

On determining the maximum runoff and flood in a river following procedures outlined in earlier chapters, an economic study has to be made regarding the relative merits of the various methods of controlling the floods. Measures of flood control essentially consist of storing water in reservoirs and regulating the flow so that the channel downstream can safely accommodate and carry it. Of all the methods, use of natural reservoirs for detention of flood is most economical. By providing control works for the regulation of outflow and increase of capacities by construction of embankments along those which increases the depth of storage the utility of such reservoirs is greatly increased.

Planning of reservoirs is closely related to their operation and both problems should be critically considered. No definite rules can be laid dawn for planning the size of multipurpose reservoirs as much depends on circumstances. However, normally it is possible to reduce the reservoir storage requirements for improved operative technique and more reliable prediction of runoff.

The main purposes of the reservoirs are for the following: (i) flood control, and (ii) conservation. The aim of flood control storage is to hold over some of the flood water of a river when the discharge rate reaches a stage likely to cause damage to the valley downstream and to release the flood water gradually at a safe rate when the flood recedes. Conservation storage is meant to store the surplus water brought down by a river during periods when the natural flow exceeds current demand, and to use this stored water during periods when demand exceeds the natural flow. Conservation may be done for any one or more of the following purposes: (i) irrigation, (ii) hydropower, (iii) regulation of low water flow for navigation, (iv) public and industrial water supply, and (v) recreation, pisciculture, etc.

The reservoir may be a single-purpose conservation reservoir or a single-purpose flood control reservoir or a multipurpose reservoir. Multipurpose reservoirs are designed for two or more purposes. The common type of such reservoirs usually have three main objectives to serve, i.e., irrigation, power generation and flood control.

The purpose of live storage in a reservoir is to guarantee a certain quantity of water usually called safe (confirm) yield with a predetermined reliability. Though sediment is distributed to some extent in the space for live storage, the capacity of live storage is generally taken as the useful storage between the full reservoir level and the minimum drawdown level in the case of power projects, and dead storage in the case of irrigation projects. Flood storage depends on the height at which the maximum water level (MWL), is fixed above the normal conservation level (NCL). The determination of the MWL involves the routing of the design flood through the reservoir and spillway. When the spillway capacity provided is low the flood storage required for moderating a particular flood will be large and vice versa. A higher MWL means greater submergence and hence this aspect has also to be kept in view while fixing the MWL and the flood storage capacity of the reservoir. Apart from above by providing additional storage volume in the reservoir for sediment accumulation over and above the live storage it is ensured that the live storage although it contains sediment will function at full efficiency for an assigned number of years. This volume of storage is referred to as dead storage and is equivalent to the volume of sediment expected to be deposited in the reservoir during the designed life of the structures.

5.2 GENERAL DESIGN FACTORS

A. For fixing-up the live storage capacity of a reservoir the following data are needed:
   (i) Stream flow data for a sufficiently long period at the site.
   (ii) Evaporation losses from the water spread area of the reservoir and seepage losses and also the recharge into the reservoir when the reservoir is depleting.
   (iii) The contemplated irrigation, power or water supply demand.
   (iv) The storage capacity curve at the site.

B. For the determination of flood storage capacity at the MWL of the storage the information needed is:
   (i) Design flood or the inflow hydrograph. (Section 2.2.4.7.1 deals with the estimation of design flood.)
   (ii) The routing of the design flood through the reservoir and the spillway. (The various methods adopted for flood routing through reservoirs are discussed in the chapter on flood routing.)

C. For fixing up the dead storage volume, it is necessary to obtain the sediment yield into a reservoir and for this purpose the information needed are:
   (i) Result of sediment load measurements of the stream, and
   (ii) Sedimentation surveys of the reservoir with similar characteristics.

5.3 STORAGE CAPACITY DETERMINATION

Whatever be the size of use of a reservoir, the main function of a reservoir is to store water and thus to stabilise the flow of water. Hence, the most important physical characteristic of a reservoir is its storage capacity. A contoured plan of the entire reservoir area would be required for estimating the capacity and water spread of the reservoir at various elevations. As the reservoir area is usually large a ground survey of the entire area requires considerable time and expenditure and aerial surveys are coming increasingly into use for this purpose. The contour interval required depends on the height of the dam and the side slopes of the basin. A contour interval of 0.5 metre for dams of moderate height would normally be sufficient. For the works area in the vicinity of the proposed dam site a very accurate triangulation survey is required with a contour interval of about 1.5 m and a scale of 1/1000 or preferably 1/500. From the contour map of the reservoir area, the water spread of the reservoir at any elevation is directly determined by measuring the area at that contour by a planimeter. The elevation vs. water spread area may then be plotted. The capacity may be determined by taking contour areas at equal intervals and totalling up by trapezoidal rule or Simpson's rule or by cone formula as given below, in the above sequence:

$$V = h/2 \ (A_1 + A_2) \tag{5.1}$$

$$V = h/6 \ (A_1 + A_2 + 4A_m) \tag{5.2}$$

$$V = h/3 \ [A_1 + A_2 + (A_1 \ A_2)]^{1/2} \tag{5.3}$$

where V = volume between adjacent contours at an interval h, and having area A_1 and A_2; and A_m = area within a contour line midway between A_1 and A_2. The capacity and area at different elevations are generally plotted in the same graph.

5.4 LIVE STORAGE CAPACITY

The live storage capacity can be determined from stream flow records at the proposed reservoir site. In the absence of such records, the records from a station located upstream or downstream of the site on the stream or on a nearby stream should be adjusted to the reservoir site. The runoff records are often too short to include a critical drought periods. In such a case the records should be extended by comparison with longer stream-flow records in the vicinity or by the use of rainfall runoff relationship.

The total evaporation losses during a period are generally worked out roughly as the reduction in the depth of storage multiplied by the mean water-spread area between the full reservoir level and the minimum drawdown level (see Appendix I for detailed estimation of evaporation loss).

The live storage capacity can also be determined from flow hydrograph and flow duration curve and mass curves as explained below. The stream flow is usually presented in the form of a hydrograph.

A hydrograph is a graph which shows the discharge at any site of a stream as ordinate and the time as abscissa. The storage capacity of the reservoir can be obtained by superimposing the hydrograph of demand on the hydrograph of inflow at the reservoir site. The demand hydrograph may be a straight line in the case of a base-load water power project or navigation project. In other cases, as in water supply or irrigation project or in a peakload water power projects, there would be a variation in different seasons of the year and the demand line would vary. The area of the flow hydrograph below the demand hydrograph gives the reservoir capacity.

When the values of a hydrological events are arranged in the order of their descending magnitude, the per cent of time for each magnitude to be equalled or exceeded can be computed. A plotting of the magnitude as ordinates against the corresponding per cent of time as abscissa results in a duration curve. This does not take into account the chronological order in which the various magnitudes may occur. If the magnitude to be plotted is the discharge of a stream, the duration curve is known as flow duration curve. In a statistical sense the duration curve is a cumulative frequency curve of a continuous time series, displaying the relative duration of various magnitudes. The flow duration curve shows the period when discharges above or below selected discharges are available for a certain percentage of total time. For reservoir design studies the time period may be a month or a year depending on the reservoir size. The plotting of the curve when monthly mean discharge for a number of years are given may be carried out in two ways, i.e., calendar year basis or on total period basis. In one case there are 12 plotting points whereas in the other the number of points will be 12 multiplies by the number of years.

An example of the use of flow duration curve is given in Fig. 5.1. On the flow duration curve the area A below the assumed demand flow of 300 m³/s and the duration curve expressed in volume represents the total natural flow up to that rate available in the total period. The full rate of 300 m³/s is available 28 per cent of the time, and less than that rate is available for the remainder of the time. The area B, below the assumed flow of 300 m³/s but above the duration curve (expressed in volume) represents the amount of flow required from storage. This latter amount is not the required stored capacity of the reservoir because the stream flow may fluctuate above or below the demand rate and the reservoir may rise and fall several times during the total period for which the duration curve is drawn. The maximum required reservoir storage is determined from a study of the mass curve or the hydrograph as explained below.

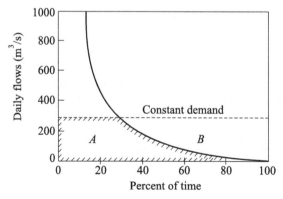

Figure 5.1 Flow duration curve.

5.4.1 Mass Curve

The mass curve [20] is a graph (Fig. 5.2) showing the cumulative net reservoir inflow, exclusive of upstream abstraction, as ordinate against time as abscissa. The ordinate may be denoted by depth in centimetres or in hectare metres or in any other unit of volume. Discharges with units of 10 days or a month may be used for cumulation in the mass curve. A segment of the mass curve is shown in Fig. 5.2. The difference in the ordinates at the ends of a segment of the mass curve gives the inflow volume during that time interval. Lines parallel to the line of uniform rate of demand are drawn at the points *b* and *c* of the mass curve. Let the line at *b* cut the mass curve at *d*.

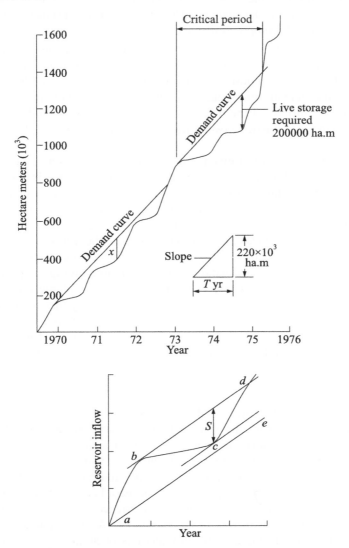

Figure 5.2 Mass curve and a segment of a mass curve.

The following information about the condition of the reservoir may be had from the mass curve:

(a) Inflow rate between *a* and *b* is more than the demand rate and the reservoir is full,

(b) Reservoir is just full at *b* as the inflow rate is equal to the demand rate,

(c) Reservoir storage is being drawn between *b* and *c* since the demand rate exceeds the inflow rate,

(d) Drawdown, *S*, is maximum at *c* due to demand rate equal to inflow rate,

(e) Reservoir is filling or in other words the drawdown is decreasing from *c* to *d* as the inflow rate is more than the demand rate, and

(f) Reservoir is full at *d* and from *d* to *b* again the reservoir is overflowing because the inflow rate exceeds the demand rate. The greatest vertical distance, *S*, at *c* is the storage required to meet the proposed demand.

The withdrawals from the reservoir to meet the irrigation demand are generally variable and in such cases the demand line becomes a curve instead of a straight line. The demand mass curve should be superimposed on the inflow mass curve on the same time-scale. When the inflow and demand mass curves intersect, the reservoir may be assumed to be full. For emptying conditions of the reservoir the demand curve would be above the inflow curve and the maximum ordinate between the two would indicate the live storage capacity required.

The live storage capacity for a given demand can be obtained as follows. Lines parallel to the demand lines are drawn at all the peak points of the mass inflow curve exclusive of upstream abstraction obtained from a long runoff record on daily or monthly or on an yearly basis (Fig. 5.2). When the demand line cuts the mass curve, the reservoir may be assumed to be full. The maximum ordinate between the demand line and the mass curve will give the live storage to meet the required demand. The vertical distance between the successive lines parallel to the demand line represents the surplus water from the reservoir.

Similarly, the estimation of demand from a given live storage capacity can be done as follows. The net inflow mass curve is plotted from the available records. The demand lines are drawn at peak points of the mass curve in such a way that the maximum ordinate between the demand line and the mass curve is equal to the specified live storage. The demand lines shall intersect the mass curve when extended forward. The slope of the flattest line indicate the firm demand that could be met by the given live storage capacity.

Comments

The capacity determination for any actual storage scheme involves a study of the seasonal storage. In regions where stream flows do not greatly vary on an yearly basis or in cases where the demand is only a small fraction of the mean annual flow, it is not necessary to go in for complete development of the catchment areas and the reservoirs generally refill within the annual hydrological cycle. From the approach indicated earlier, the seasonal storage required to maintain a given draft during, each year of record can easily be computed. From the series of storage values so obtained and from the experience with regard to drafts, the design storage for a specific purpose will have to be determined. A reasonable and uniform basis of design can be obtained by making a probability analysis of an arranged set of storage values. A reasonable frequency may be assumed and the design storage value may be selected on that basis. In underdeveloped areas the few records that are available are not as long as even 20 years. An estimation of the frequencies or of recurrence intervals requires extrapolation from available data. Probability plots may be adopted for this purpose. However, these plots must be prepared and used with considerable caution. The usual normal probability plot does not give a straight line of best fit because with a small number of observations, the frequency distribution is usually askew. For seasonal storage it is reasonable to assume that the extrapolation needed will not extend to extremely rare values of storage and as such the log probability law can be used with advantage if it is found to fit the data better.

5.5 FLOOD STORAGE

The purpose of flood storage and a reservoir is to regulate stream flow to attain flood control objectives for a give area consistent with tangible and intangible benefits. The degree of protection may be high or low depending on the objectives. Storage required during a particular flood is equal to the volume of inflow into the reservoir, minus the volume of outflow during a corresponding period. The primary objective of a flood control reservoir may be to minimize flood damages immediately below the dam site and as such the peak rate of reservoir release will usually equal to the maximum safe channel capacity through the protected area minus a small allowance for local inflow and margin of safety. Assuming the maximum rate of outflow thus established, floods of various magnitudes such as design flood, standard design flood, etc., are routed through the reservoir to determine the storage space required for effective regulation. The costs of providing various amounts of storage space are compared with average, annual flood control benefit to be expected in order to determine benefit-cost ratio. Final selection of flood control capacity is based either on economic probability analysis or based on performance study. The design storage capacities for flood control are fixed at values required to conform with the design flood condition assuming the limiting reservoir release rates and schedules specified in a reservoir regulation plan.

5.6 DEAD STORAGE

The lowest water level which has to be kept under the normal operating conditions in a reservoir is called the minimum pool level or the dead storage level. This is usually fixed after taking into account: (i) the sediment accumulation in the reservoir, (ii) elevation of the lowest outlet in the dam, and (iii) the minimum head required for efficient functioning of turbines.

In India most of the streams carry considerable quantities of sediment and any large storage reservoir will permanently store practically the entire sediment load of the stream. Sediment deposition in the reservoir causes loss of storage volume and the deposition above the reservoir causes backwater effects, swamping and flooding in the valley. The silting of reservoirs have both a long range and short range effect. In the short range the effect of silting on the economics of project development need to be considered and in the long run the loss of suitable large reservoir sites being lost forever if allowed to fill up by sediment must be taken into account. The rate of reservoir sedimentation must be estimated accurately in order to ascertain the useful life of the reservoir.

The movement of sediment into a reservoir is governed by two dynamic forces acting on the individual particle. One is a horizontal component acting in the direction of flow owing to the force of water movement and the second a vertical component due to gravitational pull and fluctuating forces in both upward and downward directions due to turbulence. On entering the reservoir, the cross-sectional area of the inflow increase and this is accompanied by a decrease in the velocity and a dampening of turbulence until both the components become ineffective in transporting the sediment and the particles begin to deposit.

The distribution pattern of sediment in the entire depth of a reservoir depends on many factors, such as slope of the valley, length of reservoir, construction in the reservoir, particle size of the suspended sediment and capacity inflow ratio. But the reservoir operation has an important control over other factors. However, a knowledge of this pattern is essential, especially in developing areas, in order to have an idea about the formation of deltas and the recreational spots and the consequent increase in backwater levels after the reservoir comes into operation.

5.6.1   Reservoir Silting (see Appendix II, IV, and V)

The important factors which affect the rate of silting are:

   (a) Quality, quantity and concentration of sediment brought down by the river,
   (b) Size of the reservoir,
   (c) Length of the reservoir,
   (d) Steepness of the thalweg,
   (e) Ratio of reservoir capacity to annual runoff,
   (f) Method of reservoir operation,
   (g) Nature of spillway,
   (h) Exposure of depleted material,
   (i) Depth and age of the sediment deposited, and
   (j) Depth and age of the head of the reservoir.

The rate of silt deposition diminishes as the reservoir is filled. With the increase of deposit lesser volume of water standing at its top is available for deposition of finer materials. This will result an increasing amount of material entering, which will be carried away by the spillway. Further, during a high flood the freshly deposited sediment may be picked up which, in turn, will flow out of the reservoir. In reservoirs whose capacity is relatively small, part of the annual flow through them which cause such effects are common.

In the estimation of reduction of capacity and in the tackling of other problems related to reservoir silting the following information is needed:
(a) rates of sediment inflow into the reservoir, (b) trap efficiency, (c) location and extent of sediment deposition, (d) unit weight of sediment deposits, and (e) control of reservoir silting. The amount of sediment Q_{td} that will be deposited in the reservoir during any given period can be written as

$$Q_{td} = Q_{ti} - (Q_{to} + Q_{ts}) \tag{5.4}$$

where

   Q_{ti} = sediment moving into the reservoir during the chosen time interval,

   Q_{to} = total sediment going out of the reservoir through sluices, head regulator and spill-way, etc.,

   Q_{ts} = change in the amount of sediment held in suspension in stored water during the same time interval. Assuming that the change in the amount of sediment held in suspension in the stored water is very small compared to the total sediment inflow, we have

$$Q_{td} \simeq Q_{ti} - Q_{to}. \tag{5.5}$$

The quantity of sediment entering into the reservoir can be ascertained following appropriate sediment transport relationship and field measurement.

5.6.1.1 Sediment outflow

The amount of sediment going out of the reservoir will depend on factors like, sediment sizes in suspension, location of outlet work, ratio of storage capacity of the reservoir to the annual runoff at the site, and quantity of water going out through the canal head regulator, the sluices and over the spillway. Reservoirs in a way can be compared to the action of sedimentation tank or settling tank but its conditions are more complicated because of inflow rate and the change of depth of the reservoir.

5.6.1.2 Trap efficiency

The efficiency of the sediment trapped in the reservoir depends on the ratio of storage capacity to inflow, age of reservoir, shape of reservoir basin, the type and location of outlets, method of reservoir operation and sediment size. If retention time is defined as the time that the water requires to pass through the reservoir, it can be seen that the trap efficiency increases with increase in retention time. Thus a storage reservoir which retains water for months will have a high trap η, whereas a smaller flood control reservoir will have a lower trap η on account of smaller period for which the silt-laden water is retained. The drainage area of the basin can be taken as an index of annual runoff and the storage capacity per hectare or square mile of drainage area may be assumed to give the relative detention time. The graphical relationship showing the trap η against storage capacity per square mile of the catchment area for some American reservoirs, indicates that the mean curve for the data can be represented by

$$T_e = 100 \left[1 - \frac{1}{1 + 0.1 \, C/W_D} \right] \tag{5.6}$$

even though there exists scatter of points where T_e = trap η, C = capacity in acft and W_D = watershed area in square miles. It will be realised that for reservoirs with large values of C/W_D, the trap η will be about 100 per cent for the first few years and will then slowly decrease. The reasons for the scatter are that C/W_D must be used as a governing parameter only within definite hydrologic regions, and not as an index of comparison over the country as a whole for with the same C/W_D ratio, the η will increase as the runoff per unit area decreases.

After analysing the data for TVA reservoirs, Churchill came to the conclusion that along with retention time the transit velocity that is the velocity with which the water flows in the reservoir governs the trap η. If the water held in the reservoir is moving fairly rapidly in the reservoir, very little sedimentation will occur because of the turbulence associated with higher velocity hinders settling even though the retention time may be high. He introduced a parameter known as sedimentation index which is the ratio of period of retention to travel velocity.

Brune [21] has analysed the records of fortyfour different reservoir in USA (41 of which were normal ponded ones) and found that the capacity inflow ratio C/I gives better correlation than the capacity to watershed area ratio. Figure 5.3 shows the Brune's plot of the trap η against C/I ratio. It can be observed that reservoirs having $C/I > 1$, T_e = 100 per cent. As suggested by Maddock [22], reservoirs with C/I greater than 1 can be classified as holdover storage reservoirs while for reservoir with $C/I < 1$ can be called as seasonal storage reservoirs. Brune found that sluicing operations can reduce the T_e by approximately 2 to 5 per cent provided the sluicing operations are timed to flush the density currents as they reach the dam. Also when a reservoir is operated as a flood control reservoir and large flows of water are allowed to pass through the dam, the T_e will be much less than that of a comparable normal ponded reservoir. Equation (5.6) is by far the best guide to the T_e of reservoirs and can be used in their planning.

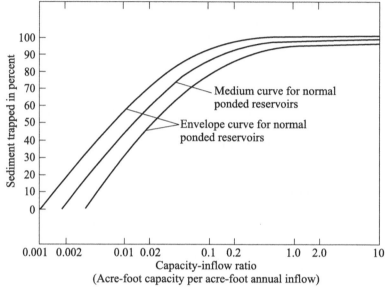

Figure 5.3 Trap efficiency vs. capacity inflow ratio after Brune.

5.6.1.3 Computation of unit weight

The initial weight is computed first by using a given set of size analysis data of suspended sediments. The unit weight is then determined for a 100-year compaction period. The equation to compute the unit weight after any given compaction period is given by

$$\gamma_T = \gamma_1 + 0.4343\, k \left[\left(\frac{1}{T-1} \right) + \log_e T - 1 \right] \tag{5.7}$$

where k = compaction factor or $\gamma_{100} = \gamma_1 + 1.588\, k$, for 100-year compaction period. The initial unit weight and compaction value of soil has been provided by Lane and Koelzer[*] and Trask[**] [23] as follows: (Table 5.l):

[*]Original reference Lane. E.W. and Koelzer, V.A., Density of sediments deposited in reservoirs, Report No. 9 of A study of methods used in measurement and analysis of sediment loads in streams: St. Paul United States Engineering District. St. Paul, Minn, Minnesota, 1953.

[**]Trask, P., Compaction of sediments, Bulletin, American Association of Petroleum Geologists, Vol. 15, Tulsa, Oklahoma, 1931 pp. 271–276.

Table 5.1 Initial Unit Weight in lb/cft and Compaction Values of Soil.

Reservoir operation	Lane and Koelzer						Trask					
	Sand		Silt		Clay		Sand		Silt		Clay	
	γ	k	γ	k	γ	k	γ	k	γ	k	γ	k
Sediment always submerged or nearly submerged	93	0	65	5.7	30	16.0	88	0	67	5.7	13	16
Normally a moderate reservoir drawdown	93	0	74	2.7	46	10.7	88	0	76	2.7	–	10.7
Normally considerable reservoir drawdown	93	0	79	1.0	60	6.0	88	0	81	1.0	–	6.0
Reservoir normally empty	93	0	82	0.0	78	0.0	88	0	84	0.0	–	0.0

As for example consider a suspended sediment having the following size analysis, i.e. 12 per cent clay, 42 per cent silt and 46 per cent sand and for a normally moderate reservoir drawdown we have,

$$\gamma_1 = 0.12 \times 46 + 0.42 \times 74 + 0.46 \times 93$$

$$= 79.4 \text{ lb/cft}$$

The unit weight after 100-year compaction is then

$$k = 0.12 \times 10.7 + 0.42 \times 2.7 + 0.46 \times 0$$

$$= 2.22$$

Therefore,

$$\gamma_{100} = \gamma_1 + 1.5888\,k = 79.4 + 1.588 \times 2.22$$

$$= 92.91 \text{ lb/cft}.$$

With the knowledge of unit specific weight of sediment after compaction in a number of years and the trap efficiency of the reservoir, it is possible to compute the total amount of sediment accumulation over the years, either in, acre-ft or in cubic metre after appropriate conversion.

5.6.1.4 Measurement of sediment yields

The sediment yield in a reservoir can be measured by any one of the following two methods:

(a) Sedimentation surveys of reservoirs with similar catchment characteristics, or

(b) Sediment load measurements of the stream.

5.6.1.4.1 *Reservoir sedimentation surveys*

The sediment yield from the catchment is determined by measuring the accumulated sediment in a reservoir for a known period by means of echo sounders and other electronic devices (Appendix III for details) since the normal sounding operations give erroneous results in large depths. The volume of sediment accumulated in a reservoir is computed as difference between the present reservoir capacity and the original capacity after the completion of the dam. The specific weight of deposit is determined in the laboratory from the representative undisturbed samples or by field determination using a calibrated density probe developed for this purpose. The total sediment volume is then converted to dry-weight of sediment on the basis of average specific weight of deposits. The total sediment yield for the period of record covered by the survey will then be equal to the total weight of the sediment deposited in the reservoir plus that which has passed out of the reservoir based on the trap efficiency. In this way, reliable records may be readily and economically obtained on a long-term basis.

5.6.1.4.2 Sediment load measurements

Periodic samples from the stream should be taken at various discharges, along with the stream gauging observations and the suspended sediment concentration should be measured as detailed in relevant Indian Standards [5]. A sediment rating curve which is a plot for sediment concentration against the discharges then prepared and used as a rough guide in the estimation of suspended sediment load. Although this is not quite reliable as the sediment concentration considerably differs for the same stage and discharge, depending upon the various parts of the monsoon.

The bed load is not generally measured but only sampled. It is estimated by analytical methods and should be added to the suspended load to get the total sediment load (Appendix II).

5.6.1.5 Distribution of sediment in a reservoir

When the sediment-laden water is impounded the sediment is distributed throughout the slope of the reservoir. At the beginning of the impounding the coarser portion of the sediment load is deposited near the head of the reservoir and the fine particles are transported by density currents towards the dam (Appendix II). As deposition continues the resulting delta front moves progressively down into the reservoir. Very little of the sediment really passes through the outlets (Fig. 5.4).

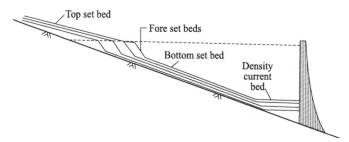

Figure 5.4 Sediment deposit profile.

By providing additional storage volume in the reservoir for sediment accumulation over and above the live storage it is ensured that the live storage will function at full efficiency for the entire life-span of the reservoir. Thus the quantity and the distribution pattern of sediment are the two most important factors to be considered in fixing up storage capacity of a reservoir. How to estimate the quantity has been dealt with earlier. As regards the distribution pattern it depends to a large extent on the reservoir operation.

As a result of detailed analysis of the deposition pattern in 30 reservoirs in USA, Borland and Miller [24] have presented six methods for predicting the distribution pattern of sediment in the reservoir. Of these, the empirical area reduction method seems to be the most rational and is described here. This is based on the adjustment of the original surface areas to reflect the decrease in the area with sedimentation. In the method the reservoirs are classified in 4 broad types depending on the approximate slope of the line of reservoir capacity vs. depth plotted on a log-log scale. The categories are listed in the (Table 5.2.)

Table 5.2 Classification of Reservoirs.

Type	Description	Slope of line of capacity vs. depth	Position of deposition	C	m	n
I	Lake	3.5 to 4.5	Top	3.42	1.5	0.20
II	Flood plain foothill	2.5 to 3.5	Upper middle	2.32	0.50	0.40
III	Hill	1.5 to 2.5	Lower middle	15.88	1.10	2.30
IV	Gorge	1.0 to 1.5	Bottom	4.23	0.10	2.50

Thus after plotting reservoir capacity against depth and determining the slope of this line for the reservoir under consideration one can refer to Table 5.2 to determine the type to which the reservoir belongs. After classifying the reservoir with capacity ranging from 4×10^4 to 3×10^7 acft and analysing the deposited sediment 4 standard type curves of per cent sediment deposit vs. per cent reservoir depth were prepared (Fig. 5.5). These curves were then converted to area design curves for use in computation. If p represents the relative depth above stream

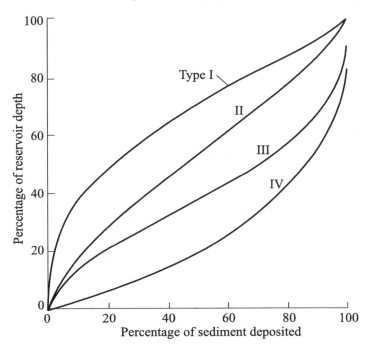

Figure 5.5 Per cent of sediment deposition as a function of reservoir depth.

bed and A_p represents dimensionless relative area p, then the relation between A_p and p for the 4 categories was of the form

$$A_p = C \, p^m \, (1 - p)^n \tag{5.8}$$

here A_p can be defined as A_s/k_1, where A_s is area of sediment deposit at p and k_1 is given by

$$k_1 = \int_0^{1.0} A_s \, dp \qquad (5.9)$$

The values of exponents m and n as well as the value of constant C are listed in the Table 5.2. The following information is useful in predicting the position of deposition, surface areas and capacities at various elevations after a known volume of sediment has been deposited. The procedure is one of trial and error and starts with a probable sediment elevation at the dam, alternately it can be found out by adopting Moody's method [5].

5.6.1.5.1 *Moody's method to find new zero elevation*

This method is used to determine the new zero elevation y_0, directly without trial and error process. Two parameters $f(p)$ and $f'(p)$ as explained below are made uses of:

$$f(p) = \frac{1 - V(p)}{a(p)} \qquad (5.10)$$

and

$$f'(p) = \frac{S - V(pH)}{H A(pH)} \qquad (5.11)$$

where

$f(p)$ = a function of the relative depth of reservoir for one of the four types of theoretical design curves,

$V(p)$ = relative volume at a given elevation,

$a(p)$ = relative area at a given elevation,

$f'(p)$ = a function of the relative depth of reservoir for a particular reservoir and its sediment storage,

S = total sediment in the reservoir in hectare metres,

$V(pH)$ = reservoir capacity at a given elevation in hectare metres,

H = the total depth of reservoir for normal water surface in metres, and

$A(pH)$ = reservoir area at a given elevation in hectares.

Table 5.3 gives the values of the function $f(p)$ for the four types of reservoirs from which the plotting of $f(p)$ against relative reservoir depth, p, for the four types of reservoirs of the empirical area method can be made.

To determine the new zero elevation $f(p)$ should equal $f'(p)$. This is done graphically by plotting the values of $f'(p)$ and superposing this over the relevant $f(p)$ curve. The intersection gives the relative depth of reservoir at new zero elevation after sedimentation. After arriving at the new zero elevation empirical area method is used for distribution of sediment along the reservoir.

Table 5.3 Values of the Function $f(p)$ for the Four Types of Reservoirs

p	Type			
	I	*II*	*III*	*IV*
(1)	*(2)*	*(3)*	*(4)*	*(5)*
0				
0.01	996.7	10.568	12.03	0.2033
0.02	277.5	3.758	5.544	0.2390
0.05	51.49	2.233	2.057	0.2796
0.1	14.53	1.495	1.013	0.2911
0.15	66.971	1.169	0.6821	0.2932
0.2	4.145	0.9706	0.5180	0.2878
0.25	2.766	0.8299	0.4178	0.2781
0.3	1.900	0.7212	0.3486	0.2556
0.35	1.495	0.3623	0.2968	0.2518
0.4	1.109	0.5565	0.2333	0.2365
0.45	0.9076	0.4900	0.2212	0.2197
0.5	0.7267	0.4303	0.1917	0.2010
0.55	0.5860	0.3768	0.1687	0.1826
0.6	0.4732	0.3253	0.1422	0.1637
0.65	0.3805	0.2780	0.1207	0.1443
0.7	0.3026	0.2333	0.1008	0.1245
0.75	0.2359	0.1907	0.08204	0.1044
0.8	0.1777	0.1500	0.06428	0.08397
0.85	0.1202	0.1107	0.04731	0.06330
0.9	0.08011	0.07276	0.03101	0.04239
0.95	0.05830	0.02698	0.01527	0.02123
0.98	0.01494	0.01425	0.006057	0.008534
0.99	0.007411	0.007109	0.003020	0.002470
1.0	0.00	0.00	0.00	0.00

5.6.1.6 Useful life of reservoirs

The density of the deposited sediment plays an important role in determining the life of a reservoir. Coarse grain sediment undergoes very little changes after deposition. Fine grade material on the other hand have a high initial porosity and considerably changes in volume due to drying or superimposed load. Studies made by Khosla [33] on a large number of rivers in the world indicates that for most rivers the upper limit of the rate of silting is 42862–47600 m^3/100 km^2 of catchment area per annum. He suggested that the rate of sediment deposition is given by $Y = 5.19/A^{0.28}$ for catchment area less than 2600 km^2 where Y = annual silt deposited in acft in the reservoir per 100 sq. miles of catchment area. For larger areas the rate is 35700 m^3/100 km^2.

Many of the reservoirs built in India have been designed to provide for silt accumulation at the rate of 75–92 acft/year per 100 sq. miles of catchment based on observations, made on small reservoirs during pre-independence period. But recent observations on completed reservoirs show that in many of the cases the sediment accumulation is almost double the above figure. The useful life of a reservoir may be considered from two points of view mainly—(i) factors that affect its span of life, and (ii) factors that affect its usefulness. As for illustration consider the following problem for the Beas river as worked out by the Bhakra Designs Directorate:

Annual silt load $= 40 \times 10^6$ tons

Average density of silt deposits $= 65$ lbs/cft

Volume of annual silt load

$$\frac{40 \times 10^6 \times 2240}{65 \times 43560} = 32000 \text{ acft,}$$

Coarse silt (33 per cent) = 11000 acft,

Colloidal silt (10 per cent) = 3000 acft,

Medium silt = 18000 acft.

It is assumed that the entire colloidal silt and 5 per cent of medium silt will flow out as density current through its low level outlets. Out of the remaining silt the entire coarse silt will deposit within the reservoir starting from the head of the lake and forming an upper delta. The medium silt will first deposit on top of the upper delta above the reservoir causing aggradation of the river slope. After a regime slope above reservoir level is attained the medium silt will travel into the reservoir ahead of the upper delta and as the reservoir depletes, it will work into the dead storage space. In the Bhakra studies it was determined that about 50 per cent of the medium silt gets deposited above the reservoir and only 50 per cent goes into the dead storage.

On the above basis the life of the reservoir with 10.5 million acft as dead storage and 5.5 m acft being live storage may be worked out as under:

Quantity of annual medium silt going out as density currents $= 5/100 \times 18000$
$$\simeq 1000 \text{ acft.}$$

Quantities of annual silt deposit in the dead storage $= \dfrac{1}{2} \times 17000$

$$= 8500 \text{ acft.}$$

Number of years required for silting up of the dead storage $= \dfrac{1.05 \times 10^7}{8500}$

Therefore, the useful life of the reservoir = 123 years

Quantities of coarse silt deposited in live storage annually = 11000 acft.

Number of years required for silting up of the live capacity $= \dfrac{5.5 \times 10^6}{11000}$

Therefore, ultimate life of the reservoir = 500 years.

After the medium silt deposit within the dead storage reaches the dam, the entire medium silt coming into the reservoir may be expected to flow out through the outlets. Thus the ultimate life of the reservoir may be taken as 500 years. Even at the ultimate stage, part of the water stored within the pore space of the silt deposit could be available by gravity flow for all time. Thus the entire live storage is not lost, which may be as high as 20 per cent of the full capacity. After about 100 years when the dead storage may be nearly full, the encroachment on live storage by coarse sediments may amount to $11000 \times 100 = 1.1$ m. acft as against live storage of 5.5 m. acft, i.e., reduction in usefulness by about 20 per cent, which few projects could sustain.

Although the sediment is distributed throughout the reservoir slope, for design of medium and minor reservoirs the sediment loads are assumed to be deposited in the lower-most part of the reservoirs. The level corresponding to the total sediment loads for the space of the reservoir is taken as dead storage level below which the water level is not allowed to fall during reservoir operation. In power projects minimum drawdown level is kept little higher than the dead storage level to prevent formation of vortex and air entrainment in the intake structures. For important reservoirs the distribution of sediment along the reservoir slope is considered while finalising the area capacity curves.

5.6.1.7   Environmental effects of dams and reservoirs

Construction of dams and reservoirs will cause a number of environmental effects to mitigate which appropriate action plans are required. The main effects are briefly mentioned or outlined herein. Goodlands et al. [40]

(i) *Land losses*: Large tracts of agricultural lands, forests or other wild lands are likely to be inundated which can be minimized by careful site selection. It is necessary to consider the value of timber and other mineral resources to be lost and the nonavailabilty of inundated land for agricultural or other purpose in the economic analysis.

(ii) *Plant and animal life*: It is necessary to carryout biotic survey, in order to prevent or minimize plant and animal life extinction by careful site selection. Loss of wild life may be mitigated by creation of a wild life management area elsewhere in the country equivalent to the inundated tract. Animal rescue, replenishment and relocation can be useful. Provision of crossing facilities over canal and others are generally essential.

(iii) *Fish and other aquatic life*: Fish migration will be effected even with the provision of fish passaging facilities. Spawning areas, aquaculture improved fishing methods and marketing may require special attention due to construction of reservoirs. A reduced supply of nutrients downstream and to estuaries may impair fishery productivity. Inter basin transfers may threaten aquatic species by introducing new predators or competitors.

(iv) *Health problems*: Some water related diseases e.g., schistomiasis, malaria, Japanese encephalitis may increase unless precautions or mitigatory measures are undertaken. Vector control, environmental modifications and education of residents may be necessary and it should be incorporated into the project. Proliferation of water hyacinth and water lettuce will impair water quality and increase disease vectors and loss of water through evapotranspiration. The clogging of water surface as a result of weeds will hamper navigation, recreation, fisheries and irrigation. The use of weeds for composting, biogas or fodder needs to be investigated.

(v) Suitability of water quality for drinking, irrigation, fisheries and other uses both within the reservoirs and downstream need to be addressed. The various issues are: salinity intrusions, water retention time, loss of flushing, increased level of nutrients in the reservoir, pollution due to agricultural leachates, pathogens, industrial effluents etc, raising or contamination of water table and salinization. Inundated vegetation on the bottom of reservoirs decomposes, thereby consuming large volume of oxygen. If thermal stratification occurs, mixing of surface and bottom water is hampered, bottom water then becoming anaerobic. Anaerobic decomposition of organic materials produces noxious gases toxic to aquatic life and harmful to machinery. When such waters are discharged downstream by the dam, fishes may be killed. Multiple level outlets in the dam can prevent discharge of anaerobic water. Conversion of forests to timber before reservoir filling reduces the projects contribution to green house gases.

(vi) *Erosion*: Upstream erosion in the catchment area leads to sedimentation or land slips which may impair storage, so catchment area management be encouraged, increased erosivity of the water on the river bed and structures below the dam including deltaic and coastal changes need to be considered during project preparation.

(vii) *Downstream hydrology*: Changes in downstream hydrology can hamper ecosystems dependent on seasonal flooding, including areas important for fisheries, flood plains, lagoons, marshes and mangroves or traditional flood recession agriculture. Hydroelectric and other developments should be concentrated on the same rivers if hydrological risks and other circumstances permit in order to preserve a representative sample of the river in the natural state and it should be considered as a part of the trade of.

(viii) *Multiple use and other considerations*: Multiple use be addressed keeping in view tourism, irrigation, fisheries, bird and other biotic sanctuaries and recreation. Water flow regulation can convert seasonal rivers into perennial waterways, reduce flooding and improve drinking and irrigation.

Other aspects that need to be examined are, involuntary settlements, tribal people, riparian rights, cultural property and safety of dams.

5.7 RESERVOIR OPERATION FUNDAMENTALS

With reference to Fig. 5.6 which furnishes a line diagram of reservoir system and various reservoir storages, the basic reservoir operation criterion can be expressed as follows, Srivastava [41]

$$S = S_t + I_t + I'_t + P_t - EI_t - O_t - O'_t \text{ for all values of time,} \qquad (5.12)$$

where,

S_t = reservoir storage at the beginning of time t,

I_t = inflow into reservoir during time t,

I'_t = local inflow to the reservoir from surrounding area in time t,

P_t = precipitation in the reservoir in time t,

EI_t = evaporation losses from the reservoir in time t,

O_t = total outflow, i.e. release from the reservoir in time t,

O'_t = release to natural channzzel from reservoir in time t, and

S = reservoir storage at the end of time t.

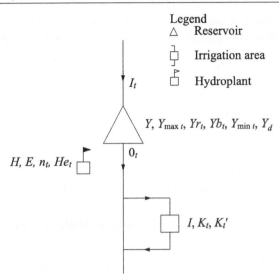

Figure 5.6(a) Line Diagram of Reservoir System.

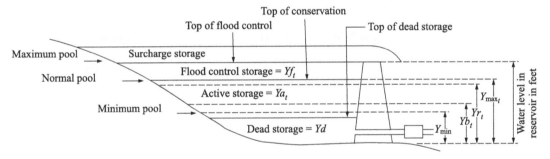

Figure 5.6(b) Various Reservoir Storage.

5.7.1 Operation of Reservoir with Conventional Rule

For reservoir operation the simplest operating rule is to supply all the water demanded, if available. The release is then independent of reservoir content and season. If there is sufficient water in the reservoir to meet the required releases the reservoir empties and this is called the conventional rule. The reservoir will operate under the following basic constraints. The volume of water released during any period cannot exceed the contents of the reservoir, at the beginning plus the flow into the reservoir during the period, i.e.

$$O_t \leq S_{t-1} - Y_{\min} + I_t + P_t + I'_t - O'_t - EI_t \text{ for all } t, \tag{5.13}$$

where

$$O_t = O'_{a_t} + S_{p_t}$$

The continuity equation for the reservoir is defined as

$$S_t = S_{t-1} + I_t + P_t + I'_t - O'_t - EI_t - O_t \text{ for all } t \tag{5.14}$$

The contents of the reservoir at any period cannot exceed the capacity of the reservoir as well as dead storage of the reservoir puts lower limit on the reservoir storage, such that

$$Y_d = Y_{\min_t} \leq S_{t-1} \leq Y \quad \text{for all } t \qquad (5.15)$$

where

O'_t = actual irrigation release from reservoir in time t,

S_{p_t} = reservoir spill in time t,

$S_{(t-1)}$ = storage at time $t-1$,

Y_d = dead storage of reservoir,

Y = total reservior capacity at max^m pool level,

$Y_{\min_t}$ = variable capacity upto min^m pool of reservoir in time t.

5.7.2 Operation of Reservoir with Single Rule Curve

Rule curve operation (vide Fig. 5.7) with constraints, (5.13) and (5.14) and a modified constraint of Eq. (5.15) may be such that as the storage of water in the reservoir decreases restrictions may be imposed in the uses so that the demand falls and releases are lowered.

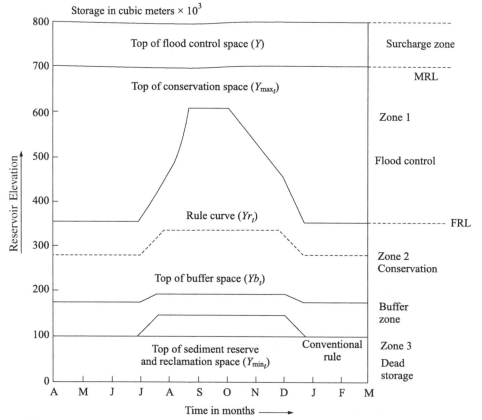

Figure 5.7 Operational zones and rule curve illustrating seasonally varying storage requirements of a multipurpose reservoir.

So the constraint is expressed as

$$Y_d \leq Y_{\min_t} \leq S_{t-1} \leq y \quad \text{for all } t \tag{5.16}$$

The important question is whether the stored water in the reservoir is to be used at present or be retained for use during possible drought years. A rule curve may be defined as a diagram showing reservoir storage requirements during the year. The persons responsible for operation of the reservoirs are expected to maintain these levels as closely as possible while generally trying to satisfy various water needs downstream. If the reservoir storage levels are above the target or desired levels, the release rates are increased, conversely the release rates are lowered if the levels are below the target. These release rates may or may not be specified but will depend in part on any maximum or minimum flow requirements and on the expected inflow.

5.7.3 Reservoir Operation with Zoning or Partitioning

In a multipurpose reservoir when the demands for water are competitive for the same time period, storage volumes in the reservoir must then be allocated to meet these competitive demands. This portioning process involves both the determination of required volumes and establishing operating rules to specify how the reservoir is to be managed. The elevations of the various zones are used as guides for operation and can vary seasonally. The five zones that are considered are as follows: (Fig. 5.7)

(i)   *Surcharge Zone*: The storage above the flood control zone associated with actual flood damage. Top of the flood control pool is used to maintain the integrity of the reservoir. Reservoir releases are usually at or near their maximum to prevent the dam from collapsing when the storage volume is within this zone.

(ii) *Flood control Zone*: A reserve for storing large inflows during periods of abnormally high runoff. The flood control zone would be evacuated of water at a time corresponding to the flood season. The reservoir would then be kept at the top of conservation space (bottom of flood control space), to provide sufficient storage to control flooding. Once the pool elevation is in this zone, the reservoir is operated to release the maximum amount of water without causing flooding ideally this would coincide with the bank full conditions downstream.

(iii) *Conservation Zone*: The zone of storage from which various water based needs are satisfied. The ideal storage volume or level is normally located within this zone. A system of priorities may be established within this zone to ensure that vital water requirement will be met.

(iv) Buffer Zone the storage beneath the conservation zone entered only in abnormally dry periods. When storage volumes within this zone, releases are restricted temporarily to satisfy high priority demands only.

(v)   *Inactive Zone*: The dead storage beneath the buffer zone which would if possible be entered only under extremely dry conditions. Reservoir withdrawal are an absolute minimum. This zone contains enough space to trap and retain sediment over the life of the project.

5.7.4 Ideal Reservoir Operation for Flood Control

The ideal operation of a reservoir for flood control is shown in Fig. 5.8 where *ABCDE*, represents the inflow hydrograph. The line *ZZ*, represents the non damaging carrying capacity of

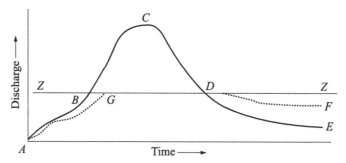

Figure 5.8 Ideal operation of a reservoir for flood control.

the reservoir. From point *B* to *D*, the natural flow in the river exceeds its safe bearing capacity. If there is no reservoir from the time corresponding to the point up to *D*, the flood water will spill out the channel banks and will cause damage. The regulation of flow for controlling the flood is given by dotted line *AGDF*. The releases is gradually increased from point *A* onwards making sure that at no point the release exceeds the safe bearing capacity. This is achieved by storing the curve *BCD* in the reservoir and after point *D*, the reservoir is gradually emptied. The ideal operation is possible when foreknowledge of hydrograph is available which is not the normal case. Therefore it is necessary to moderate the flood flows that enters the reservoir and this is achieved by storing part of the inflows in the reservoir and making releases so that damages in the downstream region is minimum. The degree of flood attenuation depends on the empty storage space available in the reservoir when the flood impinges on it. Figure 5.9 shows the normal mode of flood control operation of a reservoir which is to make releases

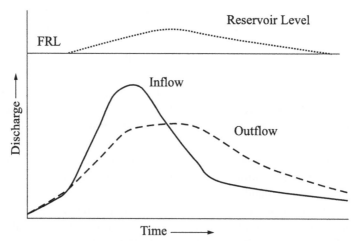

Figure 5.9 Normal mode of flood control operation of a reservoir.

equal to or less than the safe capacity of the downstream channel as long as there is empty space in the reservoir. The reservoir is allowed to rise above the *FRL* and the maximum rise in the reservoir level will be *MWL*. The zone between *MWL* and *FRL*, is exclusively reserved for flood control. After the flood has peaked the reservoir is brought back to *FRL*. When it is not desirable to allow the reservoir level to rise above *FRL*, then the outlet capacity be large, equal to the expected inflows. Alternately the flood moderation through reservoir is carried out by pre-depletion as shown in Fig. 5.10. Here the reservoir level is lowered by making prereleases in anticipation of flood and the level is brought back to *FRL* after the peak has passed, and such a strategy depends on reliable flood forecast. There again another way of flood moderation through pre-depletion and use of flood control space as shown in Fig. 5.11. Here the reservoir level is lowered by prereleases to create storage for flood moderation and the reservoir is operated such that the *RL* reaches the maximum permissible water level during the passage of the flood and this approach makes most efficient use of reservoir space.

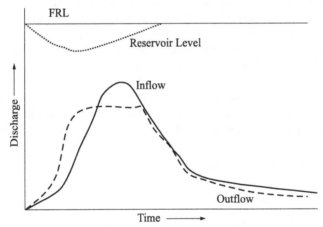

Figure 5.10 Flood moderation through reservoir pre-depletion.

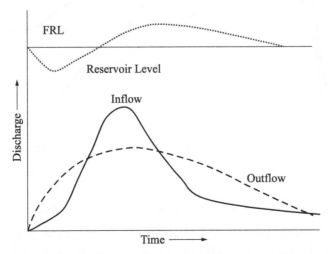

Figure 5.11 Flood moderation through pre-depletion and use of flood control space.

5.7.5 Operation Procedure of Multipurpose Reservoir

Each reservoir is operated according to the prevailing reservoir elevation and the elevation of different rule curve levels. For a reservoir with hydropower, navigation and water supply demands four rule curves can be specified a separate rule curve for each demand and one uppermost rule curve beyond which water can be spilled from the dam. The uppermost rule curve represents such water levels in the reservoir in different months such that if these are maintained throughout the year all demands from the reservoir can be met in full. Though it is desirable to fill a reservoir up to *FRL*, it is recommended that some spill should be made from the reservoir to keep the downstream river channel to avoid encroachment in the river. Keeping the upper rule curve below *FRL* (in monsoon months) can provide extra room for flood absorption, with due care not to disturb the conservation performance of the reservoir. Position of rule curves for various demands depends on their relative priority. Starting upwards from the dead storage level, first lies the rule curve for the highest priority demand i.e., water supply. This rule curve is calculated when there is very high scarcity of water and it is not possible to meet any demands except for the full highest period demand throughout the year. Below this rule curve only highest priority demands are met. Next lies the rule curve for the second highest priority demand, irrigation/hydropower. This rule curve is calculated for the case when there is scarcity of water and it is not possible to meet all demand except for the two highest priority demands throughout the year. If *RL* falls below in any month below this curve then partial demands (specified as a factor in the date of the second highest priority and full demands of the highest priority) are met. Above this lies the rule curve for the least priority

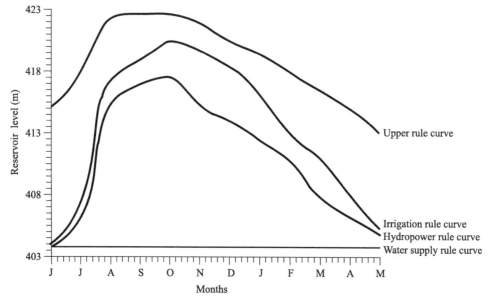

Figure 5.12 Operation rules curves for a multipurpose reservoir. (Courtesy Goel et al. [16]).

demand. This curve is calculated for the case when there is least scarcity of water and it is possible to meet all the demands in full throughout the year. If the *RL* falls in between upper and middle rule curves then partial demands of the least priority demands and other demands in full are met. Four rule curves are shown in Fig. 5.12 and it is assumed that the reservoir reaches the dead storage level at the end of the year.

5.8 RESERVOIR OPERATION FROM PRACTICAL CONSIDERATIONS

In the earlier part of this chapter, the various methods available for determining the overall capacity of the reservoirs are outlined. In actual practice, however, the relationship between supply and demand is complicated and this is taken care of by operation of the reservoir. Fundamentals of reservoir operation has been dealt with before. Additional aspects of reservoir operation is being dealt now. The operation of the reservoir can be done:

   (i) Based on annual storage capacity to the annual runoff,

   (ii) Through regulation of reservoir, and

   (iii) Through spillway gate operation schedules.

Apart from that sometimes the reservoirs are operated to ensure maximum or minimum flow.

5.8.1 Based on Annual Storage Capacity to the Annual Runoff

For the operation of the reservoirs from the consideration of annual storage capacity to the volume of annual runoff, reservoirs have been classified into two groups, i.e. within the year, and carryover or over-the-year reservoirs. Within the year reservoirs are so designed that in normal circumstances they completely fill up and even spill during the flood season and are almost completely depleted in the low-flow season. Here the storage accumulation and storage depletion period can be defined rather accurately (Fig. 5.13). As for example, in India, July to September would be the season where storage would increase and the storage would almost always decrease from November to May. For such reservoirs it would be sufficient to divide

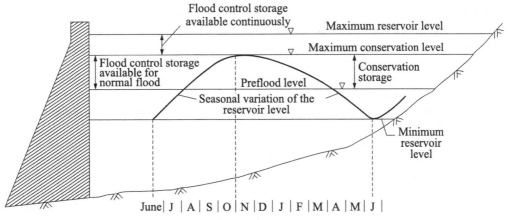

Figure 5.13 Operation of a typical reservoir.

the year in 2 parts, i.e. June to September and October to May. If, however, the flows are not so markedly cyclic or the water requirements during high-flow season are so substantial that the possibility of storage depletion during the period cannot be ignored, working tables in tabular form should be prepared on monthly basis (Table 5.4).

5.8.2 Based on Regulation of Reservoirs

From the consideration of regulation of reservoirs this can be classified into the following:

(i) Single-purpose reservoirs developed to serve a purpose, say flood control or any other use, such as irrigation, power, navigation, water supply, etc.

(ii) Multipurpose reservoirs to serve for a combination of uses with and without provision of flood control space.

(iii) A system of reservoirs in series on the main stream or tributaries consisting of a number of single or multipurpose units developed and operated for optimal utilisation of the resources of the basin.

5.8.2.1 Single-purpose reservoirs for flood control

The operation of flood control reservoirs are mainly governed by the available flood storage capacity, type of outlets, location and nature of areas, damage centres to be protected, flood characteristics, accuracy and ability of flood forecasting and size of the uncontrolled drainage area. It is difficult to evolve a plan of reservoir operation to cover all the complicated situations. Here it is generally based on one of the following criteria:

(i) Optimum use of available flood control storage during each flood event:

Under the above principle the operation aims at reducing damage, by flood stages, at the location to be protected to the maximum extent possible with the flood control storage capacity available at the time of each flood event. The releases accordingly under the plan would obviously be lower than those required for controlling the reservoir design flood. There is also a distinct possibility of having a portion of the flood space occupied upon the occurrence of a heavy subsequent flood.

(ii) Based on control of the reservoir design flood:

Releases from flood control reservoirs operated on the above concept are made so that the full storage capacity would be utilised only when the flood develops into the reservoir design flood. Since the reservoir design flood is usually an extreme event, regulation of minor and major floods which occur more frequently is less satisfactory.

If the protected area lies immediately downstream of the dam the flood control schedule would consist of passing all inflows up to the safe channel capacity and is the same for either of the above-mentioned principles. When there is appreciable uncontrolled drainage area in between the dam and the location to be protected operation under principle (i) shall consist in keeping the discharge at the damage centre within the higher permissible stage or to ensure only a minimum contribution from the controlled area when above this storage. Operation under the principle aims at reducing the damage flood stages at the location to be protected to the maximum extent possible with the flood control storage capacity available at the time of each flood event. For this it is essential to have accurate forecast of flood flows into the reservoir and the storm below it for a period of time sufficient to fill and empty reservoir.

Flood Mitigation through Planning of Reservoir Capacities and Operation of Reserviors **199**

Table 5.4 Working Table.

Month Beginning of the period						Demand during period				End of the period								
	Reservoir level	Capacity (million hectare metres)	Water spread (hectares)	Inflow (million hectare metres)	Total (million hectare metres)	For irrigation	For power	Evaporation losses	Total	Reservoir level	Capacity (million hectare metres)	Water spread area (hectares)	Spilled water	Tail Race level	Effective head (H) (in metres)	Discharge (cubic metre per second) Q	Regeneration	Power potential (P = 7.4 QH)
(1)	(2)	(3)	(4)	(5)	(6)	(7)	(8)	(9)	(10)	(11)	(12)	(13)	(14)	(15)	(16)	(17)	(18)	(19)
					(3) + (4)				(7) + (8) +(9)		(6) + (10)							

In col. (19), P = power generated in kW, Q = discharge in m^3/s and H = head in metres. In arriving at this equation, the overall efficiency for turbine, generators and the friction losses have been considered and a factor of 0.75 is taken for this purpose.

This though ideal cannot be realised in view of uncertainty in weather prediction and hence there is always the risk of having difficulty in the regulation of runoff from the subsequent storms. For reducing this risk factor an adequate network of flood forecasting stations in the upstream and downstream area of the project are absolutely necessary. For a given project, with sufficient storage capacity to control floods having a return period larger than say, 10–15 years, the operation of the reservoir is preferably aimed at the ideal.

The schedules of operation based on principles (ii) above shall consist of releases assumed for design flood conditions so that design flood could be controlled without exceeding the flood control capacity. The operation consists of discharging a fixed amount (may be subject to associated flood storage and outflow conditions) such that all excessive inflows are stored as long as flooding continues at specific locations.

When both local and remote areas are to be protected, schedules based on the combinations of principles (i) and (ii) are usually more satisfactory. In this method the principle (i), that is ideal operation can be followed to control the earlier part of the flood to achieve the maximum damage reduction during moderate floods. After the lower portion of the reservoir is filled, the operation can be based on the principle (ii) i.e. based on control of design flood so as to ensure greater control of major flood. In most cases such a combination of methods (i) and (ii) will result in the best overall operation.

In all cases procedures for releasing the stored water after the flood has passed should also be laid down in the schedule in order to empty the reservoir as quickly as possible for routing the subsequent floods.

5.8.2.2 Conservation reservoir

Conservation reservoirs meant for augmentation of river supplies during the lean period are usually operated so that they fill up as early as possible during the monsoon (June to October) when most of the annual runoff is received. All flows, in excess of the requirements of this period are impounded. No spilling of water over the spillway is allowed until the *FRL* is reached. Should any flood occur when the reservoir is at or near *FRL* releases of flood water shall be so affected as not to exceed the discharges that would have occurred had there been no reservoir.

The depletion period begins after the monsoon is over and the operation is effected to optimise the benefits, as will be indicated later. However, in case the reservoir is planned with carryover capacity its operation will provide necessary carryover at the end of the depletion period. The operation schedule of a conservation reservoir would usually consist of two parts—one for filling period and the other for the depletion period. For each project it will be necessary to prepare rule curves separately for the filling and depletion period. The rule curves for filling period can be developed from a study of the stream flow records over a long period. This will show the limits in which the reservoir levels are to be maintained during different times of the filling period for meeting the conservation design requirements. When operation is guided by such curves it will be apparent that restrictions are to be imposed on utilisation. Rule curves require periodical review and changes may be made depending on the individual background of the projects.

5.8.2.3 Multipurpose reservoir

(i) Operation of a multipurpose reservoir shall be governed by the manner in which various uses of the reservoir have been combined in the planning stage. Alterations can be made with careful analysis of improved combination of various uses. Here separate allocation of capacity has been made for each of the uses in addition to that required for flood control operation. For each of the functions the operation shall follow the principles of respective functions. The storage available for flood control, could also be utilised for generation of secondary power to the extent possible.

(ii) When joint use of some of the storage space or stored water has been envisaged the operation becomes more complicated. While flood control requires low reservoir levels, conservation interests requires, as high a level as is attainable. Thus the requirements are not compatible and a compromise will have to be effected in flood control operations by sacrificing the requirements of the function. In India major floods occur in most rivers during south-west monsoon (July to October). Some parts of the conservation storage is utilised for flood moderation at the early stages of the monsoon (Fig. 5.13). This space has to be filled up for conservation purposes towards the end of the monsoon, progressively as it will not be possible to fill up this space during the postmonsoon periods when the flows are insufficient even to meet the current requirements. This will naturally involve some sacrifice of the flood control interests toward the end of the monsoon. Where joint use of some of the stored water is envisaged operation shall be based on the priority of one use over the other, and the compatibility among demands for different uses.

5.8.2.4 System of reservoirs

In the preparation of an operational schedule for an integrated system of reservoirs, principles, applicable to the separate units are first applied to the individual ones. Modification of the schedules, so formulated, are then be considered by working out several alternative plans based on the co-ordinated operation and the best one is usually selected. The principal factors being considered are as follows:

(i) For flood control operation: The basinwise flood condition (rather than the individual sub-basin), the occupancy of flood reserves in each of the reservoirs, distribution of releases among the reservoirs, and bankful stages, at critical location shall be considered simultaneously. Thus if a reduction in outflows is required, it shall be made from the reservoir having the least capacity occupied or has the smaller flood runoff from its drainage area. If an increase in release is possible, it should be made from the reservoir where the percentage occupancy is the highest or relatively a higher flood runoff. Higher releases from reservoirs receiving excessive flood runoff may be thus counter-balanced, particularly in cases of isolated storms, by reducing releases from reservoir receiving relatively lesser runoff. The current water demands for various purposes, the available conservation storage in individual reservoirs and the distribution of releases among the reservoirs, etc. shall be considered to develop a co-ordinated plan to produce the optimum benefits and minimise water losses due to evaporation and transmission.

5.8.3 Spillway Gate Operations Schedules

It has been the practice in the operation of gated spillways to maintain the reservoir at its *FRL* until all gates are opened after which outflow would be uncontrolled as long as inflow flood exceeded the capacity at *FRL* of the spillway. Under this plan of operation where the reservoir is at *FRL* or near it, the spillway releases may be larger than the inflows since the damaging effect of valley storage within the reservoir is practically non-existent and since the flood waves originating from the upstream area travelling faster than deep water may synchronise with those from the local areas around the reservoir to produce higher peak discharge.

To investigate the damage that may occur under this mode of operation, reservoirs when full or nearly full shall be so operated that (i) peak outflow rates during damaging floods do not exceed inflow rates of the corresponding flood that would have occurred at the reservoir site before construction of the dam and (ii) the rate of increase in outflow, does not constitute a major hazard to downstream interests. The above objectives may be accomplished by one or both of the following procedures:

(i)   When the anticipated inflow is likely to raise the reservoir level above *FRL* the opening of spillway gates would be initiated before the reservoir attains the level and the increase in outflow will be gradual and limited to acceptable rate of increase. This method of operation achieves the objectives but with some risk of not attaining the *FRL* subsequently.

(ii) In the other method, the induced surcharge storage space above the *FRL* will be utilised to effect partial control of outflow rates after the reservoir has attained *FRL*. This is done by rising all gates initially by small increments thus forcing all inflows in excess of spillway capacity at those gates openings into surcharge storage which will of course be evacuated as rapidly as the prevailing conditions warrant. Upon completion of drawdown to *FRL*, the regular schedule for release of stored water shall be followed. In projects where water level is not permitted to rise above *FRL*, the method (i) is the only one applicable. In the case of a ungated spillway, there is no possibility of reservoir operation having ungated spillways except some adjustments that can be brought about by the operation of the outlets.

5.8.4   Operation to Ensure Maximum and Minimum Flow [13]

5.8.4.1   Single-purpose reservoir

Generally for flood control purpose, the storage necessary for guaranteed maximum outflow (Q_0 maximum) is necessary and for the energy and navigation purpose it is the size which guarantees minimum outflow, i.e. (Q_0 minimum). For irrigation plans, the maximum outflow might be interesting when flood plains are being reclaimed for irrigated agriculture. The minimum outflow as such is of less importance, since irrigation hardly ever requires a uniform outflow. Therefore, an outflow required for a certain irrigation project should be established, which shall result in a relationship between the size of the reservoir and frequency of failure to meet the demand. Using the records of the past, it should be possible to determine which size of reservoir is required to ascertain a certain minimum outflow. But how can it be known whether the same minimum outflow could be guaranteed for the next 10 to 30 years ? For this purpose assume, Q_0 minimum = Q_1 and determine the storage necessary (Fig. 5.14). From the mass curve read the total number of days in which the outflow would have been less than Q_1. Repeat the same for Q_0 minimum = Q_1, but with different storage value. Plot figures in probability curve, i.e.,

$S = f(n)$, where S is the storage, n = no. of days. Repeat the same procedure for Q_0 minimum = Q_2, etc. For Q_0 minimum the total duration of failure to provide Q_0 minimum is of major importance and the number of occurrences is of lesser importance, say in electricity generation. For Q_0 maximum the number of occurrences should be the criteria since the duration of each event is of lesser importance, i.e. flood control, breaking of dykes, etc. Proceeding in a manner similar to that indicated above the results are plotted in the form of frequency curve (Fig. 5.15).

$$S = F(n) \tag{5.17}$$

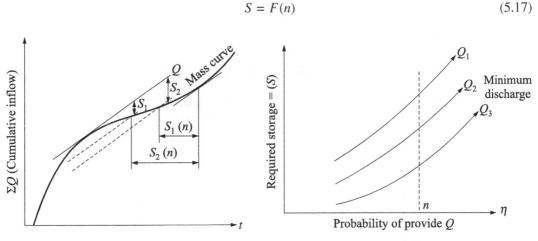

Figure 5.14 Reservoir operation for minimum guaranteed flow.

5.8.4.2 Operation of multipurpose reservoir

Consider a hydrograph and assume a wet and dry season. Say, the value of Q_0 maximum and Q_0 minimum have been determined. The problem is how to operate the reservoir when Q_0 minimum is of interest. One would start filling the reservoir as soon $Q_1 > Q_0$ minimum and keep on filling until the reservoir is full. At the point the spillway would have to be opened so that $Q_0 = Q_1$. At that moment Q_1 might be larger than Q_0 maximum. In case only Q_0 maximum is of interest, one would probably start with an empty reservoir and start filling only when $Q_1 < Q_0$ maximum. Q_0 would be kept equal to Q_0 maximum until the reservoir is empty. The programme for one of the functions would of course not be acceptable for the other function of the reservoir. Also, if for instance dotted line in Fig. 5.16 were followed both criteria Q_0 min. and Q_0 max. would have been made. The reservoir would have been full at the wet season so that Q_0 minimum for the coming dry season would be guaranteed whilst the maximum outflow would have been Q_A which is even less than Q_0 maximum. This effect is obtained by gradually filling; not so fast as for Q_0 min. alone and not so slow as for Q_0 max. However, the runoff that will occur is not known because river flow cannot be anticipated. It is not known whether when following the dotted line a very high peak will follow so that outflow will exceed Q max. and on the other hand there might not follow a very low inflow so that the reservoir would not fill. Therefore, one has to rely on statistical expectations, and this expectation will have to be calculated for each time of the year, month by month or week by week.

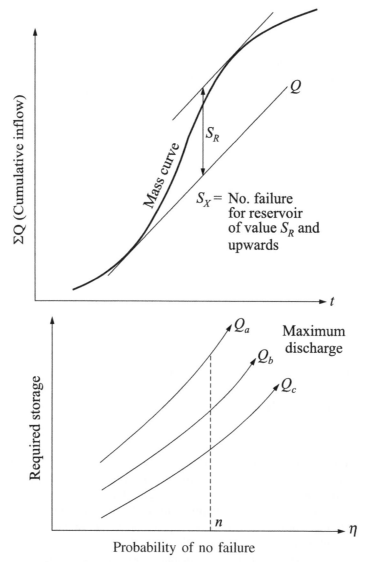

Figure 5.15 Probability operation for maximum guaranteed flow.

The storage requirements for each purpose of the reservoir have to be calculated month by month or week by week and the frequency curves have to be made for the corresponding period. These frequency curves give for each month, the required storage for the known Q_0 min. and Q_0 max.

The function for storage and outflow can then be read from the probability diagram and then plotted as rule curve as shown in Fig. 5.17. Where there is no conflict of interest, the rule curves can be obtained when the floods and droughts are very much determined by the time of the year which is true in the case of tropics because here the peak flow always comes in a specific season.

---------------------- Outflow Singlepurpose
Guaranteed Maximum

------------------------------- Outflow Single Purpose
Guaranteed Minimum

Outflow Satisfying Both
Purpose

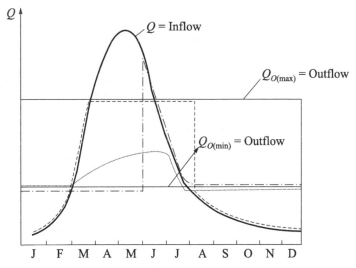

Figure 5.16 Operation of multipurpose reservoir.

Net Storage = 9 Units (Appropriate)
Guaranteed Q (min^m) Outflow = 600
Guaranteed Q (max^m) Outflow = 2000

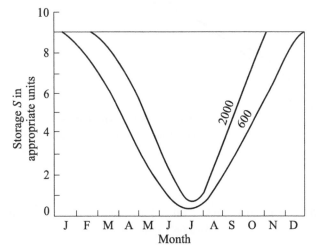

Figure 5.17 Rule curve.

Problem 5.1

It is desired to exploit the surface water of a river at a place S for various purposes. The place S is in a semi-arid zone and flood flows occur only about 5–20 times during the year, each flow lasting between 4 and 2 hours duration. Accumulation of water should normally take place during the period of high runoff. The data available are given as follows:

(i) Rainfall data on monthly basis for the years 1966–1971 being registered at the rain gauge station R (Table 5.5).

Table 5.5 Monthly Rain in mm.

	J	F	M	A	M	J	J	A	S	O	N	D
1966					8.5	10.5	68.3	301.8	180.0	67.3	60.5	
1967						34.6	37.9	112.7	189.3	95.7	3.5	
1968					5.3	5.5	169.7	178.5	134.2	59.8	34.4	27
1969					5.9	24.0	41.6	189.9	163.1	162.6	1.3	
1970			1.6		4.2	3.6	97.5	124.3	136.3	68.5	2.7	
1971						60.8	206.8	234.3	65.2	30.5	2.3	

(ii) Piche evaporation values in mm for each month in the period 1966–1971 measured at the same station (Table 5.6).

(iii) Elevation-area and elevation-storage data of the dam site being given. The watershed area is given as 130 km² (Fig. 5.18).

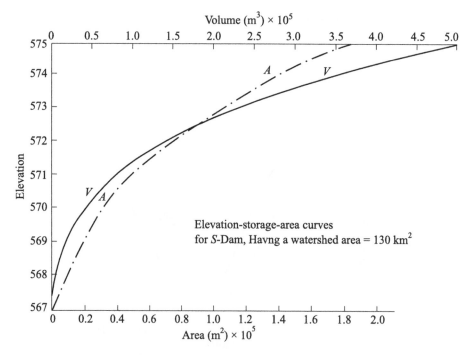

Figure 5.18 Area and storage *vs.* reservoir elevation.

Table 5.6 Piche Evaporation Values in mm.

	J	F	M	A	M	J	J	A	S	O	N	D
1966	460.0	519.7	663.0	622.2	539.9	369.0	182.1	125.6	171.2	246.2	426.0	447.0
1967	464.7	427.0	536.9	648.9	514.2	372.2	276.0	166.7	197.9	385.4	470.2	471.2
1968	494.6	467.5	703.8	607.9	553.7	341.4	191.1	156.8	225.8	348.4	435.4	470.0
1969	460.4	487.0	655.6	575.4	505.8	358.8	234.9	123.2	141.0	270.2	391.3	403.8
1970	401.9	401.9	514.8	499.3	535.1	355.1	228.0	158.7	175.0	306.3	460.5	425.2
1971	430.0	472.0	640.0	570.0	507.7	359.0	190.0	131.0	168.0	252.0	390.0	440.0

Note: The result of the calculation of storage by numerical analysis is shown in Table 5.6 and by graphical method in Fig. 5.20.

(iv) Flood data is collected during the period 1970 and 1971. Monthly runoff in percentage of monthly rains at R is given in a graphical form (Fig. 5.19).

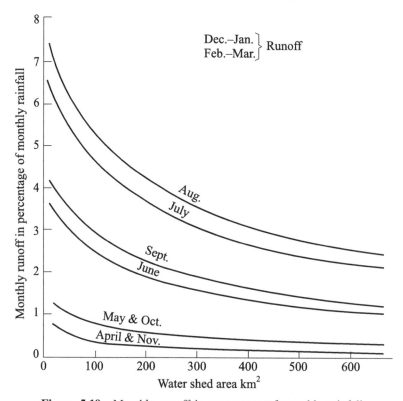

Figure 5.19 Monthly runoff in percentage of monthly rainfall.

(v) An analysis of the Piche evaporation data showed that multiplication by 0.8 yields the A class pan evaporation and multiplication by 0.56 the evaporation from the water surface.

(vi) It is assumed that the demand is at a constant rate and that in general the total annual inflow satisfies the total annual demand. No water needs to be released for downstream use since there are no prior claims to the right of doing so.

In order to obtain the actual (net available) or adjusted stream flow, the inflow at the head of the reservoir has to be corrected in several ways which are generally speaking evaporation losses from the reservoir surface, release of water for downstream use (in this case zero) precipitation on the water surface and sometimes also that of the land surface to be flooded by the reservoir. In this case the gain of a stream flow precipitation on the water surface can be taken as hundred per cent and that from the flooded area as 30 per cent.

Solution

1. Calculate the required useful storage by means of numerical analysis in Tables 5.7 and 5.8.
2. Give a graphical representation of the mass curve of net reservoir inflow and the mass curve of demand (Fig 5.20).

Table 5.7 Calculation of Storage by Numerical Method.

Month ending	Precipita-tion	Evapo-ration Piche	Runoff	Gain Pptn. on water surface	on wetted area	Loss Evapo-ration	Release	Spill-ing
		(mm/month)	Y_0	p_1	p_2	e	rl	sp
1	*2*	*3*	*4*	*5*	*6*	*7*	*8*	*9*
J—1966	0.0	460.0	0	—	—	—	—	—
F	0.0	519.7	0	—	—	—	—	—
M	0.0	663.0	0	—	—	—	—	—
A	8.5	622.2	3 204	—	—	—	—	—
M	10.5	539.9	8 872	—	—	—	—	—
J	68.3	369.0	200 665	4 371	27 375	–13 222	—	—
Ju	301.8	182.1	16 83 139	1 59 700	78 800	–53 800	—	—
A	180.0	125.6	11 30 000	1 98 000	16 200	–77 000	—	—
S	67.3	171.2	236 000	85 300	2 600	–1 21 500	—	—
O	60.5	246.2	51 000	75 000	2 900	–1 71 500	—	—
N	0	426.0	0	0	0	–2 76 000	—	—
D	0	447.0	0	0	0	–2 60 000	—	—
J—1967	0	464.7	0	0	0	–2 41 000	—	—
F	0	427.0	0	0	0	–1 91 000	—	—
M	0	536.9	0	0	0	–1 95 000	—	—
A	0	648.9	0	0	0	–1 8 000	—	—
M	34.6	514.2	30 400	12 600	11 000	–1 03 500	—	—
J	37.9	372.2	1 11 000	11 400	12 500	–62 300	—	—
Ju	112.7	276.0	6 38 000	47 400	33 100	–65 000	—	—
A	189.3	166.7	11 90 000	1 47 000	35 000	–73 000	—	—
S	95.7	197.9	3 35 000	99 800	10 300	–1 15 500	—	—
O	3.5	385.4	3 100	3 600	800	–2 18 000	—	—
N	0	470.2	0	0	0	–2 37 000	—	—
D	0	471.2	0	0	0	–2 00 000	—	—
J—1968	0	494.6	0	0	0	–1 68 000	—	—
F	0	467.5	0	0	0	–1 26 000	—	—
M	0	703.8	0	0	0	–1 38 000	—	—
A	5.3	607.9	2 060	1 200	1 900	–78 000	—	—
M	5.5	553.7	4 650	550	2 100	– 31 000	—	—

Wastershed area: 130 km^2

Evaporation: $E_{L(av)} = 0.7E_{p(av)} = 0.56 \times E_{(Piche)}$

Demand = 4500 m^3/day

Max = Water surface: A_L = approx. = 1.4×10^6 m^2

Prob: on reservoir operation

Table 5.8

Adjusted flow, f	Demand d	Reserve, γ	Mass reserve, γ (m²)	Area end of month, a (m²)	Average, α_{av}	Mass Adjusted flow $F = \Sigma f$ (m³)	Mass Demand $D = \Sigma d$ (m³)
10	11	12= (10)–(11)	13 (16)–(17)	14	15	16 = Σ10	17 = Σ(11)
—	139500	—	—	—	—	—	—
—	126000	—	—	—	—	—	—
—	139500	—	—	—	—	—	—
—	135000	—	—	—	—	—	—
—	139500	—	—	—	—	—	—
219189	135000	84189	84189	128000	64000	219189	135000
1867800	139500	1728300	1812500	930000	529000	2087000	274500
1267200	139500	1127700	2940200	1270000	1100000	3354200	414000
202400	135000	67400	3007600	1265000	1270000	3556600	549000
–42100	139500	–181600	2826000	1220000	1240000	3514500	688500
–276000	135000	–411000	2415000	1100000	1160000	3238500	823500
–260000	139500	–399500	2015500	987000	1040000	2978500	963000
–24100	139500	–380500	1635000	865000	930000	2737500	1102500
–191000	126000	–317000	1318000	730000	800000	2546500	1228500
–195000	139500	–334500	983500	570000	650000	2351500	1368000
–182000	135000	–317500	666000	420000	500000	2169500	1503000
–49500	139500	–189000	477000	310000	360000	2120000	1642500
–72600	135000	–62400	414600	290000	300000	2192600	1777500
–653500	139500	–514000	928600	555000	420000	2846100	1917000
1299000	139500	1159500	2088100	1010000	780000	4145100	2056500
329600	135000	194600	2282700	1065000	1040000	˙4474700	2191500
–210500	139500	–350000	1932700	957000	1010000	4264200	2331000
–237000	135000	–372000	1560600	840000	900000	4027200	2466000
–200000	139500	–339500	1221200	690000	760000	3827200	2605500
–168000	139500	–307500	913700	538000	610000	3659200	2745000
–126000	126000	–252000	661700	415000	480000	3533200	2371000
–138000	139500	–277500	384200	285000	350000	3395200	3010500
–73000	135000	–208000	176200	185000	230000	3322200	3145500
–23700	130500	–163200	13000	20000	100000	3298500	3285000

(4) = (2) × 10^{-3} × % × 130 × 10 = 2 × % × 130000

(5) = (2) × 10^{-3}(15) $\gamma_0 + p_1 + p_2 - e - rl - sp = f$

(6) = (2) × 10^{-3} × $[A_L - (15)]$ × some % (4) + (5) + (6) – (7) – (8) – (9) = 10

(7) = (3) × 10^{-3} × (15) × 0.56

Runoff, (4)% (different for each month) of rain

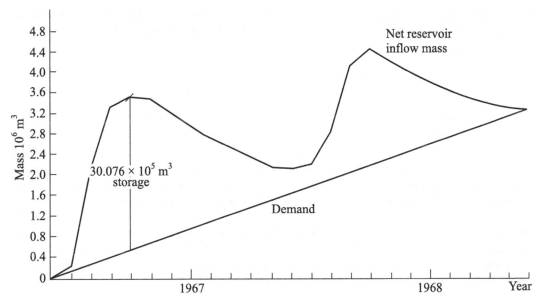

Figure 5.20 Mass curve of inflow and demand.

APPENDIX I

Evaporation

This is the transfer of water from the liquid to the vapour state. The rate of evaporation from a water surface is proportional to the difference between the vapour pressure at the surface and vapour pressure in the overlying air (Dalton's law). In still air the difference in vapour pressure becomes small and the evaporation is limited by the rate of diffusion of vapour away from the water surface. Turbulence caused by wind and thermal convection transports the vapour from the surface layer and permits the evaporation to continue. Evaporation of a gram of water at 20°C requires 585 calories, and hence without a supply of heat there cannot be any evaporation. Hence total evaporation over a period of time is controlled by the available energy.

The amount of vapour present in the air may be expressed as the pressure that the vapour would exert in the absence of any gases and in expressions for computing evaporation loss, the terms such as relative humidity, dew point temperature are used. Relative humidity is defined as the ratio of actual vapour pressure to the saturated vapour pressure expressed as a percentage. The maximum amount of saturated vapour pressure in the air depends on temperature. The higher the temperature, the more vapour it can hold. Dew point temperature is defined as the temperature to which the air must be cooled at constant pressure and with constant water-vapour content in order to reach saturation with respect to water.

In principle if one measures the humidity, temperature, and wind at two levels above a water surface it should be possible to compute the upward transport by use of turbulence theory resulting in many complex relationships. Generally simpler expression such as the following:

$$E = 0.00241 \ (e_s - e_8)V_8 \qquad (5.18A)$$

have been found to be adequate in predicting evaporation loss over a lake surface. The above one being obtained from measurements made at Lake Hefner, Oklahoma, USA.

In Eq. (5.18A)

E = evaporation in inches/day,

e_s = vapour pressure in inch of mercury at the water surface,

e_8, V_8 = vapour pressure and wind velocity at 8 m levels, wind velocity being expressed in miles per day.

Evaportation loss over land surface can be obtained either from measurements of pan evaporation loss or by application of suitable formulae. The amount of evaporation loss over a lake surface can then be obtained by multiplying annual pan evaporation loss by a suitable factor, say, around 0.7. The pan evaporation loss can be obtained by directly measuring the quantity of water evaporated from a class A, pan having the following specifications. The diameter of the pan should be 1.2 m, its depth 25 cm and its bottom raised 15 cm above the ground surface. The depth of water should be kept in a fixed range such that the water surface is at least 5 cm and never more than 7.5 cm below the top of the pan. Alternately pan evaporation loss can be determined by using the Christiansen formula.

Pan Evaporation (E_p) by Measurements

E_p can be experimentally determined by directly measuring the quantity of water evaporated from a standard class A pan (i.e. 1.2 m dia 25 cm deep of and the bottom raised 15 cm above the ground surface). The depth of water is to be kept in a fixed range such that the water surface is at least 5 cm and not more than 7.5 cm below the top of the pan. E_p can also be determined by using the Christiansen formula:

$$E_p = 0.459 \ R \ C_t \ C_w \ C_h \ C_s \ C_e \qquad (5.19A)$$

where

R = extra-terrestrial radiation in cm,

C_t = temp. coeff. = $0.393 + 0.02796 \ T_c + .0001189 I_c^2$ where T_c = mean temp. in °C,

C_w = coeff. for wind velocity and is given by

$0.708 + 0.0034 \ W - 0.00000 \ 38 \ W^2$, where W is the mean wind velocity at 0.5 metre above the ground in km/day.

C_h = coefficient for relative humidity, and is given by

$1.250 - 0.0087 \ H + 0.75 \times 10^{-4} \ H^2 - 0.85 \times 10^{-8} \ H^4$ where H is the mean percentage relative humidity at noon or average relative humidity for 11 and 18 hours.

C_s = coefficient for per cent of possible sunshine, and is given by

$0.542 + 0.0085 - 0.78 \times 10^{-4} \ S^2 + 0.62 \times 10^{-6} \ S^3$, where S is the mean sunshine percentage.

C_e = coefficient of elevation

= $0.97 + 0.00987 \ E$, where E is the elevation in 100 metres.

Values for R for different latitudes are tabulated in Table 5.9.

Problem 5.2

Determine the pan evaporation from the following data for the month of April, using Christiansen method.

Table 5.9 Mean Monthly Values of Extra-Terrestrial Radiation R in cm.

Latitudes	Jan.	Feb.	Mar.	Apr.	May	June
in degrees	15.621	19.990	31.963	41.012	50.317	52.146
North 45	July	Aug.	Sept.	Oct.	Nov.	Dec.
	52.420	46.101	35.204	25.730	16.981	13.843

Latitude 15°09′ N, Elevation + 449 metres

Month April

Mean temperature 31.8°C

Mean wind velocity at 0.5 m above the ground = 183 kilometres per day

Mean relative humidity = 40 per cent

Mean sunshine per cent = 80 per cent

Use Table for extra-terrestrial radiation.

Solution

Find the value of R from Table 5.7 for the month of April and for a latitude of 15°09′ N. It comes out to be about 47.3 cm. Using Eq. (5.19A). to compute coefficients we get,

$$C_t = 0.393 + 0.02796 \times 31.8 + 0.0001189 \, (31.8)^2$$

$$= 1.403$$

$$C_w = 0.708 + 0.0034 \times 183 - 0.0000038 \, (183)^2$$

$$= 1.200$$

$$C_h = 1.250 - 0.0087 \times 40 + 0.75 \times 104 \, (40)^2 - 0.85 \times 10^{-8} \, (40)^{-4}$$

$$= 1.000$$

$$C_s = 0.542 + 0.0085 \times 89 - 0.78 \times 10^{-4} \, (89)^2 + 0.62 \times 10^{-6} \, (89)^3$$

$$= 1.073$$

$$C_e = 0.97 + 0.00984 \times 4.49$$

$$= 1.014$$

$$E_p = 0.459 \times 47.3 \times 1.403 \times 1.200 \times 1.00 \times 1.073 \times 1.01$$

$$= 39.8 \text{ cm.}$$

Problem 5.3

The monthly flows of a stream over the recorded driest year are furnished below

Month	Jan	Feb	Mar	Apr	May	Jun	Jul	Aug	Sept	Oct	Nov	Dec
Flow in 10^6 m^3	5.00	3.00	6.00	2.50	1.50	1.75	1.50	1.80	2.00	2.25	6.00	7.00

Assuming each month is 30 days, estimate the maximum possible uniform draw off from the stream. Using the following methods: tabulation, a hydrograph and a cumulative flow diagram, find the reservoir capacity required to achieve uniform draw off and the minimum initial storage to maintain the demand.

Solution

The reservoir water levels for the driest year on record along with outflow hydrograph are indicated schematically in Fig. 5.21 and Table 5.10.

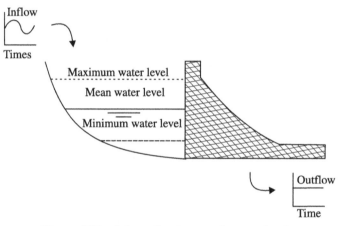

Figure 5.21 Schematic of reservoir water levels.

Table 5.10 Tabulation of inflow and outflow data

Month	Inflow (10^6 m^3)	Outflow (10^6 m^3)	Required storage volume (10^6 m^3)
January	5.00	3.36	0.00
February	3.00	3.36	0.36
March	6.00	3.36	0.00
April	2.50	3.36	0.86
May	1.50	3.36	1.86
June	1.75	3.36	1.61
July	1.50	3.36	1.86
August	1.80	3.36	1.56
September	2.50	3.36	1.36
October	2.25	3.36	1.11
November	6.00	3.36	0.00
December	7.00	3.36	0.00
Total	40.30	40.3	10.58

The inflow, outflow and required storage data are tabulated Table 5.10 Examination of Table indicates that the required total storage between maximum and minimum water levels is 10.58×10^6 m^3. Assuming that the reservoir commences service in the month of January, the amount of water that must be initially stored in the reservoir to maintain the uniform demand is 6.3×10^6 m^3. The figure is calculated by deducting the fill volumes in month of January (1.64×10^6 m^3) and March (2.64×10^6 m^3) from the required total storage of 10.58×10^6 m^3.

Hydrograph

The inflow and outflow hydrographs are graphically represented in Fig. 5.22. The reservoir storage volume to maintain a uniform demand is equal to the shaded area shown in the figure, that is 10.58×10^6 m^3. The fill volumes for the months of January and March equal the area between the inflow and outflow hydrographs for these months, that is, 1.64×10^6 m^3 and 2.64×10^6 m^3 respectively. The amount of water that must be initially tored in the reservoir to maintain the uniform demand is therefore 6.3×10^6 m^3.

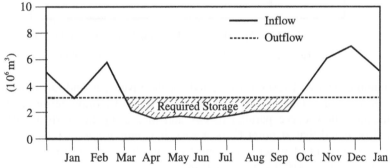

Figure 5.22 Inflow and Outflow Hydrograph.

Cumulative Flow Diagram

The monthly flows and the cumulative flows from the start of the year are presented in Table 5.11 and graphically illustrated in Fig. 5.23.

Table 5.11 Cumulative reservoir inflow and outflows

Month	Inflow (10^6 m^3)	Cumulative Inflow (10^6 m^3)	Outflow (10^6 m^3)	Cumulative Outflow (10^6 m^3)
December	0.00	0.00	0.00	0.00
January	5.00	5.00	3.36	3.36
February	3.00	8.00	3.36	6.72
March	6.00	14.00	3.36	10.08
April	2.50	16.50	3.36	13.38
May	1.50	18.00	3.36	16.74
June	1.75	19.75	3.36	20.10
July	1.50	21.25	3.36	23.46
August	1.80	23.05	3.36	26.82
September	2.00	25.05	3.36	30.18
October	2.25	27.30	3.36	33.54
November	6.00	33.30	3.36	37.00
December	7.00	40.30	3.36	40.36

The cumulative mass diagram shown in Fig. 5.23 below can be interpreted as follows: The uniform draw-off rate is the gradient of line OA = $3.36 \times 10^6/(30 \times 24 \times 60 \times 60) = 1.30$ m³/s. The vertical ordinate from the x-axis to point D represents the area under the inflow curve (and the volume of inflow) for the corresponding time duration. Similarly, the vertical ordinate to point E represents the area under the outflow curve (and the volume drawn down). The distance [DE] therefore represents the increase in volume stored in the reservoir from the start of the year to (approximately) the month of March. When the outflow rate exceeds the inflow rate (March to October) the water level in the reservoir is drawn down.

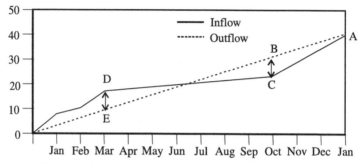

Figure 5.23   Cumulative reservoir inflow and outflow.

The minimum capacity of reservoir required for the given demand OA is represented [DE] + [BC] = 10.58×10^6 m³. This figure is conveniently found by drawing the tangents at C and D to the cumulative flow curve parallel to OA and measuring the vertical distance between them. It should be noted that at point C the reservoir is empty and at point D the reservoir is full. Therefore, the amount of water which must be initially stored in the reservoir to maintain the uniform demand OA is [BC] = 6.3×10^6 m³.

APPENDIX II

Sediment Transport

Disintegration, erosion, transportation and sedimentation are the different stages leading to siltation of reservoirs. In the catchment areas of streams certain forms of rock and soil formations are more susceptible to weathering than others. In the case of arid climate, excessive weathering action and torrential rains carry the weathered materials which results in sporadic and heavy flow of sediment into the stream. The erosive power of streams varies with the square of the velocity while its ability to transport varies as the sixth power of velocity.

Principles of Measurement

During transport the sediment is not uniformly distributed but varies more or less throughout the cross section of the stream, it being more in the vertical direction than in the lateral. For an accurate assessment measurement of both suspended and bed loads are required. However, the suspended load constitutes the predominant mode of transport and can be more easily measured than the bed load. In a stream because of the variation in the vertical distribution of the sediment with particle size the concentration increases more rapidly from the surface towards the bed of the coarse particles than the finer particles. The finer the sediment the more gradual the concentration gradient with respect to depth. Conversely, the gradient assumes a steeper slope as the particle gets coarser.

The manner in which sediment is transported is important from the design considerations of engineering structures. Generally speaking, total sediment load is divided between bed material load and wash load. Bed material load is referred to as the sediment which is picked up from the river bed and moves either along the bed termed as bed load or in suspension designated as suspended load. The wash load refers to the sediment load which originates from other areas of the catchment and is finer than the bed material. This is also carried in suspension by the river and there is no relation between it and the water discharge, the rate being determined by the amount in which it becomes available in the upper catchment area. However, there is a relation between the water discharge and the bed material load and accordingly there exist formulae to estimate this type of transport. There exists extensive treatment of the the theoretical aspect of the subject matter by different authors [7, 25, 26]. Herein only a brief account regarding estimation of the amount of sediment transported in streams based on field measurements is furnished.

Sediment concentrations are usually expressed in parts per million (p.p.m), i.e. the ratio of dry weight to total weight of the water sediment mixture. Sediment transport is usually expressed as (a) dry weight or volume per unit time or (b) bulk volume including pores in the settled volume per unit of time.

Measurement of Suspended Sediment Load [5]

Principles of measurement

The concentration of suspended sediment C_i and the velocity of flow v_i are measured practically simultaneously at a large number of points m in the sampling area of a cross-section. Each concentration and velocity represents small area a_i of the sampling cross section the summation of all being the whole sampling area A. The average static concentration C_s is given by

$$C_s = \sum_0^m \frac{C_i}{m}$$ and the mean concentration of suspended sediment load in motion being

$$\overline{C}_m = \frac{\sum C_i \, v_i \, a_i}{\sum a_i \, v_i} \text{ or } \frac{\sum C_i, \, v_i, \, a_i}{Q} \tag{5.20A}$$

where $Q = \sum a_i v_i$ is the discharge in the sampling area. The suspended sediment load through the sampling area is the product of the mean concentration in motion and the discharge, i.e.

$$\overline{C}_m \times Q \tag{5.21A}$$

The site for the measurement of suspended sediment is the same as the one used for measurement of discharge. The concentration of suspended load not only changes from point to point in a cross-section but also fluctuates from moment to moment at a fixed point. It can be obtained with the help of sediment samples or utilizing the principle of isokinetic withdrawal. One such instrument point integrating sample known as USP-61 is developed in USA. The equipment is made of cast bronze and equipped with tail vanes. A removable bottle is located in a cavity of sampler. An intake nozzle of 0.47 cm points into the flow and air exhaust leaves the body of the sampler on the side. Operations of intake and exhaust are controlled by an electrically-operated valve system. There is also a pressure equalising chamber where the air pressure in the container is equalised with the external hydrostatic head in the nozzle at any depth. The sampling period is restricted but in principle all particles in suspension are kept in the bottle. Suspended load and the velocity are measured simultaneously at each of a number of points in the vertical like 0.1, 0.2, ... 0.9 depth or the lowest practicable point. Alternately an integrating depth sampler can be used. This is essentially similar to the point integrating sampler but without a valve. In this case the sampler begins to collect the sample when it enters the stream and stops on leaving it. During this time and over the distance travelled it collects the average concentration of suspended solids. The sampler is required to cover the distance with a constant speed. For a depth of 5 m or less the depth integration is performed on a round trip basis, i.e., it is lowered and raised in an open position and for a depth of 10 m or less it is performed on a single trip. This means the sampler is lowered in an open position but on contact with the bed a foot trigger closes the intake and exhaust line and is raised in a closed position. The concentration thus obtained represents the mean value. Knowing the mean concentration and discharge, the sediment discharge can be ascertained.

The suspended sediment concentration as well as the grade of sediment in a flowing stream decreases from bottom to top and it also varies transversely across the section. The variation depends on the size and the shape of the cross-section, the stage of flow and other channel characteristics.

The mean sediment discharge per unit and the mean sediment concentration in motion at the vertical, may be obtained by drawing the velocity and sediment concentration curves and then obtaining the products of concentration $C \times v$ (velocity) at corresponding points which give the rate of sediment discharge.

Measurement of bed load

Bed load can be measured by collecting the sediment moving along the bed with the help of an instrument. For this purpose the instrument is usually placed on the river bed for a fixed interval of time and the volume or the mass of the material thus collected is measured. The transport can then be determined using a calibration curve. This is because of the fact that the presence of the instrument in the stream causes increased resistance to flow compared to undisturbed conditions which results in a smaller flow velocity and smaller rate of bed load transport. Hence, only 40–60 per cent of the material moving towards the instrument is actually trapped on it. So to convert the results of measurements into the true transport, the efficiency of the instrument has to be determined by calibration.

Of the various types of bed load samplers the so-called pressure-difference type is generally employed. This type of instrument has a section which diverges in a downstream direction causing a pressure drop at the exit of the instrument to compensate for the energy due to the presence of the instrument. One such type is known as VUV sampler initially designed by Karolyi and improved by Novak has been found to be satisfactory [25].

COMMENTS

The total sediment transport can, therefore, be estimated from the knowledge of suspended and bed load. When the bed load component is small it may be neglected and the suspended load transport can be considered to approximate to the total sediment transport.

Depending on the purpose for which sediment measurement results are to be used there exists the necessity of separating wash load from the suspended load measurement as for example in the morphological prediction of river bed. On the other hand, for the purpose of reservoir sedimentation no such separation is needed. In many rivers the grain size of wash load is much smaller than the bed material used. Usually a diameter lying between 50 and 70 μm is taken as a practical distinction between wash load and bed material load in rivers sediment of the above size is uniformly distributed over the vertical.

APPENDIX III

Measurement of Depth and Cross-section of a River by Echosounder [2]

The echo sounder works on the principle of sending out a pressure wave for a very short duration, receiving back the echo, i.e. the reflected wave, and measuring and translating the interval of time between start of the sound impulse and receiving the return echo in terms of depth of flow. All these steps are repeated speedily and automatically so that in shallow water, soundings may be made at a rate of 100 or more per minute. The nature of operations for obtaining a sounding can be summarised as follows. A strong and short duration electrical impulse is produced by the signal generator which is then amplified and sent to the transducer which converts it into an acoustic signal. This acoustic signal is sent to the river bed which after reflection from the bed is received by the transducer and converted into any electrical impulse which is again amplified. The true lapse between the transmission and reception of the signal is measured and converted into the depth of water under the transducer. The depths are shown on a suitable indicator and recorded. Alternately, it can be had as a digital output. Because of the high frequency of transmission, the individual depths appear on the recording paper as an almost continuous bed profile. To obtain the total depth of water the known depth to the transducer below the water surface is added to it. When cross-sections are sounded and the measurement of distance to a reference point is made electronically. The speed of the recording paper is adjusted automatically to the variable speed with which the survey boat moves. Such cross-sections will be in horizontal scale also. Usually in practice 80 kHz frequency sounder is used for routine soundings of alluvial sandy river beds.

The operating frequency of the transducer also needs to be properly chosen in order to— (i) obviate the effect of ordinary water noises, (ii) minimise loss of energy due to absorption in eddies and swirls in turbulent flow, (iii) minimise error on account of wide sound cone, and to (iv) provide a measurable time lag between incident and echoed waves.

Water noises are avoided by choosing a frequency beyond the audible range. Loss of energy in turbulent flow is minimum with longer wavelengths and hence with smaller frequency. This is, however, of little concern as long as depths are limited to less than 150 m. Another factor to be considered is the angle of the sound cone since appreciable error in recording depths would be produced with sloping bottoms. This is because when the transducer gives out pressure waves they travel in the form of a cone of which the central angle A varies with the radius of the transmitter plate and the frequency of the sound signal.

The sound cone can be expressed as:

$$\sin (A/2) = 0.183 \ L/R \qquad\qquad (5.22A)$$

where

L = wavelength of sound signal in water.

As $L \propto \dfrac{1}{f}$, where f = frequency of sound signal, it follows from the relationship that the sound cone angle would reduce by adopting higher frequency with a given size of transmitter plate of radius R. As for example, for $R = 0.15$ m and frequency of 8000 cycles/sec the sound cone angle works out to 94° while for the same R, the angle is reduced to 43° (Fig. 5.24) by doubling the frequency to 16000 cycles/s. With that angle, ratio of sound cone diameter to depth of flow works out to be 0.79 which means if the echo sounder is vertically above the toe of the sloping river bank, for a side slope of $2\dfrac{1}{2}$: 1, depth recorded by the echo sounder should be about 0.30 m less than the true depth. In order to avoid that higher frequency, supersonic range is preferred.

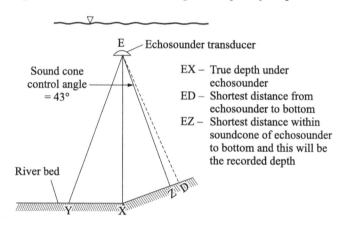

Figure 5.24   Influence of cone angle on accuracy.

Radiolog System

Electronic instruments are used particularly when the distances to be measured are considerably large. They are, however, costly but are nevertheless easier to handle. They provide fast consecutive measurements and can be adjusted to produce a direct digital input for computer processing.

In this system a portable shore-based relay station with antennae is placed at a point from which the distance to the zero mark of the cross-section is known. The second station with antennae and necessary equipment is placed on the survey boat. The on-board station transmits a radio wave of a certain frequency which is received by the shore-based station and then transmitted after amplification at twice the original frequency. On board the frequency of the original wave is also doubled to avoid confusion between transmitted and received waves. The shift $\delta\varphi$ of the two waves is measured electronically and converted into a distance, i.e.,

$$\delta L = \frac{1}{2} \frac{\delta \varphi}{2\pi} \lambda \qquad (5.23A)$$

where
  λ = length of a wave of particular frequency,
  δL = increase or decrease of the distance between on-board and shore-based station.

The measurement is continuous and the increases and decreases are totalled automatically. Then

$$L = \frac{1}{2} \lambda \frac{\Sigma \delta \varphi}{2\pi} + x \qquad (5.24A)$$

where x = known distance measured with a tape or range finder. Accordingly, the counter is adjusted and L can then be read directly from the counter. The relationship between distance measured and simultaneous sounding by an echo sounder is plotted on an echogram

APPENDIX IV

Density Current [11]

In man-made reservoirs stable density stratification is caused due to variation of temperature with depth as well as due to a variable concentration of dissolve and suspended solids. The dense water as a natural process sinks to the bottom where if there exists a slope it may continue to flow until its progress is checked by an obstacle, say in the form of a dam. A moving stratum of this kind is called a density current because its slightly higher density gives it the power of motion. The interface separating the two fluids is assumed to be a streamline and coincident with the density discontinuity. A steady uniform flow of a lower layer of fluid will occur along and incline when the driving gravity force for unit area is in equilibrium with the shear stresses exerted by the fixed boundary and the moving interfacial boundary.

Theoretical Development

Considering the induced velocity in the upper layer to be negligible and for a fully established flow at a distance far enough from the origin of the current, the equilibrium equation can be written as:

$$\tau_0 + \tau_i = \Delta \rho g h_2 \, S \qquad (5.25A)$$

under the assumption that the interface is smooth and direct use may be made of an analogy of flow between parallel boundaries in which the lower boundary is stationary and the upper boundary, i.e., the interface has a velocity u_i. In such a flow the shear stress varies linearly from τ_0 at the bottom to zero at the point of maximum velocity to τ_i at the interface (Fig. 5.25). The interfacial shear stress τ_i is taken to be a constant proportion of the bottom shear and it depends only on the vertical location of the maximum velocity. Or in other words,

$$\tau_i = \alpha \, \tau_0$$

and

$$\alpha = \left[\frac{1 - 2 \, y_m / h_2}{1 + 2 y_m / h_2} \right] \qquad (5.26A)$$

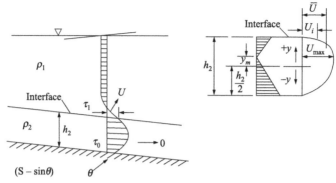

Figure 5.25   Velocity profile in a two-layered flow.

Eliminating τ_i from Eqs. (5.25A) and (5.26A) we have,

$$\tau_0 = \Delta \rho g \left(\frac{h_2}{1 + \alpha} \right) S \tag{5.27A}$$

Again the shear stress may be expressed in terms of the friction factor f or

$$\tau_0 = \left[\frac{f}{4} \rho_2 \frac{\bar{U}^2}{2} \right] \tag{5.28A}$$

Now equating Eqs. (5.27A) and (5.28A) the expression for $\bar{U}$, the average velocity of the lower layer becomes

$$\bar{U} = \left(8g' \frac{h_2 \, S}{f \, (1 + \alpha)} \right)^{\frac{1}{2}} \tag{5.29A}$$

where

$$g' = \left[\frac{g \, \Delta \rho}{\rho_2} \right]$$

It may be remarked here that the uniform underflow in turbulent flow is not subject to exact analysis and hence Eq. (5.29A) can be adopted for estimating the order of magnitude for velocities in turbulent flows. The factor $(1 + a)$ for two-dimensional flows represents the factor by which the resistance coefficient f is increased by the presence of interface. Since $\alpha = 0$, for free surface flows f may be obtained from Moody's diagram using $(4h_2)$ as hydraulic radius. The value of f thus obtained has to be increased by the factor $(1 + \alpha)$. Experimental investigations indicate that in the lower-layered turbulent flow the maximum velocity on the average occurs at $0.7h_2$ which gives the corresponding values of $\alpha = 0.43$, for turbulent flow and it has been found to be fairly constant for values of Reynolds

number $\dfrac{4 \, U \, h_2}{\upsilon} < 10^5$.

Problem 5.4

A density current is created in a reservoir as a result of entry of sediment-laden water from the river. The density difference $(\Delta \rho/\rho_2)$ between the two layers is given as 0.0008. If the average bottom slopes of the reservoir is taken as 10×10^{-4} and the depth of the moving current as 5 m, estimate the total sediment transport due to density current when the percentage concentration of sediment is given as 0.128.

Solution

Given

$$S = 0.0010; \quad \frac{\Delta\rho}{\rho_2} = 0.0008$$

$$h_2 = 5 \text{ m}, \ \alpha = 0.43, \text{ assume}$$

$$f = 0.010 \text{ then}$$

$$\bar{U} = \left[\frac{8 \times .0008 \times 9.81 \times 5 .0010}{.010 \ (1.43)} \right]^{\frac{1}{2}}$$

$$= 0.148 \text{ m/s}$$

Reynolds No. $= 4 \ \bar{U} \ h_2/\upsilon = 4 \times .148 \times 5/10^{-6} = 2.96 \times 10^6$

The f value is within 10 per cent of assumed value, for above Reynolds number. Hence $\bar{U}$ is taken as 0.148 m/s.

Hence, the sediment transport rate $= \dfrac{0.1280 \times 0.148 \times 5}{100}$

$$= 9.47 \times 10^{-4} \text{ m}^3/\text{s/m}.$$

APPENDIX V

Reservoir Surveys

Reservoir surveys are carried out to estimate the amount of material deposited and its distribution over the reservoir bed. This enables the engineer to:

   (i) Know the prevailing rate of sedimentation
  (ii) Estimate the probable future sedimentation damages to the reservoir
 (iii) Estimate the total useful life of the reservoir
 (iv) Estimate the time that will elapse before any of the purposes are interfered with
  (v) Modify the area-capacity curve to assume efficient operation of the reservoir.

The capacity surveys of reservoirs are being carried out using conventional equipment. The conventional method of survey is cumbersome and time consuming, and it takes about two to three years to conduct a survey of a big reservoir. Survey techniques have not yet been developed by which it is possible to conduct surveys within two to three months.

Conventional Method

Usual topographical methods are used to determine the spot levels at different locations in the empty portion of the reservoir area. An echo-sounder and boat are used for hydrographic surveys in the submerged portion. From the survey data obtained, the changes in the storage volume that has occurred between two successive surveys are computed.

Modern Techniques

(i) HYDAC System: This has been developed by the Canada Water Authority. The HYDAC System consists of:

    (a) Positioning system which includes Trisponder and remote sensing units.

    (b) Depth measuring unit which consists of echo-sounder and transducer.

    (c) Computer system which includes plotter, printer, precision digital electronic clock, data coupler, and magnetic tape recorder.

For the actual hydrographic surveys, the HYDAC system (less remote sensing units) is mounted on a high speed jet boat having speed of the order of 35 to 40 km/hr. The data acquisition is fully automatic and is done by using a computer software called HYDAC (Hydrographic Data Acquisition). The analysis of data is carried out with the help of another software called HYDRA (Hydrographic Data reduction and Analysis).

The number of cross-sections or range lines and their spacing have an important bearing on the accuracy in measurement of capacity of a reservoir. More the number of ranges, more the accuracy. More ranges, however, involve more cost. Data on the minimum number of ranges required to be adopted in sedimentation survey in order to restrict the error or uncertainty to the desired limit is accordingly important. Such information is however lacking at present.

In India, the ranges are spaced about 0.8 km (0.5 mile) apart.

In USA, the practice followed by the Tennessee Valley Authority is to space the range lines 1.6 to 3.2 km (1 to 2 miles) apart for measurement of sediment deposition. Closer spacing of 0.8 km (0.5 mile) is preferred at the upstream and downstream ends of the reservoir and at the junctions with the large tributaries. According to the US Army Corps of Engineers the number of ranges depends on the objectives applicable to the individual project. The relationship between the number of range lines and reservoir surface area was developed by regression analysis of data of 57 reservoirs and is given by

$$\text{Number of ranges} = 2.942 \, (\text{Reservoir area in acres})^{0.28}$$

Use of Satellite Imageries

When measuring sediment deposition in a reservoir, hydrographic surveys of underwater areas and ground surveys for areas above low water are normally used. Now that the satellite imageries can be obtained easily, the possibility of their adoption for measurement of sediment deposits between minimum water level and full reservoir level needs to be examined. Instead of the ground survey, satellite imageries at various water levels of the reservoir water spread area can be used to prepare area-elevation curve using the contour method, from which elevation-capacity relationship can be obtained. Main advantages of the use of satellite imageries are economy and saving in time.

Information about satellite sensing techniques, interpretation of data and calculation of water-spread area is now available. The difference between areas of water-spread measured by ground survey and from satellite imageries was within 10 per cent.

The large the water-spread area of a reservoir, the greater the accuracy of measurement from satellite imagery. The minimum size of the reservoir for which the imagery can permit measurement of area with accuracy comparable with or more than that with the ground survey is yet an unanswered question. This study can be made on basis of available data. Sedimentation surveys have been carried out on several reservoirs and it should be possible to work out capacity survey using satellite imageries and ground survey, and their comparative costs.

Sedimentation surveys are currently undertaken on many reservoirs, and on some others such surveys are planned. The authorities concerned can collect all the necessary data and make comparative studies to determine the scope of use of the images.

6

Flood Mitigation through River Protection and Improvement Works

6.1 INTRODUCTION

From time immemorial human habitats are centred around river courses and with that man has been blessed alternately by the life sustaining flow in the rivers when under control and then plagued by their destructive might when out of control. With the increase in population and to sustain higher standards of life it has become imperative for man to adopt river control measures so as not to destroy their cities, agricultural lands, transportation system, etc.

This, however, is not at all an easy affair because of the unpredictable nature of the river behaviour which is primarily because it usually carries huge amount of silt load. The interaction between water discharge, the quantity and the character of the silt load, the composition of the bed material gives each river its own characteristic geometrical features such as meandering, straight and braided in plan from and various bed configurations such as ripples, dunes, bars, etc. in elevation. In the course of last several years, the subject of sediment transport and river mechanics has advanced considerably through sustained research efforts, backed by field study and quite a good number of textbooks exist dealing with various aspects of the subject matter [7, 23, 25, 26].

6.1.1 Types of Rivers

Rivers are formed along more or less well-defined channels to drain from land all the waters received by way of precipitation and melting of snow. Mainly two sources of water flow into the rivers, i.e. tidal and freshwater discharges. Tidal water enters at the lower end of a river and is derived from the tidal wave of the ocean. This water is available throughout the year and its variation depends on tides and freshwater discharges.

River reaches can be divided according to the topography of the river basin, i.e. upper hilly regions and the lower alluvial plains. Further subdivision of lower and upper reaches can be made as under:

Upper regions—(1) Incised rivers, and (2) Boulder rivers. Under (1) the channel is formed by the process of degradation. The sediment that it transports is often dissimilar in character to that of river bed since most of it comes from catchment due to soil erosion. The bed and banks of the reach are normally highly resistant to erosion. As bed conditions do not determine the sediment load the transportation rate cannot be determined as is normally done as a function of bed characteristics. These rivers are further characterised by steep slope, supercritical flow and the formations of rapids along their courses. They do not present a regular meander pattern because of varying resistance to bed and bank erosion which vary along the banks.

The rivers under (2) are characterised by the steepness of their slopes and their beds consist of a mixture of boulders, gravel, shingle and sand. They, moreover, differ considerably from that carrying sand and silt—here in place of regular meandering courses deep well-defined beds and wide flood plains, boulder rivers tend to have straight courses with wide shallow beds and shifting, braided and interlaced channels. During the flood time the high velocity flow transports boulders, shingle, gravel downstream. As soon as flow subsides the flow of material is checked and the materials are dumped in heaps. The flow with reduced flow velocity finds it difficult to go around them. The channel thus wanders often attacking the banks thereby widening its dimension.

Rivers in the lower reaches have the characteristics of meandering freely from one bank to another. It carries material which is similar to those of the bed. Here material is eroded from the concave banks and then it is deposited either along convex sides of successive bends or in between successive bends to form a bar. The erosion of banks, the path of bed load movement and the location of deposits significantly change with the changes in stages. The cross-sectional shape and the slope of the river are determined depending on erodibility of bed and bank along with relative quantum of sediment. Generally they inundate very large areas during flood time causing severe changes. Such rivers are further subclassified as—(i) aggrading, (ii) degrading, (iii) stable, and (iv) tidal and deltaic rivers.

Aggrading river is one which builds up its bed-because of a variety of reasons, i.e. heavy sediment load, obstruction to flow due to barrage, dam across it thereby raising the level of water and flattening the slope, extension of the delta at the river mouth, sudden intrusion of sediment from a tributary. Such a river is usually straight with wide reaches having shoals in the middle which shifts with flood and the flow being divided into a number of braided channels. Degrading type, on the other hand, is found either below a cut-off or below a dam or barrage. These result in the sudden lowering of the water surface upstream of the cut-off which increases slope of flow and in the sudden diminution of sediment load. A stable river is one which mostly carries the sediment load brought into it from upstream into the sea. Here there is stability of the channel alignment and slope as well as its regime. There may be changes within a year. However, there exists little variation from year to year. Here also the changes such as scouring and silting of the bed, delta advancement to sea and changes in bed and water surface slope over a long period of time also takes place which, however, are small. Seldom does a river remain of a single type. It may have all the characteristics from source to mouth depending on the amount and size of sediment entering the river, the load-carrying capacity and other factors. Even the same reach in river may pass through different types depending on variations of sediment load and discharge with time.

The river reach is called tidal where periodic changes in water level occur due to tides. During its last journey to the sea before becoming tidal, a river may split into branches and from a delta. This portion is known as deltaic river. In this last reach the river water receives tidal water derived from the tidal waves of the ocean. The ocean waters enter the river with flood tide and the process is reversed during ebb tide. The amount of water that enters during flood goes out during ebb, thereby undergoing a periodical rise and fall in the level depending on the nature of tide. The distance up to which the tidal effects are felt depends on slope of the river, tidal range, freshet discharge and configuration of the river.

The delta river is a stage of a river where the river comes closer to its outfall with the sea and is characterised by several branches which multiply in number as it approaches the sea. Due to low velocity the channels get silted resulting in splits and eventual formations of new channels.

In this chapter, however, only a brief treatment of river hydraulics is made. This is with a view to making the reader familiar with the necessary theoretical background before he decides on the methods of flood control and embarks on the design of various river improvement works.

The various methods can be categorised as:

(a) Construction of flood retaining levees to protect the adjoining areas from flooding.

(b) Improvement of the discharge capacity of a river reach. This can be achieved by reducing its roughness, enlarging the conveying cross-section, or shortening the river channel and thus steepening its slope.

(c) Increase of the storage capacity of the river and or diversion of flood water from the river. The storage capacity can be increased by reservoirs (already discussed) or by using parts of reclaimed areas for flood water storage. The use of diversion canals will directly reduce the discharge downstream resulting in lower water levels.

6.2 THEORETICAL BACKGROUND IN RIVER ENGINEERING [14]

6.2.1 Types of Flow

The various types of flow are as follows:

Case (i) two-dimensional: Steady uniform flow with no acceleration or deceleration which means rate of change of velocity with respect to either distance or time equals zero, i.e.,

$$\frac{\partial v}{\partial x} = 0, \ \frac{\partial v}{\partial t} = 0$$

Case (ii) two-dimensional: Steady flow along symmetrical stream lines but with acceleration or deceleration with respect to distance i.e.,

$$\frac{\partial v}{\partial t} = 0$$

Case (iii) two-dimensional: Steady uniform flow along curved streamline with no exchange between vertical planes, i.e.,

$$\frac{\partial v}{\partial s} = 0 \text{ and } \frac{\partial v}{\partial t} = 0$$

Case (iv): Steady along curved lines but with exchange between vertical planes such type of flow is termed as spiral flow and the flow is considered three-dimensional, here also $\partial u/\partial t = 0$ (Fig. 6.1).

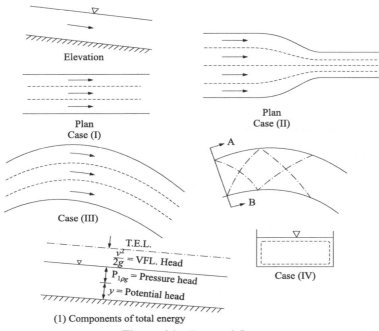

Figure 6.1 Types of flow.

Case (v): Non-steady flow, either uniform or nonuniform, i.e.,

$$\frac{\partial v}{\partial t} \neq 0, \quad \frac{\partial u}{\partial x} = 0 \text{ or } \neq 0.$$

Under nonuniform type unsteady flow may be gradually varied or rapidly varied. In the first case, the curvature of the wave profile is small and the change in depth is gradual and in the second type the curvature is very large and so the surface of the profile may be discontinuous. Eg: various kinds of flood waves and surges of various types.

6.2.2 Resistance Laws (Simplified Treatment)

The two most well-known equations developed and used by hydraulic engineers are the Chezy's and Manning's equation for simplest type of flow, viz., steady and uniform, Chezy formula can be written as:

$$V = C \sqrt{hS} \tag{6.1}$$

which is obtained by equating the propelling force which is due to the weight component to the retarding force due to the roughness of the bed where,

V = mean or average velocity,

h = depth of flow,

S = energy slope.

Chezy's formula can be combined with Cole-Brook-White's formula which is based on Karman-Prandtl equation of velocity distribution and on Nikuradse test data and finally can be expressed in the form:

$$C = 18 \log \frac{12h}{k_s}$$

or,

$$V = \left[18 \log \frac{12h}{k_s} \right] \sqrt{hS} \tag{6.2}$$

where

k_s = Nikuradse equivalent roughness height.

The Manning's foumula is empirical in nature, viz.,

$$V = \frac{1}{n} h^{2/3} S^{1/2} \tag{6.3}$$

and it can be combined with Strickler's formula also empirical, viz., $n = \dfrac{k_s^{1/6}}{26}$ or $0.0342\, k_s^{1/6}$ to give

$$V = \left[0.0342 \left(\frac{h}{k_s} \right)^{1/6} \right] \sqrt{hS}$$

The two formulae are thus closely related and they differ only in the constant before ($\sqrt{hS}$) term. Generally speaking, the ratio (h/k_s) remains nearly constant for a particular alluvial river which is because of the tendency of the ripples on the bed to grow larger as the water level rises. The value of C can be calculated from an assumed roughness k_s which can be estimated from measurements in actual rivers or by consideration of sediment transport formulae. The normal range for C is from 30 m$^{1/2}$/s for ripples to 70 m$^{1/2}$/s for a deep channel having a smooth clayey bed. Usually in a river the width is very much greater than depth but in the case of narrow river the side friction should be considered. Accordingly,

$$V = C \sqrt{RS}$$

where

R = hydraulic radius.

Further, the rivers have an irregular cross-section and accordingly the discharge is given by

$$Q = \Sigma C h^{2/3} S^{1/2} (dB) \tag{6.4}$$

where dB is an increment of breadth with constant C and S. Hence, $Q = CS^{1/2} \Sigma h^{3/2} dB$. In practice summation can be done graphically or numerically.

6.2.3 Energy Slope

Consider first three components of total energy of a flowing water, viz., potential, pressure and kinetic energy (Fig. 6.1). Potential energy is due to the position a particle of water has with respect to a fixed datum. Pressure energy is due to the weight of the column of water above the particle and finally kinetic energy is due to the velocity of the water particles. The components of total energy are to some extent interchangeable, viz., potential energy may be converted to kinetic and vice versa. Consider the case of frictionless flow between bridge piers (Fig. 6.2). Usually, however, friction will cause loss of energy, i.e. the total energy line will fall with gradient depending on the rate of loss of energy. Hence, at equilibrium

(S_b), Bed slope = energy slope, (S_e) – water surface slope, (S_w).

Using Chezy's formula,

$$S_e = \frac{Q^2}{C^2 \, B^2 \, h^3} \tag{6.5}$$

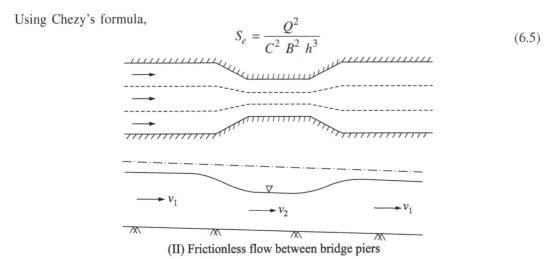

(II) Frictionless flow between bridge piers

Figure 6.2   Total energy components and flow between bridge piers.

Consequently, the water surface line is $(v^2/2g)$ below the energy line. As a corollary to above the equilibrium depth h_e for a given discharge per unit width, i.e., $q = Q/B$, and a given (S_b) can be calculated from

$$h_e = 3\sqrt{q^2/C^2 S_b} \tag{6.6}$$

Apart from energy loss due to friction other losses almost always occur more so when the flow is decelerating (Fig. 6.3). Usually accelerating flow readily changes direction without a great loss of energy except in the very extreme case of abrupt contraction. When eddies are formed, most of the kinetic energy will be transferred to heat which represent loss to the hydraulic system; the worst case will be when all kinetic energy differences between upstream

and downstream of the given section would be lost. Now energy loss = $\alpha\left(\dfrac{v_2^2 - v_1^2}{2g}\right)$, where

$\alpha = 0$, ideal condition, $\alpha = 0.6$, fair condition and $\alpha = 0.8$, poor condition. In the computation of energy losses in a river (Fig. 6.3) both the type of energy losses should be accounted for; sometimes friction predominates, at other times, deceleration predominates. The river-bed configuration is normally comprised not only individual grain roughness but also ripples, dunes, bars and other irregularities. Accordingly, the water surface is affected by regions of acceleration and deceleration and show large variation in level. Usually when the upstream gradient is measured between two gauging stations some distance apart only the average gradient between two parts is obtained and a straight line is drawn. Consequently, both the effects of friction and declaration losses are included, which means the measured values of S are greater and, therefore, the actual C is smaller than the C value of the bed. Same considerations apply when flow takes place over high water bed, because of hedge, wall, groynes, etc. which greatly increase the resistance. Consider the flow in a meander with cutoff as shown in Fig. 6.4. Here the highwater flow along two channels must be balanced to give the same head loss when

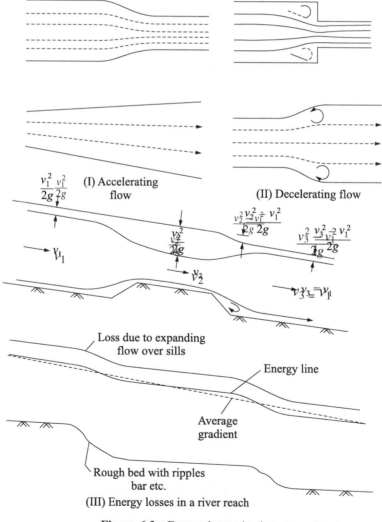

Figure 6.3 Energy losses in river.

flowing through two different paths, viz. meander and straight cut-off portion. However, the nature of the losses is different for the two paths. Accordingly, when river improvements are executed the value of C for various water levels will have to be calculated from measurements beforehand. In new situations the same values may be used except when the roughness has been appreciably changed due to cutting of bushes, construction of smooth reaches, trimming of the bed. In such a case new values would have to be selected by theoretical calculation or by experience of comparable situations.

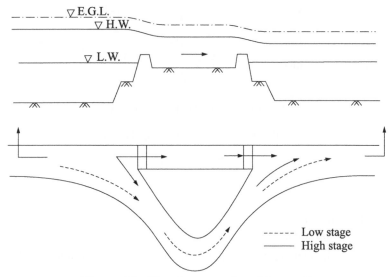

Figure 6.4   Flow in a meander with cut-off.

6.2.4   Gradually Varied Flow

6.2.4.1   Backwater effect

The general treatment of gradually varied flow is furnished in textbooks of open channel flow [27]. However, in river engineering problem, they need not be treated on rigorous theoretical basis. Usually a simplified analysis is adequate. Consider a river having a rectangular cross-section discharging into a lake (Fig. 6.5). It is desired to know the backwater profile given the gradient of this river reach, discharge and the level of the lake. The steps involved in the analysis are:

    (i) The normal depth h_e of the undisturbed flow, a great distance upstream from lake, can be determined from Eq. (6.6), i.e.,

$$h_e = 3\sqrt{q^2/C^2 S_b}$$

   (ii) The influence of back water starts from lake and here the conditions are known, i.e., h_1 at the section is known.

  (iii) The problem now reduces in determining the distance Σdl from the section at lake where depth is h_1 to depth h_e by taking suitable increments in h_1, i.e. by several steps.

(iv) This is accomplished in tabular form, Table 6.1 given below using the relationship

$$dl\ [S_b - (S_e)_m\ ] \simeq dh_0 \tag{6.7}$$

where dh_0 change in specific head for an arbitrary increment or decrement of depth. $(S_e)_m$ average value of energy slope for the depth h and $(h + dh)$ which is assumed to represent the mean value over the distance dl.

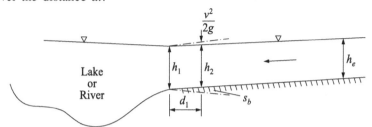

Figure 6.5 Gradually varied flow.

Table 6.1 Backwater Computation.

h	A	v	$\dfrac{v^2}{2g}$	h_0	P	R	C	S_e	$\dfrac{v^2}{C^2 R}$ or $\dfrac{v^2}{C^2 h}$	$(S_e)_m\, dh_0\, dl$

where
> P = wetted perimeter
> A = cross-sectional area.

In the case of wide rectangular section with discharge q per unit width, $v = q/h$ and $R = h$.

6.2.5 Three-dimensional Flow

6.2.5.1 Uniform steady flow along curved streamlines

Case (i): Consider the flow during low water in a river as shown in Fig. 6.6, where it is entirely confined in a definite channel reach. One way of the calculation procedure is to divide the cross-section into lanes of equal discharge and then applying the formula relating to two-dimensional flow.

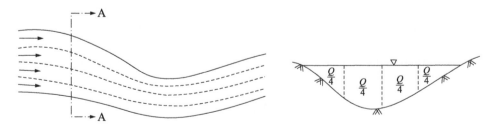

Figure 6.6 River flow during low water.

Case (ii): Where the riverbed configuration is more complicated, large differences in gradients, transverse gradients and intermediate flow across sand banks are some of the difficulties that can occur. In such a case the problem is essentially three dimensional. In order to change the problem into more amenable two-dimensional network it is necessary to draw equivalent streamlines which are not given by nature.

Consider the forces which act to cause water to flow round a bend. It is obvious that such a force must act at right angles to the direction of flow giving an acceleration towards the centre of the bend. When the flow takes place in a straight course, the water surface is horizontal in the cross-section, but in curved flow a certain cross-gradient I_p (transverse slope) is set up (Fig. 6.7). Consider an element of width db. The difference of pressure between outside and inside of the element is

$$dp = \rho g \, I_p \, db, \quad \text{or} \quad \frac{dp}{db} = \rho g \, I_p \tag{6.8}$$

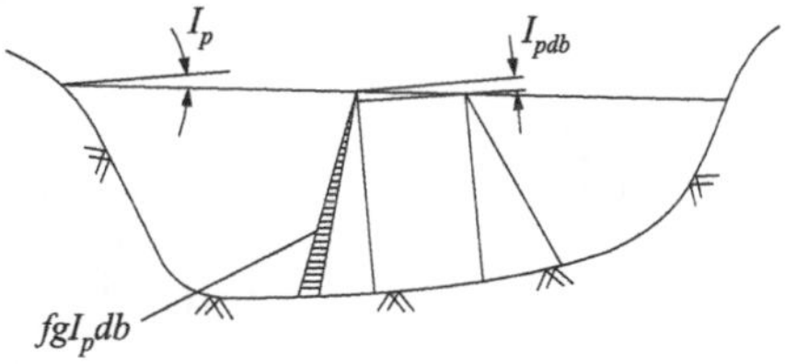

Figure 6.7 Transverse gradient.

i.e., pressure difference per unit width which is equal to force per unit volume acting normal to the flow. If v, represents the flow velocity in the bend, then the acceleration normal to flow is (v^2/r). The force per unit volume acting normal to flow is $\rho v^2 / r$. Hence,

$$\frac{\rho v^2}{r} = \rho g \, I_p \quad \text{or} \quad I_p = \left[\frac{v^2}{gr} \right] \tag{6.9}$$

As an example consider the flow in a bend (Fig. 6.8) forming a semicircle. Ignoring the effects of friction and deceleration, the flow can then be considered as potential.

In the inside of the bend from (A-A) to (B-B), the velocity increasing, i.e., $v^2/2g$ increases. As there are no losses the energy line is horizontal and the water surface is, therefore, lowered. Conversely, from section (B-B) to (C-C) the water surface returns to the initial position. The process is exactly the opposite in the outside of the bend. Consider a streamline carrying a discharge equal to q. If the velocity increases and the depth is changed only slightly (second order of magnitude) then the width of the lane must decrease the streamlines, therefore, get closer to the inside of the bend.

Since the velocity and radius both changes across any section, the surface slope also changes. The water flows in such a way that the forces are always balanced. Usually in practice the flow is unsteady. Accordingly, a stream lane receives too little water at the one moment too much next as such the water surface fluctuates continuously over a mean. Since full centrifugal force must be available at the start of the bend, there will be a transition section upstream of the bend where the water already has a transverse slope. Sometimes the flow can be seen to be hugging the inside bank of the bend at a distance of 1 to 2 km upstream. Next, let us discuss the complication arising out of consideration of frictional resistance and deceleration for the case of flow in a river bend (Fig. 6.9). In such a case, the streamlines may be drawn as indicated in Fig. 6.9.

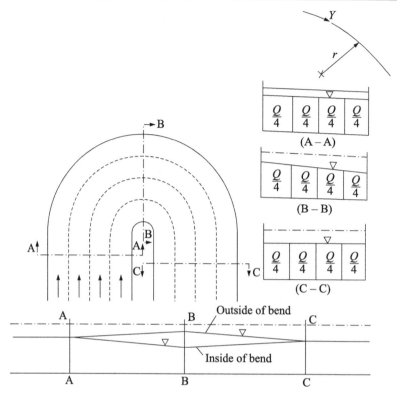

Figure 6.8 Flow in a river bend.

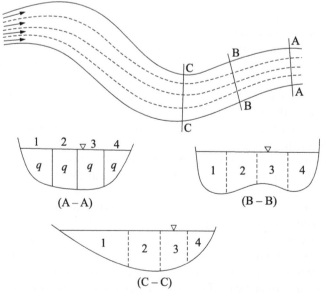

Figure 6.9 River flow with irregular bed features.

For this purpose all the data such as levels, velocity or discharge at the starting section (A-A) and the cross-section at other points along the river course should be available i.e., (B-B), (C-C). With the above data it is possible to obtain the relevant informations such as streamlines, velocity, water surface elevations, etc. at other sections.

The steps to be followed are:

(i) Always start calculating from the upstream direction.

(ii) Sketch estimated streamlines, i.e., say four, lanes with $Q/4 = q$.

(iii) Using Chezy's formula, calculate the longitudinal slope in each stream lane for each section. This will give the water level at the respective lanes.

(iv) Check the transverse slope at each cross-section conforming with the equation

$$I_p = \frac{v^2}{gr}$$

where

v = velocity in the stream lanes considered;

r = radius of the lane;

I_p = average gradient across stream lane considered and obtained by consideration of the bends in adjacent stream lanes.

(v) If the gradients and levels thus obtained are incompatible the streamlines must be adjusted and the values recalculated until satisfactory agreement is reached.

6.2.5.2 Circular and helicoidal flow

Decelerating flow generally causes additional energy losses. Flowing water cannot decelerate quickly and abrupt changes in direction of flow cannot occur. Kinetic energy is not readily converted to potential energy. Only with very gradually divergent angles will the flow follow the wall; normally there will be separation and eddies formed between the wall and the first regular streamlines. However, it is not possible to calculate the exact dimension or location of eddies. As in normal flow, the eddies are unstable having oscillation in strength and direction about a mean. The kinetic energy of eddies is supplied by the main stream and used up in extra frictional losses on bed and banks. The velocity of the main stream and its length of contact with the eddies are parameters for the measurement of the energy supply to an eddy. Eddies usually have an elliptical shape with the major and minor axes of similar length. If the ratio between the axes gets too great the eddies split up into several smaller ones. Eddies may be compared to ball bearings. They support the flowing water when there are no fixed walls. Instead of a frictionless hard ball there is a flexible ball, absorbing energy and transferring it to the nearest boundaries. The centrifugal force of the eddies which is proportional to v^2/r, provides the support. If r becomes too large, the force becomes too small, the eddies break up into smaller units of r values. Also when the outer streamline of the eddies has a certain radius then the main stream, must be curved also. The eddy grows out of its fixed walls and causes a narrower section of the main stream. The main stream must, therefore, accelerate, the water surface will fall and a transverse gradient will be established. The streamlines thus become curved and a gradual deceleration is possible.

Eddies frequently occur in river engineering problems such as in groynes used for protection of embankments and river training, mouth of a tributary or canal entering on to a river, in bifurcations of rivers.

In the treatment so far made it has been assumed that the flow takes in parallel streamlines in the horizontal and vertical planes, i.e., along stream tubes. Actually, however, internal flow takes place due to exchange between adjacents stream tubes both horizontally and vertically. This is the reason for sedimentation and scour in river. As has been mentioned earlier in curved sections the water surface has a transverse slope. The force resulting from this gradient is acting on all the water particles and below it the vertical. Now the actual water velocity varies throughout the vertical with a higher than average velocity at the surface and a lower one near the bed. To compensate, therefore, the radius the water particle takes must be greater on the surface and smaller near the bed. This results in a component of velocity across the main direction of flow which is outwards at the surface and inward along the bed thus setting up a spiral movement. Approximate calculation of spiral flow can be made with the help of following relations:

(a) At the surface:

$$\frac{v_y}{v_x} = 0.025 \frac{V^2}{gr\,S}$$

where

v_x = velocity along axis of river flow,
v_y = velocity at right angles to river flow.

Replacing S from Chezy's relationship, the above can be written as

$$\frac{v_y}{v_x} = 0.025 \frac{C^2 h}{gr} \tag{6.10}$$

(b) At the bottom:

$$\frac{F_y}{F_x} = 0.04 \frac{V^2}{gr\,S} = 0.04 \frac{C^2 h}{gr} \tag{6.11}$$

where

F_x = force acting on the bed material along the average curve,
F_y = force at right angles to F_x.

For a circular bend with fixed bend having a constant discharge the mobile bed will eventually assume a lateral slope such that the force on a grain at the bed will be zero in the radial direction. This means that the resultant shear force due to current and friction at bed is balanced by the component of weight force (Fig. 6.10).

Accordingly, the component of weight force on a spherical particle of diameter d is

$$G_y = (\rho_s - \rho)\,g \sin \beta_y \, \frac{1}{6} \, \pi d^3 \tag{6.12}$$

and

$$F_y = 0.04 \frac{C^2 h}{gr}, \; F_x = 0.04 \frac{C^2 h}{gr} \left(\frac{u_*^2 \, \rho \pi d^2}{4} \right)$$

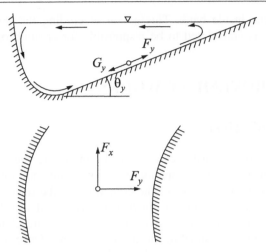

Figure 6.10 Spiral flow in a river.

Equating G_y and F_y, we have

$$(\rho_s - \rho)\, g \sin \beta_y \, \frac{1}{6} \pi d^3 = 0.04 \, \frac{C^2 h}{gr} \cdot u_*^2 \, \frac{\rho \pi d^2}{4}$$

or,

$$\sin \beta_y = \frac{0.04 \times 6}{4} \cdot \frac{h u_*^2 \, C^2}{g^2 r \, \dfrac{(\rho_s - \rho)}{\rho} d} = \frac{0.6 h u_*^2 \, C^2}{g^2 \, r \Delta d}$$

where

$\Delta = \dfrac{\rho_s - \rho}{\rho}$, again, $u_*^2 = ghS$, and considering $C^2 = 2000$ m/sec^2,

$$\sin \beta_y = 0.06 \times 2000 \, \frac{h^2 S}{\Delta r d g} = \left[\frac{12 \, h^2 S}{\Delta r d} \right] \tag{6.13}$$

The lateral bed slope is small and hence $\sin \beta_y \simeq \beta_y$. However, the bed slope is still large compared to the lateral surface slope and so $\sin \beta_y \simeq \beta_y \simeq \dfrac{dh}{dr}$ or $\dfrac{dh}{dr} = \dfrac{12 h^2 \, S}{\Delta r d}$ or integrating

$$\left[\frac{1}{h} - \frac{1}{h_0} \right] = \left[\frac{1}{r} - \frac{1}{r_0} \right] \frac{12 r_0 \, S_0}{\Delta d}$$

where,

$$S_0 = \frac{S_x \, r_x}{r_0} \tag{6.14}$$

i.e., gradient of water surface the bend. If the outer bank is erodible, there will be no equilibrium since there will be a continuous supply of material from the outer bank, therefore, $F_y > G_y$

causing deposition on the inside bank. This action of eroding the outside bank and causing deposition on the inside is considered to be responsible for creation of meanders.

6.3   RIVER IMPROVEMENT WORKS

6.3.1   Behaviour of River

The silt carried by flow is primarily responsible for moulding the behaviour of rivers. The available energy is balanced by the dissipation due to bed, side and internal friction as well as energy required to transport the sediment. The conditions in the river such as stage, discharge, silt charge change all along the course from day to day and it is seldom that river sections attain equilibrium. Every river tends to develop bends which are characterised by scour and erosion of the concave side and shoaling on the convex end. Once a bend is established the flow tends to make its curvature longer. Studies indicated that particles move from the convex shore to another crossing the intermediate shoals diagonally. Generally, at high water excessive deepening of the pools occurs with marked shoaling at the crossings. The river gradient during the period is more or less uniform resulting in uniform sediment movement. With the drop of high water level, slope variation occurs being flatter in the pools and steeper in the crossing as it acts like a weir. Scouring and transporting action decreases at a rate relatively greater in pools than in shoals. The shoals thus scour, however, the rate of scouring is much smaller than the rate of fall of water surface. Hence, at low water the river shows deterioration when it is to be used for navigation.

6.3.2   River and Hydrographic Surveying

Before any scheme can be undertaken it is necessary to conduct survey. Usually an arbitrary point representing the start of a river mouth or confluence is chosen and the distance is scaled on the map along the centre line of the river. A line is drawn through each mile or kilometre point at right angles to the river and permanent marker posts are placed on this line on each bank of the river. Further, it is necessary to take cross-sections of the river as well as the land adjoining the river at regular intervals. The maximum interval usually is of the order of 300 m, depending on whether the river is regular or irregular in shape. Cross-sections are taken at right angles to the river at the point of sectioning and distances between sections are measured along the centre line of the river. The intervals between sections as well as the intervals between soundings should be constant say about 2 m for smaller to 4 m for bigger rivers. Soundings are taken from the water level. The water level at the last section taken on one day should be taken first the next morning, to obtain the surface fall throughout the length. Cables or pipelines crossing above or below the river should be noted. The sections are required for calculation purpose and should be plotted to the same horizontal and vertical scale in order to determine wetted perimeter. There are two conventions for plotting river sections looking (i) upstream, or (ii) downstream.

Under (i), the further upstream section is plotted looking downstream and is placed at the top of the sheet so that the right bank is on the right-hand side of the paper. The other sections are plotted successively under each other.

Under (ii), the furthest downstream section is plotted looking upstream and is placed at the bottom of the sheet with the right bank to the left of the paper. The sections are plotted successively above one another.

For survey of the land adjoining the river, usually they extend from each bank to the land which is 3 m or more above flood level. In addition to field level at regular intervals across the section, levels should be taken when the line cuts drains, road and railways. Inverts of bridges and culverts in the area are also required. Land section are usually plotted in suitable scale for easy calculation but different vertical and horizontal sections are used to show up the relatively small changes in level.

6.3.2.1 Hydrographic Surveying

The hydrographic survey is required for the assessment of short- and long-term morphological changes of river. Further, it is essential for navigational requirement and safety of vessels playing in the river. For carrying out hydrographic survey the first requirement is fixing up position offshore. They vary depending on complexities and cost such as simple optical Sextant to electronic positioning system. Generally, optical methods are commonly adopted aboard most of the survey launches for river hydrographic survey. The methodology of fixing up position to carry out survey along a cross-section is described below:

(a) The survey vessel is anchored first on the river bank near a cross-section.

(b) The anchored point in most of the cases will be either upstream or downstream of cross-section, seldom it will be on the actual line itself.

(c) To locate the anchoring position survey personal make use of Sextant for measuring the two angles between any three shore objects or mark or pillars as the case may be so that the angle subtended between the right and left object with respect to central object is almost equal.

(d) The two angles are then plotted on the survey plan (field sheet) with the help of station pointer and the anchored position is found out.

(e) In case, the position is downstream or upstream of the cross-section the launch is moved towards upstream or downstream of the initial position and the above process is repeated till the position of the cross section has been correctly located on the bank.

(f) Having fixed up the anchor point the alignment of the cross-section is given with an angle to any one of the shore objects. The alignment is prefixed in such a manner so that the cross-section is normal to the direction of flow. Generally, two flags are placed across the alignments of the cross-section on one bank. This alignment is also plotted on the survey plan.

(g) In the upper deck of the launch two surveyors with Sextant in hand stand together at a very close distance. The launch fitted with an echosounder then is moved at a constant speed from one bank to the other along the alignment.

(h) While crossing over the river along the cross-section the position for different points on the cross-sections are measured by Sextant. At the same time respective depths for those points are also recorded from the echosounder for each of such points which are then plotted on the survey maps.

(i) In the shallow regions the launch will not be able to move because of draft problems. In such a situation country boats are used for survey. The different locations on the cross-section in the shallow-reaches are also fixed by Sextant and the corresponding depths are measured from the boat using the lead line.

(j) For gradually sloping bank the survey is done only up to the water line on the day of the survey. The bank elevations above the waterline are measured with the help of Engineer's level and wooden staff. The bank line and their elevations are thus evaluated and plotted on the survey plan.

(k) Bench mark pillars having values of local elevations with respect to Mean Sea Level exist on both the banks of the rivers at suitable interval of length. Water levels on the day of survey w.r.t., M.S.L. are measured by connecting two consecutive B.M. Pillars with the help of the Engineers level and a wooden staff. These water levels are distributed for each of the cross-section according to the calculated slope for the reach in between two B.M. pillars.

(l) On the survey plan the cross-section is completed with soundings reduced to local datum (with respect to values given in terms of GTS) called local zero.

(m) The depths are shown in metres and decimetres as per conventions. As for example sounding 5_2 indicates that the position is 5.2 m below local zero. Thus the hydrographic survey for one cross-section is duly completed in all respect. Following the above procedure the entire river reach can be surveyed.

6.3.3   River Training for Flood Protection

6.3.3.1   Embankment

River training broadly covers all engineering works constructed in a river to guide and confine the flow to the river channel and to control and regulate the river bed configuration for effective and safe movement of floods and sediment. River training for flood protection is generally referred to as high water training. Its main objective is to provide sufficient cross-sectional area for the safe passage of the maximum flood. Embankment for flood protection is one of the earliest engineering achievements of man. Historical records reveal embanking of river Nile by Egyptians. Similarly, the city of Babylon was protected by levees. Other rivers like Euphrates and Tigris in Iraq and Ganges in India present early attempts in leveeing. In India and other countries generally retired embankments were constructed across the spill channels.

A levee or dyke may be defined as an earthen embankment extending generally parallel to the stream course and designed to protect the area behind it from overflow by flood waters. In general, they should be aligned on the high ridge of the natural banks of a river where land is high and the soil is suitable for the construction of embankments. Their alignment has to be determined in such a way that the high velocity flow is sufficiently away from them. In case it is not possible to set back the embankment to avoid high velocity of flow, protection in the form of spurs and revetments is necessary. Sometimes a ring bund has to be provided to protect a town or urban property. In such a situation the river-side of the ring bund may need to be protected by short permeable or impermeable spur or stone revetments. Also pumps have to be provided for drainage and rainwater when the river is in flood. The success of an embankment depends on vigilant and continuous supervision during floods. Where the embankments are likely to be attacked immediate protection is required. In case it is found that the protective measures will not give desired security to the embankment a loop behind the existing embankment should be constructed which will act as a second line of defence should the original embankment fail.

Barriers generally parallel to the stream course in the form of levees, floodwalls or highway or railroad embankments are widely adopted for the protection of flood plains associated with a particular stream. This method is a compromise with nature. Here the normal flow is confined to the river and flood flows are allowed to spread to a limit determined by flood embankments. If the flood banks can be set back at a reasonable distance from the river and the area between the flood banks can be kept free of obstructions, the method is good and full use can be made of the whole cross-sectional area between the flood banks to carry the flood discharge. Because of the reduction in velocity as the water flows over the natural banks most of the suspended load is deposited on the flood plain. When levees are constructed the width of the stream is reduced and hence the velocity is increased. Therefore, the rate of deposition of the flood plain is reduced. The material that should have been deposited on the flood plain in the absence of levees is now carried downstream and deposited either in the unleveed portion of the stream or, in the sea. If the sea is deep enough the outlet of the stream is unaffected, otherwise extension of the delta takes place thereby flattening the stream slope. Other effects of confining the flood water of the stream by a system of embankment are: (a) increase in the rate of travel of flood wave in the downstream direction, (b) rise in the water surface elevation of the river during the flood, (c) reduction of the storage and thus an increase of the maximum discharge downstream, and (d) decrease in the water surface slope of the stream above the levees portion as a result aggradation will occur upstream of the leveed reach.

From the point of view of desirability of high discharge capacity for a given stage levees should be located far apart. Since alluvial streams meander the levees should be at least on the edge of meander belt so that the levees will not be washed away on account of the stream changing its course. They should also follow the natural curvature of the stream so that they are unlikely to be attacked by the stream current.

6.3.3.2　Discharge capacity

Earlier it has been explained how to obtain the cross-sectional shape of river inclusive of flood for river survey. Usually the irregular cross-sectional shape can be approximated into a double expanded channel with side slope as shown in Fig. 6.11. By adopting a suitable height of embankment, the capacity of the double channel can be estimated with reasonable accuracy by considering it, to be made up of two channels a deep one and a shallow one.

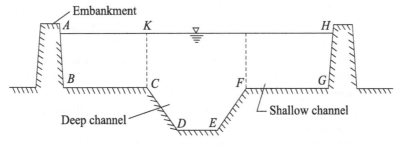

Figure 6.11　Flow in a river with flood plain.

The shallow channels are usually found to be rougher than the main channel, so the mean velocity in the main channel is greater than the mean velocities in the side channels. In such a case the Manning's formula may be applied separately to each subsection in determining the mean velocity of the subsection. Then the discharges in the subsections can be computed. The total discharge is, therefore, equal to the sum of these discharges. The mean velocity for the whole channel section is equal to the total discharges divided by the total water area. Ignoring the effect of nonuniform velocity distribution in the subsection and considering only the mean velocity V_1, V_2, V_3 corresponding to subsection, viz., *ABCK*, *KCDEFI* and *IFGH* having areas equal to ΔA_1, ΔA_2, ΔA_3, etc. then as per Manning's formula:

$$Q_1 = V_1 \, \Delta A_1 = \frac{1}{n_1} \, R_1^{2/3} \, S_e^{1/2} \, \Delta A_1 \qquad (6.15)$$

or,
$$Q_1 = K_1 S_e^{1/2}$$

where

$$K_1 = \text{conveyance} = \left[\frac{\Delta A_1}{n_1} \, R_1^{2/3} \right]$$

or
$$Q_2 = K_2 S_e^{1/2} \quad \text{and} \quad Q_3 = K_3 S_e^{1/2}$$

where R_1, R_2, R_3, etc. are the hydraulic radius.

$$R_1 = \frac{\Delta A_1}{(AB + BC) = \Delta P_1}, \, R_2 = \frac{\Delta A_2}{(CD + DE + EF) = \Delta P_2} \text{ a n d } R_3 = \frac{\Delta A_3}{(FG + GH) = \Delta P_3} \, ,$$

ΔP_1, ΔP_2, ΔP_3 being the wetted perimeters n_1, n_2, n_3, etc. being the roughness coefficient, S_e = energy slope. Hence, total $Q = [Q_1 + Q_2 + Q_3]$.

6.3.4 Design of River Dyke or Embankment

Generally, development and construction of various types of embankments has been largely as a result of experience. The essential conditions governing embankment design and construction are provision of an adequate freeboard to prevent overtopping at a cross-section, sufficiently massive for severity against seepage through the body of the embankment. Normally they have to be designed to keep the seepage gradient inside the body of the embankment with a minimum cover around 1 m. As regards the seepage gradient it varies from about 1 : 4–1 : 5 according to the character of the soil. The width at the top of the embankment should be sufficient to provide a road for supervision and transportation of materials during flood. The road should preferably be supplemented by a berm on the landside slope to avoid reconstruction of the road in case it is desired to raise the height of dyke subsequently. The alternative approach based on theory for design is discussed below.

Great care should be taken in determining the height of levees. Levees must be designed to contain a flood of reasonable return period, i.e., say of the order of 500 years. Such a flood can be routed through the stream and flood stages can be determined at desired locations along the reach. The level of the top of the levee at any place can be fixed after providing a free board of 1–2 m. The freeboard is a provision against waves and the possibility of occurrence of a flood of higher return period than assumed. One should also consider the probable settlement of the levee after it is constructed.

An example of the design of a levee is shown in Fig. 6.12. Consider the area stretching from the railway station at location Y up to the high ground at location X on the side of a river. It is required to design a suitable section of the dyke. From the record of the stream flow it has been ascertained that once in a thousand years the flood discharge in the river is 2400 m³/s and based on this discharge the relevant backwater computation gives a river level of 6.5 m at location X and 5.9 m at location Y and almost a linear variation in between.

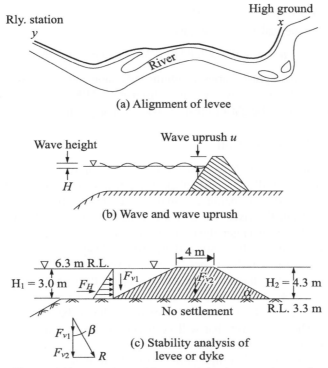

(a) Alignment of levee

(b) Wave and wave uprush

(c) Stability analysis of levee or dyke

Figure 6.12 Locations of levee and the forces acting on it.

6.3.4.1 Computation of wave heights

Assume a maximum of 20 m/s of long duration wind velocity and consider fetch F in the river influencing wave height to be about 1 km. In f.p.s. units the wave height is given as

$$H = 0.17 \sqrt{v \cdot F} + 2.5 - \sqrt[4]{F} \qquad (6.16)$$

where

H = wave height in ft,
F = fetch in miles,
v = wind velocity in ft/s.

Now in the present problem $v = 20$ m/s = 65 ft/s and 1 km = 0.62 miles or $H = 0.17\sqrt{65 \times 0.62}$ $+ 2.5 - \sqrt[4]{0.62} = 2.7$ ft or 0.82 m. Now at other positions of the dyke where the fetch is much less than 1 km, the wave height would still be close to 0.8 m so we will consider the wave height of 0.82 m to be constant along the dyke. After hitting the dyke water will rush up along the slope of the dyke and the distance is known as wave uprush. It can be estimated from the relationship

$$u = 8 f H \tan \alpha \cos \beta \qquad (6.17)$$

where

α = dyke slope,
β = angle wave makes with the dyke, the minimum is about 30°,
f = roughness of dyke surface = 0.7.

By substituting the numerical values we have

$$u = 8 \times 0.7 \times 0.82 \times 0.33 \times 0.866 = 1.3 \text{ m}$$

The above value of wave uprush varies insignificantly along the dyke so level of dyke crest is about 1.3 m above the water level when it discharges 2400 m³/s. Now at location X, the river level is about 6.3 m so height of crest of dyke at that position is 7.6 m above the datum.

6.3.4.2 Design of dyke section

In designing a dyke section (Fig. 6.13) suitable clayey soils have to be obtained from a source and if it is not available locally the cost of excavation, transport ploughing, etc. of the given clay would prove prohibitive. Generally, it may prove cheaper to dredge the dyke material, say, sand and gravel from the river bed and use a smaller amount of clay just for the cover of the dyke. This will need a larger dyke cross-sectional area since the side slope will have to be gentler but it should cost less. Since there would always occur seepage through a river dyke made from dredged material it is necessary to have a gentler backslope in order that the backslope will be wetted at high river stages due to seepage through the dyke. Backslope should be less than or equal to $\phi/2$, where ϕ is the angle of repose of the material having a value of about 36° for sandy gravel. Hence, backslope of dyke should be 18° or 3 : 1 and thus can also be front slope. The clay cover that will be needed is about 1 m in thickness at the river side and crest, and about 0.5 m thick on the backslope plus a 2 m deep clay cutoff. Because the levees consist of gravels with a high permeability there is no danger of uplift pressures to there is no need to provide a filter at the toe of the dyke but a clay cutoff is needed at the toe to protect it from damage due to erosion and increased seepage path under the dyke. Also if stone is readily available in the area the seaward slope can be protected with dumped stones. Generally, for such small wave height the pitching is of the order of 1.5 t/m² on the river banks with stone size of 0.25 m dia placed to as low a depth as possible. On the inside slope and

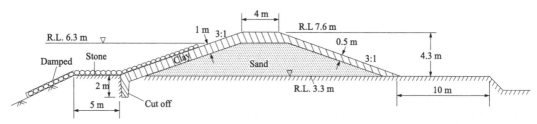

Figure 6.13 Design of a dyke section.

crest grass is sown and a collection ditch 2.5 m is dug from about 10 m inside the toe of the dyke, the material from which can be used in the construction of the dyke.

6.3.4.3 Stability analysis of the dyke

With reference to Fig. 6.12, since α is less than ϕ, for stability it can be shown that $\tan \beta < 0.23 \tan \phi$.

Now

$$F_H = \frac{1}{2} \gamma H_1^2; \; F_{V_1} = \frac{1}{2} \gamma H_1^2 \cot \alpha,$$

$$F_{V_2} = \gamma_s \, H_2 \, (H_2 \cot \alpha + 4) \text{ if } \gamma_s = 1.6$$

$$\tan \beta = \frac{F_H}{F_{V_1} + F_{V_2}}$$

or

$$\tan \beta = \frac{\dfrac{1}{2} \times 1 \times 3^2}{\left(\dfrac{1}{2} \times 1 \times 3^2 \times 3 \right) + 1.6 \times 4.3 \, (4.3 \times 3.0 + 4)}$$

$$= \frac{4.5}{129.5} \text{ or } \tan \beta = 0.035.$$

Now $\phi = 36°$; $\tan \phi = 0.726$; or $0.23 \tan \phi = 0.167$. Hence, since $\tan \beta < 0.23 \tan \phi$, the dyke is stable.

6.3.5 Bank Protection

The lower bank acts as the foundation for supporting the upper bank and is more susceptible to erosion. Recession of bank is caused by the erosion of the lower bank, particularly at the toe. The recession is fast especially when there is sandy substratum below, i.e. sand is washed away by a strong current and the overhanging bank collapse. The upper bank is the portion between the low water and the high water. Action on this is more severe when the current impinges normal to the bank. When the water level rises with increase in the flood stage banks become partially or fully saturated with water. For silt or silty sand the angle of shearing resistance may be as low as 50 per cent of its original values before saturation. If the angle of the sloping surface is steeper than the reduced angle of shearing resistance sloughing may result.

In the case of flood control measure, protective works only along rivers will be considered. Because of the variation of the force of water acting at different parts of a cross-section of a river and because of the variation of duration for which the various parts of a cross-section are in contact with water, one can divide the river bank in three parts (i) embankment, i.e. the sloping surface of embankments facing the river, (ii) upper bank or the portion of the river bank that is located above the low water level and below the foreshore level, and (iii) lower bank, i.e. the portion of bank below low water level extending quite a distance horizontally to the river bed which is usually called the toe. Embankments are usually set back away from the main channel of the river and are, therefore, usually not subject to very strong currents. As embankments are constructed on the flood plain or foreshore of a river they come into contact with water only during floods. Problems encountered in the protection of embankments are the protection of the surface against currents of relatively low velocity and the maintenance of stability as a result of infiltration of water into the embankment as already discussed under design. When embankments are constructed very near to the main channels of rivers they become a part of the upper bank. In torrential rivers where embankments are constructed to confine the flood flow and where the foreshore is subject to a strong erosion during floods, special protection is required to be provided at the toe of the embankment as well as on its sloping surface.

6.3.5.1   Causes of bank recession

A bank may fail owing to any or a combination of following causes:

(i) Washing away of soil particles of the bank by current or waves and this is called erosion.

(ii) Undermining of the toe of the lower bank by currents waves swirls or eddies followed by collapse of overhanging material deprived of support called scour.

(iii) Sliding due to the increase of the slope of the bank as a result of erosion and scour.

(iv) Sloughing or sliding of slope when saturated with water; this is usually the case during floods of long duration.

(v) Sliding due to seepage of water flowing back into the river after the receding of flood; the internal shearing strength is further decreased by the pressure of seepage flow.

(vi) Piping in a sublayer due to movement of ground water to the river which carried away sufficient material with it.

The way banks fail under (i), (ii) and (iii) may be attributed to direct erosion or scour. Failure under (iv) and (v) may be said to be due to reduction of internal strength, and those under (vi) due to foundation failure. The last three are the result of saturation of seepage of water.

Generally speaking, the stability of a sloping bank under the action of water is subject firstly to the static head of water which induces seepage and saturation and secondly to the erosive force of running water which changes the form of the slope and in turn affects its stability.

6.3.5.2   Classification of bank protection works

Bank protection works may be broadly classified into two groups—direct protection and indirect protection.

Direct protection includes works done directly on the bank itself such as slope protection of embankment and upper bank and the toe protection of lower bank against erosion, and grading of sloping surface or provision of drainage layers to ensure stability against seepage and saturation. As such works cover continuously a certain length of the banks, they are also called continuous protection.

Indirect protection includes such kind of works as are not constructed directly on the banks but in front of them with a view to reducing the erosive force of the current by deflecting the current away from the banks or by inducing deposition in front of them. This is usually done by means of groynes spaced at a certain distance apart and is, therefore, also called non-continuous protection.

Under both direct and indirect protection, some differentiation can be made according to the durability of the works among (i) temporary protection to check erosion at times of emergency,

(ii) semi-permanent protection using brush wood, bamboo, wire mesh, etc., that will last a comparatively short time, and (iii) permanent protection including the use of more permanent materials like stone, concrete and also timber or brushwood mattresses if kept always under water.

6.3.5.2.1 *Direct protection*

(a) *Slope protection of embankments and upper banks*

River-side slope of embankments and upper banks of rivers if not subject to strong current can be well protected from erosion by a vegetable cover, either by turfing or by sodding or by other natural growth (Fig. 6.14). Low growth of shrubs and willows is by far the most effective cover, but they require a longer period to grow. In such cases, a temporary cover with mattresses of woven willow brushes is often necessary for the initial period of one or two years. After the natural growth has taken place, it just needs cutting once every year to keep the brushwood from growing into tall trees.

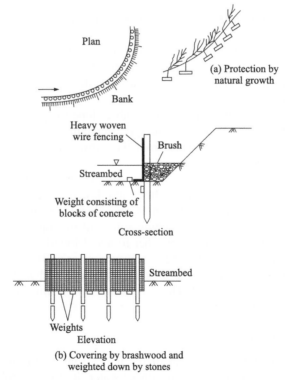

Figure 6.14   Slope protection through vegetable cover or turfing.

If the current near the upper bank and embankment is quite strong particularly during floods, paving of slopes with materials that can resist erosion is necessary. Figure 6.15 gives an idea of the size of stone to resist a given current. Temporary covering by brushwood and weighted down by stones is used for emergency purposes or by brush used everywhere. All pavings must have a firm connection with or be made continuous with the toe protection (Fig. 6.14).

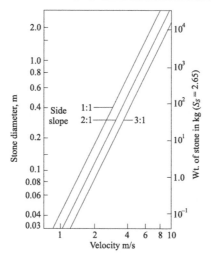

Figure 6.15   Size of stone to resist a given current.

All sodded, vegetable covered, or broken stone pavement must be graded beforehand to a slope at least equal to or flatter than the angle or shearing resistance of the soil under water. Drainage behind the paving is sometimes necessary in order to prevent the return seepage flow from carrying away fine particles of soil.

All protection work covering earthen banks with a slope steeper than the angle of shearing resistance of earth may be classified under the type 'retaining wall'. Thus the protection work has to be strong enough to resist pressure from the earth behind it and at the same time to resist the erosive force of water in front.

Temporary retaining walls may be built of closely driven piles, piles and timber boards or sheeting, piles and brushwood packing. All these may last for one to a few years and may be replaced by more permanent structures later on. Permeable pile fencing or rib work is sometimes built along the bank for the purpose of catching some silt to stop erosion.

Masonry retaining walls built on pile or other forms of foundation are heavy and expensive but are effective in protecting the banks against wave upwash. Besides being heavy, masonry retaining walls for banks protection have another drawback in that they are not feasible for deep foundation below low water level. The daily construction time is also shortened to a few hours only, particularly on tidal rivers.

(b) *Toe protection of lower bank*

Protection of lower bank differs from that of the upper bank in the respect that, in addition to the fact that the lower bank is constantly under water, its toe is liable to scour. Protection of lower bank starts somewhere near the low water level and extends towards the river up to a distance corresponding to the possible depth of scour. Toe protection must be extended to possible depth of scour, even if such scour may not be realised for years to come. The protection must be as flexible as possible so that it will cover up the bed when any scour does take place.

In general, there are four different ways now used in practice. The most common method is to provide a flexible apron extending to the river bed as far as needed. This may include all kinds of mattresses, as well as the falling apron (Fig. 6.16). The second method is to provide enough material, usually stone, as a reserve, which may be dumped into the hole to stop further scouring. Where the current is specially strong, stones packed in different kinds of cages such as brushwood or bamboo sausages, wire mesh sausages or mats, etc. may be used. This second method may be cheaper in first cost but is less reliable in time of emergency, especially when extraordinary scour occurs during times of great floods and storms, as no one can predict how far and how deep the scour would extend. It is too hazardous to take the chance of saving a little in first cost, with the probability of exhausting the material provided and the entire loss of an important structure.

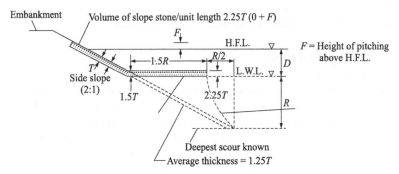

Figure 6.16 Toe protection through falling apron.

The third method of toe protection is by the provision of an impregnable curtain of retaining walls or sheet piles, etc. This is quite expensive, especially where deep scouring is expected. Steel sheet piles can answer the purpose best. Reinforced concrete sheet piles or wooden sheet piles are usually not watertight and the flow of water through sheet piles may weaken the foundation for the superstructure. Closely driven round piles can perform the work where scouring is moderate. Retaining walls built at the foot of a slope serve both as foundation for the pitching above and as toe protection below; but they cannot penetrate deep below water and another form of toe protection in front of them will be necessary.

For protection of embankment exposed to wave and tidal action the following relationship may be used to determine, the weight of stones required.

Zone I Defined as lying from 6.0 m below MSL to 2.0 m below MSL. In this zone the design maximum current velocity of the areas around Sunderban delta of West Bengal say will be of the order of 1.35 m/s. The formula for obtaining the weight of stone size can be expressed as:

$$W_1 = \frac{\gamma_s}{(\gamma_s - 1)} V^6 A,$$

where A = coefficient $\simeq 1$ with $V = 1.35$ m/s and $\gamma_s = 2.72$ the weight of stones comes to 9.5 kg.

Normally two layers of such stones over appropriately laid filter bed or over geotextile fabric need to be placed for less damage.

Zone II It is defined as 2.0 m below MSL to 3.0 m above MSL. In this zone the design maximum velocity of the area around Sunderban Delta say will be of the order of 3.90 m/s. Using the above relationship the weight of stone comes to 5500 kg.

For protection of slope above the zone the size of stone required may be calculated from the following relationship:

$$W_2 = \frac{\gamma_s H^3}{K_D (S_r - 1) \cot \alpha} \text{ in lbs}$$

where

H = wave height (significant),

K_D = coefficient,

S_r = sp. gr. relative to water,

α = slope of the embankment exposed to the river side.

6.3.5.2.2 *Indirect protection*

Groynes (Fig. 6.17) are employed as an indirect way for protection of banks and this is in general cheaper than direct or continuous protection. The object of using groynes is to reduce the net width so that the depth of the net cross-sectional area is increased. The net width, the distance between outer streamlines is always changing with alternate reaches of acceleration and deceleration. The greatest width is just upstream from each groyne and this portion plays the dominant role, since the smallest water depth will occur here.

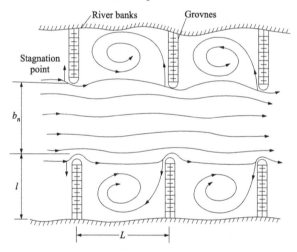

Figure 6.17 Groynes and the flow pattern.

The optimum distance between groynes cannot be exactly calculated, too short a distance is expensive, too long will cause the effective width to be larger than the normalised width with consequent shoaling upstream of the groyne, as also there may be deep-scoured attack of the head of the groyne. From experience the following rules have been found to be helpful, i.e.

Case (a): $L < b_n$ (if $L > b_n$ the outer streamline will spread outside the normalised river width).

Case (b): $L < (3 \text{ to } 4)\, l$ (if the component is too long and narrow the eddies will break up into smaller ones) l = length of groynes where L = spacing between groynes, b_n = transverse clear width between groynes.

The direction at which groynes should be placed relative to the high water river flow is subject of much discussion. If it is pointed downstream also known as attracting groyne, it may help to maintain normalised width immediately downstream of the groyne and it is liable to cause excessive attack on the bank. If it is pointed upstream also known as repelling groyne it may help in keeping the maximum flow in the net cross-section, but in this case the head of the groyne will be subject to heavy scour. Repelling groynes have the main purpose of deflecting currents away from the bank or of shielding the bank with a strong arm in front of it. Hence, sediment may be caught above or below the groynes. These groynes must have a strong head to resist the direct attack of a swirling current and are usually built of stone or stone pitched earth embankment, with paving or other form of protection at the toe of the slope. The scour gradually diminishes from the head to the bank and the protection of the slope and the apron can be reduced accordingly. As the heads are always under attack, much repair may be required from year to year. Repelling groynes may be of permanent, semi-permanent or temporary construction. Recent practice tends to favour the upstream inclination making an angle of 15–30 degrees with the line normal to the flow. Repelling groynes are usually constructed in a group. The first or uppermost groyne should either be just a bankhead attached to the bank, or it must be very strongly constructed, as the attack on this groyne is usually the severest.

Sedimenting groynes: Sedimenting groynes is a name coined just to distinguish these from the repelling type. Most of the permeable types of construction work fall in this category. Permeable groynes are not very well adopted for clear water rivers. As sediment is accumulated between the groynes, the foreshore is more or less permanent so that there is no need of using too durable a material. Sometimes, especially for toe-protection, these groynes may be constructed low, and after the accumulation of silt to that level they can be raised to a higher level. These groynes may be of the solid or permeable type depending upon the condition of the current and the waves. If of solid construction, groynes must be made as smooth as possible with the current so as not to induce turbulence of action.

The requirements of indirect protection are effective location and a form of structure that would answer the purpose of deflecting or slackening the current or silting up of the foreshore with the material available without unduly high cost. If works of great extent are projected, model testing as well as construction of experimental structures is considered desirable to begin with.

Bank protection, owing to the extensive nature of the work, is always bulky and expensive. Unless the property or the land to be protected is extremely valuable, such as the water-front of an industrial city, engineers are prone to choose the cheapest type rather than a stronger and expensive one. Moreover, there is yet no such criterion as a factor of safety for use in designing structures that can be used for bank protection, and cost becomes the dominating factor in choosing the type. The cost should include the initial outlay as well as the annual maintenance. In the following paragraphs, an attempt is made to indicate only the general preference of one method over another, in the light of experience gathered.

(i) When river banks are subject to erosion, it is highly desirable that protective measures be taken as early as possible rather than to tackle the problem when erosion has become serious. The cost of bank protection, if deep scour has already developed will be many times more expensive than if the problem has been tackled its initial stage. On some occasions the river, after showing signs of erosion for a short or long period without causing serious scour, may change the direction of attack to some other place; while on other occasions erosion may be persistent and may develop serious trouble. It is for the engineer to make a careful study of the changes that have occurred and use his knowledge and judgement to decide whether action should be taken.

(ii) If a deep scour has already developed, which usually attracts the current towards it, particularly in large rivers, it is preferable to divert or deflect the attacking current somewhere upstream rather than to protect the place of scour directly. Bank heads, repelling groynes or Denehy's groynes are examples of such constructions and on occasions some dredging of the shoals upstream may be necessary with a view to changing the direction of the attacking current. The layout of such structure should be tested in models before actual work is commenced.

(iii) Regarding the relative merits of continuous and non-continuous protection, experience seems to show that their use depends on the sediment load of the river and on whether the river is of the aggrading or the meandering type. For non-continuous protection such as a series of groynes, best results would be obtained if sediment could accumulate between the groynes and a permanent foreshore could thus be developed. For solid groynes, the accumulation of sediment deposits would enable the gradual raising of the height of groynes in stages over the deposit between them, thus greatly reducing the cost. Therefore, the use of non-continuous protection is most suitable for the aggrading type of river where sediment load is relatively high in respect to the carrying capacity of the river. The usual wide and shallow cross-section of aggrading rivers are particularly adopted to the use of permeable construction in which piles or posts of ordinary length can serve the purpose. The successful use of such structures in aggrading rivers is demonstrated by the construction of bandals in deltaic rivers in Bengal.

(iv) For meandering type of rivers, where the water course is well-defined, deep and winding, it appears that continuous protection by revetment is preferable to non-continuous protection. On account of the great depths encountered at the concave banks where protection is deemed necessary, the use of non-continuous protection such as bank heads or short groynes, which have to be spread quite closely and have to be brought up to a certain height in order to make them effective, is much too expensive. Revetment (Fig. 6.18) has the advantage of achieving the object of protection without adverse effect on the flow in other parts of the river and does not create undesirable obstacles to navigation as groynes may do especially the low groynes when submerged at higher stages of flow.

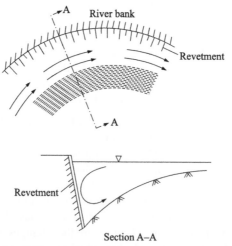

Section A–A

Figure 6.18 Bank protection through revetment.

6.4 CHANNEL IMPROVEMENT

Introduction   Enlargement of the channel by dredging of bars, by removing obstructions to flow, by straightening of bends or complete lining would improve the hydraulic capacity of stream and permit the more rapid escape of flood waters without overspill. Natural river channels are formed by the runoff from the catchment area. The size of the channel is dependent upon the formative discharge. There is some uncertainty of the magnitude of this discharge but it is generally accepted that it is never the maximum discharge. It is, however, often observed that the bankful discharge of a natural river channel is about one-half to one-fourth of the maximum discharge.

All rivers have or have had a main channel and side channel. The side channel can be traced in rural areas and it may vary in width from the river several miles either way. In natural form the flood discharges are accommodated in this double channel. Fundamentally, it is only because of man's use of the land lying near the side channel, either for agricultural or urban development purpose, that flooding occurs. Flood discharges last only several days in a year as the river channel is sufficient to accommodate average flows.

Widening and deepening   To prevent overspill, it is necessary to enlarge the channel two to four times its natural size unless continually dredged of deposited silts. During low-flow periods the depth of water in overlarge channels is shallow which favours weed growth and creates a costly maintenance task. Dredging as a flood abatement measure is often a losing struggle. There is also a spoil disposal problem. Sometimes it may be convenient to form the excavated spoil into flood banks on either side of the river. Channel enlargement is practised at many places particularly where navigable channels must be maintained. The effect of such improvements on flood crests are computed by usual hydraulic procedures.

6.5 CUTOFFS

A cutoff is essentially a characteristic of alluvial rivers and signifies development of river meander to acute conditions in the form of hairpin bend. These bends become large loops with narrow necks under favourable conditions. After a limit is reached in the narrowing, breakthrough occurs resulting in the formation of a chute channel known as a cutoff form of the neck. Cutoffs result in violent changes in river regime. The river slope upstream of the cutoff steepens and flood levels are lowered. This increases the velocity resulting in appreciable bank erosion and bed scour in the upstream reaches of the river. At the same time due to reduction in channel storage upstream the inrushing floods threaten the lower reaches also. Due to excessive erosion upstream the channel carries excessive sediment and deteriorates the lower channel also. Sometimes the beneficial effects of uniformity in cross-section also results. It is seldom possible to have a full-scale diversion of the river through a cutoff, although it has beneficial effects in the reduction of flood heights, flood periods and shortening of navigational course. Hence, it is a river training measure. However, it becomes necessary to carry out extensive work between cutoffs to improve upon the alignment, width and depth of the channel by supplementary training works. Such works involve directing the flow and closing pockets found at unduly wide parts of the channel by training groynes of dredged sandfill. Revetment at places of severe erosion is also required. The objective is the creation of a uniform river width and a central river channel deep enough to maintain itself by normal scour action.

Within recent years, man-made cutoffs have been constructed to straighten streams in order to increase their hydraulic efficiency. Figure 6.4 indicates a situation where the high water flow along two channels must be balanced to give the same head loss over the section but the nature of the losses is different for two paths. When all the losses are known, the distribution of discharges can be calculated as per Kirchoff's law. When river improvement works are executed the value of resistance coefficient, viz., *C* for various water levels will have to be calculated from measurements before hand. In new situation the same values may be used except when the roughness has been appreciably changed, i.e. by cutting of bushes, construction of smooth revetments, triming of the bed, etc. In such cases new values would have to be selected by theoretical calculation or by experience of comparable situation. By decreasing the length of the stream, the cutoffs increase the gradient. Due to increased gradient and velocity, flood crests reach downstream more rapidly because of the shorter distance the water has to travel. Above the cutoff for some distance there is an increase in velocity resulting in a lowering of flood heights. Consequently, the flood danger above a cutoff is diminished because of the increased efficiency of the stream. Those portions of the flood plains below the cutoff may be more liable to flood danger, for not only is more water passed downstream more quickly than before, but also the storage capacity of the old channel at bend is no longer utilized. This may result in a closer synchronisation at some point downstream with peak inflows from a tributary causing the flood. One effect of cutoffs that may reach far downstream is a result of the change in direction of the current as it comes out of the cutoff as compared with its direction in the natural condition, when the bulk of the stream flowed around the bend. Deflections of currents may give rise to a whole new series of downstream impingement points initiating new meanders. Some areas previously safe are subjected to erosion and some areas previously being eroded are no longer scoured. Cutoffs are, therefore, to be stabilised against the characteristic stream pattern of meandering in alluvial valleys, and hence cutoff maintenance is a continuous battle. Again, any uncompensated lengthening or shortening that changes gradient disturbs the profile of equilibrium which has been established with reference to the average size of the bed-load particles. As a result the stream starts building up (aggrading) or scouring until the profile of equilibrium is re-established. A stream is said to be 'graded' (should not be confused with gradient stated in m per km) or to have reached its profile of equilibrium when its slope and volume are in equilibrium with the sediment load transported.

A series of cutoffs and straightening of bends had been completed on the Mississippi river between Cairo, Illinois, and the mouth in a total river length of about 1275 km. In this reach about 497 km of river channel were shortened by 179 km during 1933–36 between the mouth of White river and Angola. Comparison of stages reached during the flood of June 1929 with those reached during the flood of January 1937 at equal rates of flow indicated lowering during the latter flood that ranged between 0.67 and 4 m in the 497 km reach. However, because of the difficulty of comparing floods, it is not possible to state quantitatively how effective the cutoffs have been or what their effect has been on the regimen of the river.

The enlargement of the discharge capacity by channel improvements may be efficacious but it is fairly limited in value and is essentially a local protection measure. There is a practicable limit to the amount of straightening, widening and deepening that can be maintained. Once this limit is reached, the benefits end. These measures are generally used in conjunction with and to supplement flood embankment systems. Any channel improvement is a technique which must be used with caution and with consideration for possible adverse effects upon downstream reaches which may not have been improved to pass the greater flows produced by upstream channel improvement. Channel improvements should, therefore, be considered as a portion in the overall development plan for the entire stream and most be so planned and executed as not to lead to unfortunate consequences and that their benefits at one point are not offset by increased damages elsewhere.

6.6   DIVERSION

Introduction   Diversion as a flood control measure opens up a new exit channel which carries the excess flow safely round the area to be protected. Sometimes, the flood water of one river may be diverted to another river system. For example, the flood waters of Missouri river and other Mississippi tributaries were diverted into the Rio Grande and other southwestern rivers in USA where water is needed. The Babylonians diverted flood water of the Euphrates into natural depressions at Habbania and Abu Dibis to protect the ancient city of Babylon, but the first flood control works were probably those of Amenenhat, King of Egypt, who constructed flood banks on both sides of the River Nile and diverted flood waters into Lake Noeris. After the flood had receded in the main river, the water stored in these depressions was permitted to flow back to the stream. Diversion as a flood control measure is, however, for the most part of limited application. Diversion of flood is also practised on some Indian rivers such as Mahanadi and the Cauvery. Waters of the Mahanadi at Naraj are distributed approximately equally between the main river and the Katjuri, however, there is no control of the distribution. There are also a number of escapes in the leveed reach of the Mahanadi where the gaps in the banks are stone-pitched and water flows in this escape as soon as the water rises above a certain level. The Cauvery divides into two branches in the delta, the right arm being called the Cauvery and the left the Coleroon. There is a barrage with lift-shutters across the Coleroon which passes down the flood discharges that are not required for irrigation in the Cauvery. The effect of diversion where the diverted water returns to the main channel some distance downstream of the protected area depends on the distance between the point of diversion and point of return. In case it is small, the backwater effect may somewhat negate the effect of diversion. However, if this is large the backwater effects exists over a relatively small distance and diversion will be effective in reducing the flood stage. Another consideration is the availability of adequate head for the diversion channel to develop the necessary velocity. If it is low the cross-section diversion channel will be large. Depending on local condition a dam can be built to raise the water level and create adequate head for diversion. Depending on the terrain through which the artificial diversion channel has to pass it can either be complete cutting or partly cutting and partly in embankment.

6.7   FLOOD RELIEF OR BY-PASS CHANNEL

These channels have particular application in highly developed areas where it is impossible to increase the size of existing rivers. For economy it is obvious that the existing river should be used to carry the maximum possible flow without flooding and the by-pass channel should be designed to carry on the balance. Theoretically the existing river should carry all the discharge up to that which causes flooding then the by-pass should carry any further increase. This, however, is hardly practically possible. The detail design of a by-pass entrance is not easy to calculate with any accuracy but the problem can be solved without difficulty by a model, for example:

A river having a slope of 1 in 2000, of trapezoidal cross-section having base width equal to 30.5 m, side slope 2:1, has a maximum discharge of 198.22 m^3/s where the flow depth has been found to be 3.05 m. At some point, lower down flooding occurs when the river depth is 2.44 m. By calculation when the depth is 2.44 m, the discharge will be 141.5 m^3/s. Hence, the by-pass will have to carry 56.6 m^3/s of the maximum discharge. The slope of the by-pass will approximate to that of the river, i.e. 1 in 2000. The channel required has 12.19 m bottom width 2 : 1 side slope and 2.44 m depth. If the by-pass channel takes off from the river at 30° then the length of the weir possible will be about 33.55 m. The head to discharge 56.60 m^3/s over this weir is about 0.91 m. There will be a fall in the height of water surface along the length of the weir. Assume that the fall is straight uniform from 3.05 to 2.44 m and the average water level is 2.74 m. Therefore, the crest of the weir will be 1.83 m. The weir will operate when the river level reaches 1.83 m, i.e. when the discharge passing in this river is 79.24 m^3/s. If it is determined that this happens too often then the shape of the outlet could be altered to give a longer weir or sluices could be added to the crest, opening at some predetermined level, say 2.29 m.

Care must be taken that the by-pass channel is of uniform carrying capacity along its length, otherwise flooding may occur in some areas along its banks not previously affected. At normal flows the by-pass will be dry which will lead to the growth of the weeds. If, however, arrangements are made to drain adequately the bottom of the by-pass, cattle will be able to graze and keep down growth. A by-pass channel is costly and often necessitates construction of a bridge, etc. Such channels may also be constructed entirely of concrete in order to keep their dimensions to a minimum and to economise on land.

6.8 INTERCEPTING OF CUTOFF CHANNEL

Intercepting of cutoff channel provides an entirely alternative system and diverts either all or part of the flow away from the area subject to flooding. It is commonly used in storm drainage scheme in low-lying areas. When it is not possible to enlarge the channels and heighten the embankments sufficiently, the highland waters of the streams can get intercepted by a cutoff channel which may discharge into a relief channel running parallel to main stream for some length before joining it finally.

6.9 FLOODWAYS

Floodways serve two functions in flood abatement, first by providing an additional channel storage for storing a portion of the flood water and second by opening up a new channel to carry a part of the main stream flow. Non-availability of low cost land and unfavourable topographic condition limit the opportunities for construction of floodways. These are ordinarily utilized only when there is a major flood, however, periodic floodings are necessary to scour any deposit so that the channel does not aggrade. The land in the floodways may be used for agriculture but extensive or costly development should never be encouraged. Flood waters are admitted at critical times to the floodway by constructing. (i) a low section of the embankment with poor material which when overtopped will wash out rapidly providing full discharge capacity, (ii) a gap in the embankment line, (iii) a concrete weir, or (iv) a spillway with gates or stop logs. Weirs and gated spillways provide better control over the flow but are costly and their construction can be justified only when overflow is expected to occur quite frequently and when neighbouring important cities are to be protected.

The hydraulic design of the diversion works for a floodway is rather complicated and the actual design is done largely by trial. At the point of diversion, the depth of flow in the stream required to pass a given flood will be smaller than before, due to increased channel capacity equal to the flow through the diversion channel. The water-surface slope above the diversion is, therefore, increased due to drop-off resulting in higher velocity and lowering of river stage in upstream reach for some distance. The influence of the drop-off curve becomes more pronounced when channel slope is flatter. Similarly, water surface slope below the point of diversion is decreased. This effect on a channel with flatter slope will cause a reduction in the discharge capacity at a given stage and thus will offset the benefits derived from the diversion of flow. The lowering of the flood height will not, therefore, be as large as would be expected corresponding to reduced flow, because the prediversion stage discharge relationship will not be valid. However, with a substantial amount of water diverted, flood-height may be

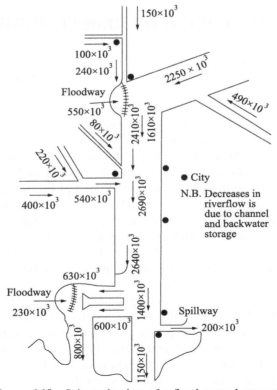

Figure 6.19 Schematic view of a flood control measure.

expected to be smaller for some distance downstream from the point of diversion. The point of diversion is fixed from the hydraulic behaviour of flow. If the channel is relatively steeper, the diversion work should be constructed on the stream above the area to be protected and conversely below the area subject to flooding if the channel is relatively flatter. A trial weir length and crest elevation are assumed and the water-surface profiles above and below the diversion weir at design discharge are computed based on the consideration that for a given stage at the weir the sum of the weir discharge and the flow in the downstream channel below the weir must be equal to the given inflow. At the head of the diversion channel velocity is decreased because of the divided flow. A loss of volume causes deposition, if the stream with reduced volume cannot transport the sediment load. Hence, below the point of diversion there is danger of silting of the channel. The engineering problem is to plan the diversion in such a way that the capacity of the main channel is not choked up with sediment and flooding hazards are not made greater than before. The diversion channel must have a regimen of balance between velocity and sediment load supplied so that it will continue to be effective.

A comprehensive flood control plan involves a judicious blending of carefully selected quantities of number of features each of which is intended to perform a contributing function. After its execution it should protect a large part of the alluvial valley from the project flood. A typical schematic view is given in Fig. 6.19 of a flood control measure showing the discharges of tributaries into the system plus the distribution of flood water.

6.10   FLOOD-PLAIN ZONING OR REDEVELOPMENT

In the industrial revolution indiscriminate development took place in the flood plains of many rivers, because the lands are fertile and level, water supply is close at hand, waste disposal may be easy, etc.

Many flood plains are densely populated. Perhaps, between a quarter and a third of world's population lives on flood plains. A direct consequence of the increasing intrusion upon river banks and river valleys by cities, industrial plants, and highways may be found in the mounting figures of damages by flooding. Most of the flooding problems of today are, therefore, man-made and are due to the extraordinary lack of foresight and misunderstanding of the behaviour of rivers. The invasion and growing use of flood plains disregard the basic functions inherent in the flood plain as a part of the river. Despite the repeated flood damages and many other lessons of the consequences of foolhardy development, flood plains are not yet sacrosanct and ill-advised encroachment still continues. Use of flood plains is essential to economic life and people want unhindered development of the flood plains through flood-control works that reduce the risk of floods. To safeguard life and property, it has become, therefore, necessary to keep the rivers within prescribed levels.

A scheme favouring the return of the flood plain to the river and to stay out of flood plain is a logical solution. But demolishing all existing installations and rebuilding them on high ground clear of the flood area or moving people out of the flood plain are usually impracticable or unjustified, but to prevent new development may be practical or economical. Flood-plain zoning (Fig. 6.20) to restrict the use of flood plain land would meet with considerable local opposition. Zoning must pass the test of economic justification and may be adopted if the benefits from zoning will be greater than those from an investment in flood control.

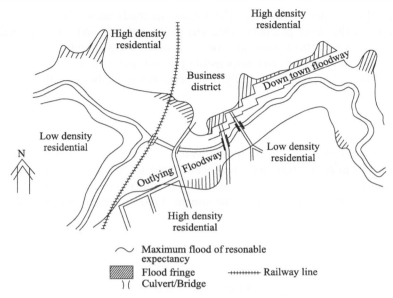

Figure 6.20 Flood-plain zoning layout.

Emergency evacuation of the threatened area with timely warning to flood should be undertaken whenever possible to prevent loss of life and to permit removal or protection of property. Isolated units of high value or special importance like water-works, power-stations, may be individually flood-proofed if these cannot be shifted to safer areas.

6.11 SPREADING GROUNDS

Flood flows are sometimes diverted into large flat land areas that are capable of absorbing water at relatively higher rates. Spreading grounds serve to reduce downstream flood peaks and replenish underground reservoirs which provide water by pumping for irrigation purpose or other water supply uses. Similar arrangements are used to raise ground water levels along coastal areas for general water supply purpose and to suppress salinity intrusion from the sea.

6.12 SOIL CONSERVATION METHODS

Recent studies have shown that there exists quantitative effect of soil conservation methods on flood runoff and erosion. Improved vegetable cover has a significant influence on the reduction of floods of small magnitude. However, its effect is less pronounced in the case of larger floods which cause serious damage.

Soil conservation methods also results in substantial reduction of the rate of erosion. Rate of erosion depends on the character and density of vegetation cover, besides rainfall, slope and soil character. It is of prime importance to determine accurately as far as practicable focal points in the catchment which contribute most of sediment load. This can best be done by aerial photography or satellite imagery. The various methods are— (i) afforestation, grassing, contour bunding, cultivation practices, etc., (ii) gully control, and (iii) prevention of bank erosion. For gully control check dams may be constructed across a gully to control the gradient of the channel and the high velocity responsible for bed and bank erosion. Prolonged flow of runoff and percolation into banks and gully floor promotes vegetable growth through controlling erosion.

Problem 6.1

A stream channel whose shape can be approximated as a rectangle of base width 100 m carries a discharge of 450 m³/s. Its bed slope has been found to be 18×10^{-4} and its roughness in terms of Manning's n equals 0.02. The stream is a tributary to a river where the existing flood stage is 3 m above the normal depth in the stream. Find the approximate backwater curve in the tributary.

Solution

Given

$$Q = 450 \text{ m}^3/\text{s}, \ S_b = 0.0018$$

$$n = 0.02, \ B = 100 \text{ m}$$

For a wide rectangular channel hydraulic radius can be approximated by the depth in the calculation. Now equilibrium depth in the tributary can be calculated from

$$h_e = \sqrt[3]{\frac{q^2}{C^2 \, S_b}}$$

when

$$C = \text{Chezy coefficient} = \frac{h_e^{1/6}}{n} \text{ and } q = \frac{450}{100} = 4.5, \text{ and}$$

by substitution the values

$$h_e = \sqrt[3]{\frac{4.5 \times 4.5 \times n^2}{h_e^{1/3} \cdot 0.0018}}$$

or

$$h_e^{10/9} = \sqrt[3]{\frac{45 \times 45 \times 4 \times 10^{-6}}{1.8 \times 10^{-3}}} = 10^{-1} \times 16.46 = 1.646$$

or

$$h_e = [1.646]^{9/10} = 1.566 \simeq 1.57 \text{ m}.$$

With the above information in view the backwater computation can be carried out as per the Table 6.2. Since the flood stage is 2 m above the h_e, hence $h_1 = 3 + 1.57 \simeq 4.57$ m.

Table 6.2 Backwater Computation.

h	A	V	$V^2/2G$	$h_0 = h + \dfrac{v^2}{2G}$	$C = \dfrac{h^{1/6}}{n}$	$S_e = \dfrac{V^2}{C^2 h}$	$(S_e)_m$	dh_0	$dl = \dfrac{dh_0}{S_b - (S_0)}$	Σdl
(m)	(m²)	(m/s)	(m)	(m)				(m)	(m)	(m)
4.57	457	0.98	0.05	4.62	64.4	5.06×10^{-5}	5.91×10^{-5}	−0.36	−206.8	0
4.20	420	1.07	0.06	4.26	63.5	6.75×10^{-5}		−0.19	−109.8	−206.80
4.00	400	1.13	0.07	4.07	63.0	8.04×10^{-5}	6.98×10^{-5}	−0.49	−288.7	−316.60
3.50	350	1.29	0.08	3.58	61.6	12.52×10^{-5}	10.28×10^{-5}	−0.47	−287.7	−605.30
3.00	300	1.50	0.11	3.11	60.1	20.79×10^{-5}	16.66×10^{-5}	−0.44	−292.3	−893.00
2.50	250	1.80	0.17	2.67	58.3	38.17×10^{-5}	29.48×10^{-5}		−339.5	−1185.30
2.00	200	2.25	0.26	2.26	56.1	80.33×10^{-5}	59.25×10^{-5}	−0.41	−468.7	−1524.80
1.60	160	2.81	0.40	2.00	54.1	168.72×10^{-5}	124.53×10^{-5}	−0.26		−1993.50

Note: Starting section is the junction of tributary with the river where h_1 is known.
$S_b = 18 \times 10^{-4}$ or 180×10^{-5}. Negative sign indicates distance upstream of the junction.

Problem 6.2

The following data concerning flow in a semicircular bend are given

$$r = 490 \text{ m}, \ C = 40 \text{ m}^{1/2}/\text{s}, \ h = 5 \text{ m}$$
$$v_x = 1.1 \text{ m/s}$$

Find the deviation in the path of a particle starting along the centre line at the beginning of the curve and finishing at its end.

Solution

Since $v_x = 1.1$ m/s the time to compute the journey of the bend by the fluid particle

$$= \frac{\pi r}{1.1} = \frac{22}{7} \left(\frac{490}{1.1} \right)$$
$$= 20 \times 70 = 1400 \text{ sec}$$

Again
$$v_y = \frac{0.025 \times 40^2 \times 5}{9.81 \times 490}$$
$$= 0.0416 \text{ m/s}$$

Hence the deviation from the centreline at the end of the curve equals

$$0.0416 \times 1440 = 59.91 \text{ m.}$$

The practical significance of it will be apparent when one considers a ship travelling around such a curve. The ship must steer itself towards the centre to avoid being moved towards the outerbank.

Problem 6.3

Compute the transverse bed slope for a rectangular river section having a fixed outer bank from the following data:

Discharge $Q = 3750 \text{ m}^3/\text{s}$,
Outer-radius $r_0 = 2500 \text{ m}$;
$$S_0 = 11 \times 10^{-5},$$
$$C = 40 \text{ m}^{1/2}/\text{s},$$
$$B = 1000 \text{ m};$$
$$\Delta = 1.5,$$
$$d = 0.5 \times 10^{-3} \text{ m}.$$

Solution

The average $\bar{h}$ for rectangular section

$$\sqrt[3]{\left(\frac{3750}{1000} \right)^2 \frac{1}{40^2 \times 11 \times 10^{-5}}} = 4.3 \text{ m}$$

As a first approximation assume

$$h_0 = 2\bar{h} = 2 \times 4.3 = 8.6 \text{ or say } 9.0 \text{ m}$$

and
$$\frac{12\,S_0\,r_0}{\Delta d} = \frac{12 \times 11 \times 10^{-2} \times 2500}{1.5 \times 0.5} = 4400$$

Calculation of transverse bed slope is done in the following tabular manner (Table 6.3) from equation, viz.

$$\left[\frac{1}{h} - \frac{1}{h_0}\right] = \left[\frac{1}{r} - \frac{1}{r_0}\right]\left[\frac{12\,S_0\,r_0}{\Delta d}\right]$$

$$= \left[\frac{1}{r} - \frac{1}{r_0}\right]4.4 \times 10^3,$$

where

$r_0 = 2500$ m, $h_0 = 9$ m.

It is now necessary to check the discharge which is

$$Q = C\sum bh^{3/2}\sqrt{S}$$

In this case, the river width is 1000 m and it is divided into segments of 200 m each. Accordingly,

$$\sum bh^{3/2} = 200 \times \sum h^{3/2}$$

$$= 200\,[27 + 8.4 + 4.2 + 1.85 + 1.07 + 0.67]$$

Table 6.3 Calculation of Transverse Slope.

r	$1/r$	$(1/r - 1/r_0)$	$(1/h - 1/h_0)$	$1/h$	h	$h^{3/2}$
2500	0.04×10^{-2}	0	0	0.11	9.0	27.0
2300	0.043×10^{-2}	0.003×10^{-2}	0.132	0.242	4.13	8.4
2100	0.048×10^{-2}	0.08×10^{-2}	0.352	0.462	2.16	4.2
1900	0.053×10^{-2}	0.013×10^{-2}	0.550	0.66	1.52	1.85
1700	0.059×10^{-2}	0.019×10^{-2}	0.840	0.95	1.05	1.07
1500	0.067×10^{-2}	0.027×10^{-2}	1.190	1.30	0.77	0.67
1300	0.071×10^{-2}	0.027×10^{-2}	1.190	1.630	0.58	0.44
1100	0.091×10^{-2}	0.051×10^{-2}	2.25	2.36	0.42	0.27
900	0.110×10^{-2}	0.070×10^{-2}	3.08	3.19	0.32	0.18

i.e. $r = 2500$ to 1500 which gives $b = 1000$ m

or $\sum bh^{3/2} = 2.00\,[43.19] = 8638$

Now S varies across the width of the river so that a value slightly higher than S_0 can be taken. In this case consider, $S = 12 \times 10^{-5}$.

Accordingly,

$$Q = 40 \times 8638\,\sqrt{12 \times 10^{-5}}$$

$$= 345520 \times 10^{-2}\,\sqrt{1.2}$$

$$= 3784.98 \simeq 3785 \text{ m}^3/\text{s}$$

which is almost equal to the discharge passing, i.e., 3750 m³/s. For more accurate calculation S can be calculated for each increment of width.

CASE STUDY 1

Design and layout of a solid spur for strengthening an embankment:
The layout of the embankment and the proposed location of the spur is shown in Fig. 6.21. The data pertaining to the design is as follows:

> Normal discharge of river = 20000 cumec.
> Catchment area = 246 km^2
> Lacey's silt factor = 0.9
> Velocity of approach = 2.44 m/sec.

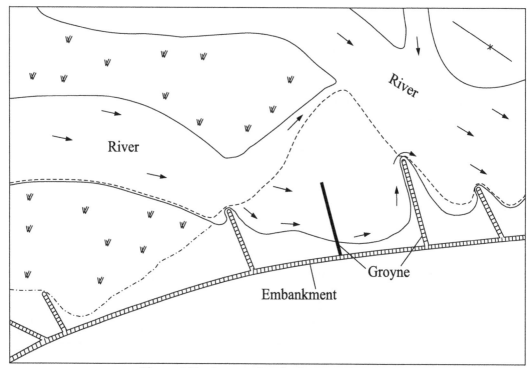

Figure 6.21 Layout of embankment and groyne.

Design calculations

(a) *Design of spur:* Maximum discharge = Normal river discharge + Flood discharge,

$$Q = 20000 + 12(246.0)^{3/4}$$

Assuming,	$C = 12$, from Dickens formula, $Q = CA^{3/4}$
where	A = catchment area in sq. km
	Q = 20746 cumec.
Say	Q = 20800 cumec.

Stable waterway, $L = 4.75 \ (20800)^{1/2} = 685$ m, from Lacey's regime relations.

Discharge per metre run (q)

$$= \frac{20800}{685} = 30.4 \text{ cumecs}$$

Lacey's scour depth

$$R = 1.35 \left(\frac{q^2}{f} \right)^{1/3}$$

taking $f = 0.9$

$$R = 1.35 \left(\frac{30.4^2}{0.9} \right)^{1/3}$$

$$R = 13.65 \text{ m (say)}.$$

(b) *Armouring of spur*:

(i) Apron at nose of the spur:

Effective depth of scour at nose = $2R$ = 2×12.65 = 27.3 m.

Apron is proposed to be laid at LWL which is 3.7 m below HFL

Depth of scour below LWL = 27.3 − 3.7 = 23.6 m.

Slope length of launching apron = $23.6\sqrt{5}$ = 52.8 m.

Width of apron of depth 1.2 m with minimum cover of 0.6 m

$$= \frac{52.8 \times 0.6}{1.2} = 26.4 \text{ m} \simeq 26.5 \text{ m (say)}.$$

Provide 26.5 m × 1.2 m loose boulder apron with 1.2 × 0.6 m rectangular sausage over 0.6 m loose boulder apron at 30° interval.

(ii) At 15 m U/S of Nose:

Effective scour depth is considered = $2R$

Hence, 26.5 m × 1.2 m loose boulder apron is provided.

(iii) At, 45 m U/S of Nose:

Effective scour is 1.2 R, i.e., 1.2 × 13.65 = 16.38 m.

Depth of scour below LWL = 16.38 − 3.7 = 12.68 m.

Slope length of the launched apron = $12.68\sqrt{5}$ = 28.4 m with depth of apron, 1.2 m, width of apron required

$$= \frac{28.4 \times 0.6}{1.2} = 14.2 \text{ m} \simeq 14.5 \text{ m (say)}.$$

(iv) Remaining portion of U/S:

In this zone nominal apron will be transitioned from 14.5 m × 1.2 m to 5 m × 0.9 m.

(v) Effective scour depth is considered = 1.5 R, i.e., 1.5 × 13.65 = 20.5 m or 21 m (say)

Depth of scour below LWL = 21 − 3.7 = 17.3 m

Slope length of lauched apron = $17.3\sqrt{5}$ = 38.7 m

With depth of apron 1.2 m, width of apron required

$$= \frac{38.7 \times 0.6}{1.2} = 19.4 \text{ m} = 19.5 \text{ m (say).}$$

Provide 19.5 m × 1.2 m boulder apron.

(vi) At 14 m D/S of Nose:

In this zone apron will be transitioned from 19.5 × 1.2 m to 5 m × 1.2 m.

(c) Side slope pitching:

A 0.45 m thick loose boulder pitching over 0.10 m shingles is provided.

(d) Height to be raised for velocity of approach (Afflux).

$$h = \frac{U^2}{2 \times g} = \frac{(2.44)^2}{2 \times 9.80} = 0.3 \text{ m} = 30 \text{ cm}$$

The formation level at the junction of the spur with main embankment is to be raised by 30 cm. The crest level of the spur will be gradually increased from 30 cm at nose to 50 cm at anchoring point. The complete design drawing of the spur is shown in Fig. 6.22.

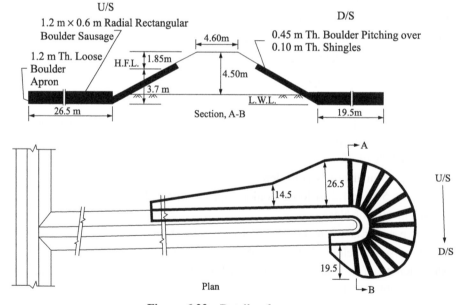

Figure 6.22 Details of groynes.

Problem 6.4

Design and layout of a screen for the protection of river bank from erosion

The layout drawing of the protection scheme is shown in Fig. 6.23. The data pertaining to the design is as follows:

$$\text{River discharge} = 7250 \text{ cumec}$$
$$\text{Catchment area} = 1476 \text{ km}^2$$

Average HFL = 86 m

Lacey's silt factor f = 0.75

Average LWL = 82.5

Design calculations

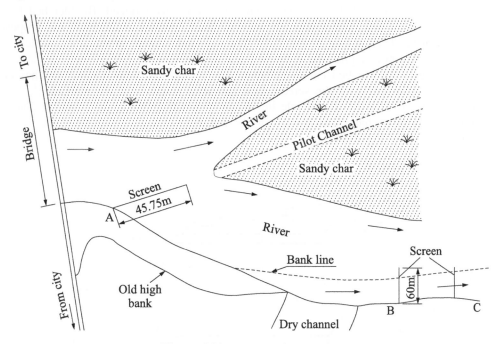

Figure 6.23 Layout of the screen.

Solution

(i) Considering C = 11

Adding flood discharge = 11 × $(1476)^{3/4}$ = 2620 cumec

Probable max. discharge = (7250 + 2620) cumec

= 9870 cumec

(ii) Scour depth:

Stable waterway = 4.75 $(9870)^{1/2}$ = 475 m

Discharge per metre run = $\dfrac{9870}{475}$ = 20.75 cumec.

Lacey's scour depth = 1.35 $\left(\dfrac{20.75^2}{0.75} \right)^{1/3}$ = 11.2 m

Lacey's silt factor = 0.75

Average high flood level = 86 m

RL of the bottom of pile at nose should reach the level = 86 – 11.2 = 74.8 m

RL of top of the pile = Av. LWL + 0.6 m = 82.5 + 0.6 = 83.1 m

Length of the pile = 83.1 – 74.8 = 8.3, say 8.5 m.

Hence, provide 8.5 m length of each pile. Length of each pile should be 8.5 m of which 4 m to 6 m should be driven below the river bed level. The pile top should be gradually raised towards bank end and driven length may be accordingly reduced. Boulder bed bar of crest 3 m, height 1 m and side slope 1 : 1 with rounded nose up to the bank has proposed. The arrangement has been shown in Fig. 6.24. The salbullah screens at the zone marked B–C will consist of two rows of pile line restricting the flow not more than 20 per cent and 1 m of boulder filling to form a bed bar in between. The spacing between the screens has been kept at 2.5 times the length of the screen. One sulbullah screen at point marked '*A*' length 45.75 m will consist of single row of pile line to be started from the guide bund for diverting the flow through a pilot channel as marked on the Fig. 6.23. The complete design drawing of the screen is shown in Fig. 6.24.

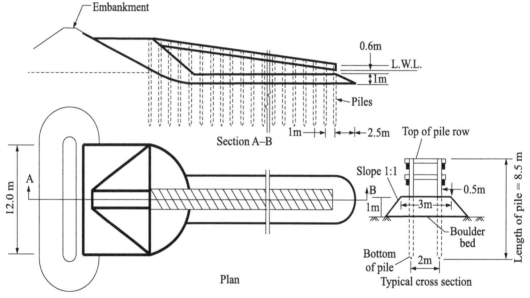

Figure 6.24 Details of screen.

CASE STUDY 2

Design of armouring for the strengthening of an embankment. The typical layout drawing for the protection scheme is shown in Fig. 6.25. The data pertaining to the design are as follows:

Discharge = 7360 m³/s, Lacey's silt factor f = 1

Low water level = 2 m below high flood level,

Height of pitching above HFL = 0.6 m.

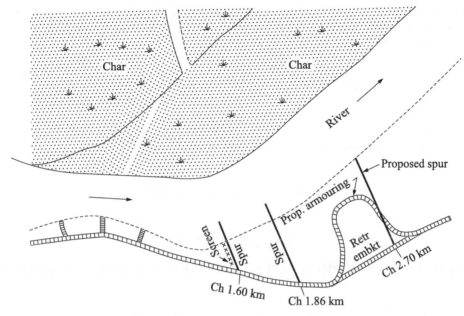

Figure 6.25 Layout of the embankment.

Design calculations

The armour design for the strengthening of the embankment is based on the concept of design of falling apron. The design calculations are as follows:

Stable waterway = $4.75(Q)^{1/2}$ = $4.75 \times (7360)^{1/2}$ = 407.5 m.

Discharge per metre run

$$q = \frac{7360}{407.5} = 18 \text{ m}^3/\text{s/m}$$

Depth of scour as per Lacey

$$1.35 \left(\frac{q^2}{f} \right)^{1/3}$$

$$= 1.35 \left(\frac{18^2}{1} \right)^{1/3}$$

$$= 9.3 \text{ m}$$

As the reach is a protected one the intensity factor of scour for design is taken as 1. Hence, maximum scour depth below HFL = 9.3 m. With reference to Fig. 6.16, the numerical values of D, F and R for the present problem are 2 m, 0.6 m and 7.3 m, respectively. Accordingly, volume of slope stone required per metre length will be $2.25T(D + F)$ or $2.25T(2 + 0.6)$ m³. Considering the thickness T = 0.60 m of the volume of stone comes $2.25 \times 0.6 \times 2.6$ m³ = 3.5 m³. The volume of apron stone per unit length will be 2.81 RT. Now $R \times 7.3$ m and considering average $T = 0.90$ m, the volume comes to $2.81 \times 7.3 \times 0.90$ m³ or 18.46 m³. The length apron will be equal to $1.5\,R = 1.5 \times 7.3 = 10.95$ m. The complete design drawing is shown in Fig. 6.26.

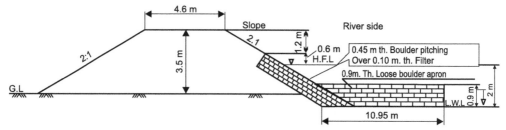

Figure 6.26   Details of armouring.

CASE STUDY 1

A TYPICAL BANK PROTECTION SCHEME USING GEOFABRICS

River bank geometry

Typical cross-section at the reach is shown in the below.

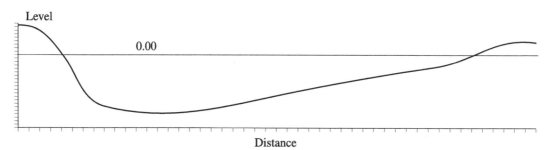

Figure 6.27   A typical cross-section of river.

The typical cross section at the site snows a steep slope of 1:1.5 to 1.75 from high bank to '0' contour (14.62 m, above MSL) and thereafter a comparatively flatter slope of 1:3 towards the river. The steep slope at bank area should be trimmed back at a minimum slope of 1:3.

The scour depth calculated is given below:

Scour depth $\hspace{3cm} (D_s) = 1.34\ (q^2/f)1/3$

where

> q = design discharge intensity per meter width
> f = silt factor = 1.76 (m)$^{1/2}$
> m = mean diameter of the bed material in mm, taking weighted average upto the anticipated scour depth.
> $f = 0.7$
> v = maximum velocity of the flow area = 2.5 m/s

So,
$$q = \frac{Q}{w} = \frac{2316}{260} = 8.9 \text{ m}^2/\text{s}$$

Scour depth
$$(D_s) = 1.34 \ [(8.9)^2/0.7]^{1/3} = 6.48 \text{ m}$$

Therefore the maximum depth of scour from HFL, considering moderate bend = 1.5 × 6.48 = 9.47 m. In the present case HFL = 16.57 m,

Therefore the level at expected maximum scour = 16.57–9.47 = 7.1 m and scour level with respect to local zero = (–) 7.52 m

Considering an average width of protection of 35 m from high bank, the suggested measure is stated below

The suggested measure:

Selection of materials:

As discussed earlier, under cutting, caving and mass failure cause the bank- failure under study. Typically the protection of the bank in such case is achieved by (1) providing a stable bank slope not steeper than 1:3 and (2) retaining the soil particle in place and at the same time allowing the water flow, to prevent significant pore- pressure development. A well-designed filter- layer placed over the graded bank slope can serve the purpose. To secure the filter- layer in place and protecting against various forces, hydrodynamic/ mechanical and ultra violet ray, an armour layer, conventionally made of stone, is used.

It is proposed that upper part of the bank up to 1 m below LWL be protected with a non-woven geo- fabric filter- layer with a layer of sand bag on it and a top layer of 0.6m thick stone boulder over the poly bags filled with sand. The lower part of the protection should be of a woven geo-fabric filter- layer, with a 0.60 m thick layer of geo-bags over it.

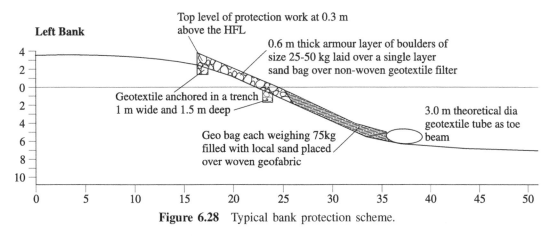

Figure 6.28 Typical bank protection scheme.

Designing with geo-textiles for filtration is essentially the same as designing graded granular filters. A geo-textile is similar to a soil-layer,, only in that it has voids (pores) and particles (filaments, and fibers). However, because of shape and arrangement of the filaments and the compressibility of the structure with geo-textiles, the geometric relationships between filaments and voids are more complex than in soils. In geo-textiles, pore size is measured directly, rather than using particle size as an estimate of pore size, as is done with soils. Since pore size can be directly measured, relatively simple relationships between the pore size and particle size of the soil to be retained can be developed. Three simple filtration concepts are used in the design process:

If the size of the largest pore in the geo-textile filter is smaller than the larger particles of the soil, the soil will be retained by the filter. As with graded granular filters, the larger particle of soil will form a filter bridge over the hole, which in turn, filters smaller particles of soil, which then retain the soil and prevent piping.

If the smaller openings in the geo-textile are sufficiently large enough to allow smaller particles of soil to pass through the filter, then the geo-textiles will not blind or clog.

A large number of openings should be present in the geo-textiles, so that proper flow can be maintained even if some of the openings later become plugged.

To perform effectively, the geo-textile must also survive the installation process (surviability criterion).

7

Flood Forecasting, Warning and Flood Fighting

7.1 GENERAL

Flood forecasting is the prediction of water levels, areas and depths of flooding in rivers and flood plains. Flood warning is the preparation of forecasts in a meaningful format.

Forecasts in hydrological phenomenon along with the weather forecasts are of great significance in almost all countries in the world specially for their economic advancement. In water resources engineering, forecasts are of still greater value for the proper utilization of river energy, inland transport, irrigation, water supply and flood control. Forecasts of highly dangerous phenomenon on rivers such as the occurrence of a flood are of great relevance for all the sectors of a country's economy and to the rural and urban populations around places which are prone to flooding. If timely warnings are given in advance it is possible to minimize the amount of damage by taking precautionary measures like evacuation of persons and property to safer places. Flood forecast is also very important for efficient operation of existing reservoirs. Advance warning of an approaching flood permits the reservoir to be suitably operated for moderating its intensity and also at the same time ensuring full storage for flood relief purposes. If a flood forecasting service is at hand, the initial high reservoir level can be drawn down to the lowest level possible before the arrival of the flood wave, thereby providing enough storage for its moderation and also necessary adjustment of its operation so as to pass the flood and at the same time have required storage in the end.

7.2 BASIC DATA

For the formulation of the forecasting service, adequate data is needed. The development of the river forecasting procedure requires historical hydrological data and for the preparation of operational forecasts sufficient current information is required. Generally, it can be said that it is necessary to have a minimum of 10 years of basic hydraulic data, available to develop adequate river forecasting procedures, the primary requirement being that the period of record should contain a representative range of peak flows. Records of rainfall should be adequate to provide a reasonable estimate of the precipitation over the area under study. The density of rainfall reports required varies with basin topography and meteorological factors. As for example, areas of extremely spotty precipitation necessitates greater demands than those areas where precipitation is of a uniform nature. For formulation of a forecast system for the stream flow, the forecaster must have reliable information as to the nature of hydrological conditions in the basin including the amount and aerial distribution of rainfall and the water equivalent programme of snow if present. Further information regarding snow conditions, duration of rainfall, are also needed. Also reports of the river stage, at key positions, are necessary to serve both as aids in forecasting as well as for verification of data used in forecasts.

7.3 COMMUNICATION NETWORK

Quick flow of data is necessary to avoid loss in warning time. Telegraphic and telephonic communication have been found to fail sometimes during storms and has not been considered dependable. It is, therefore, advisable to install wireless network connecting all the observation stations to the forecast control room. Usually high frequency link (day frequency 6.9545 megacycles and night time 3.954 megacycles) and very high frequency link (82.9 megacycles) are usually used for hilly and plain portions respectively. The VHF sets are operated by 12 V, batteries, whereas HF sets are operated by 230 V AC. Hence mains-operated battery chargers are necessary at selected centres for recharging the batteries and diesel engine generators are usually provided as a standby for 230 V AC sets. For every major river system a comprehensive communication network is planned connecting the central control room with the rain gauges and gauges discharge stations of the river basin.

7.4 FORECASTING TECHNIQUES AND PROCEDURES

The standard and the normal river forecasting principles that are encountered in usual practice are dealt with here. The basic techniques that are used in river forecasting are rainfall runoff relationship, unit hydrographs, synthetic unit hydrographs, flood routing methods, stage discharge relationships, etc. Essentially forecasting involves two main components:

  (i) prediction of the volume of runoff resulting from the storm rainfall and
  (ii) prediction of the distribution of nunoff with time at preassigned locations in the course of the river. In the practical procedure it amounts to the estimation of:
    (a) the magnitude of the stage or level of water in the stream,
    (b) the time of occurrence of the stage and the corresponding peak discharges,
    (c) the magnitude of the peak discharges, and
    (d) the volume of the flood runoff.

Hence, it will be evident that forecasting involves the following: (a) the method of determination of runoff from the rainfall data and (b) the method of calculating the flood hydograph from the runoff volumes, and finally the method of forecasting the discharge stages, based on the regulation of runoff movement and the transportation in the river channel.

Since there is a need to formulate the forecast in the shortest possible time it is necessary to streamline the forecasting organisations and techniques. For this purpose special forms for logging the incoming reports, entering them in rainfall plotting maps and preparation of continuous hydrograph charts are necessary. Also correlation charts giving relationship between runoff and rainfall taking various factors including antecedent precipitation index, time of the year, etc. into consideration should be prepared on the basis of past data. Sheets for computing forecast should be available, dated and levelled in advance.

In cases near the upstream reach of the river it is usually sufficient to forecast crest height and time. The rapid rate of rise and fall makes the duration above flood stage so short that there is little value in forecasting the entire hydrograph. In such cases only the empirical charts relating the flow at a point to stream flow events at an upstream station, and relations between rainfall and crest stage, are enough for forecasting. For stations in the lower reaches of large rivers where rate of rise are slow, it is important to forecast the time when various critical heights will be reached on the rising and falling stages, as well as its time. Here unit hydrographs, basin runoff routing and flood wave routing techniques will be used. For forecasting purpose, in the downstream stations the entire hydrograph at the upper stations needs to be predetermined, since efficient operation of flood control reservoirs, requires its complete prediction to ensure maximum storage space for reducing flood intensity and at the same time ensuring complete filling in at the end of flood period. Where the available data are sufficient flood forecasting by unit hydrograph method can be adopted. Rainfall-runoff relationship can be adopted to estimate the amount of water expected in the stream while the unit hydrograph and stream-flow routing procedures in one form or other are utilised to determine the time distribution of water at a forecast point. Stage discharge relations are then adopted to connect the flow to stages.

7.5 DETERMINATION OF RUNOFF FROM RAINFALL DATA

Due to the complex physical processes involved a direct physical or analytical approach to the problem of forecasting runoff from rainfall is very difficult. A graphical approach is often used in the problem of rainfall and runoff correlation. Herein analysis of storms covering a wide range of conditions is done for some relatively small drainage basins. Rainfall and runoff from these storms are evaluated and procedures developed to correlate them. The same procedures are then applied to the surrounding drainage basins having similar hydrometeorological characteristics.

The rainfall-runoff relationship correlates storm rainfall, antecedent basin conditions, storm duration and the resulting runoff usually expressed as an average value in cm of depth over the basin. The basic techniques is the coaxial graphical method (Fig. 7.1). Herein the antecedent basin conditions are represented by two variables, i.e., API (antecedent precipitation index) and period of the year, i.e., week of the year. The API which is an index to moisture conditions with the basin, is essentially the summation of the precipitation amount occurring prior to the

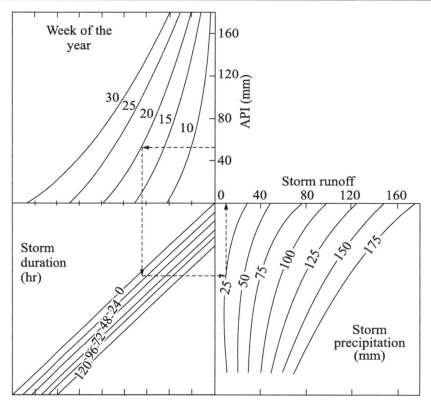

Figure 7.1 Co-axial rainfall runoff correlation.

storm, weighted according to the time of occurrence. It is generally defined by an equation of the type

$$P_a = k^1 p_1 + k^2 p_2 + \cdots\cdots + k^t p_t, \tag{7.1}$$

where

P_a = antecedent precipitation index,

p_t = amount of precipitation which occurred t-days, prior to the storm under consideration and

k = recession factor.

Thus the index for any day is equal to k times the index of the day before. If rain occurs in any day then the index after the rain is equal to that before the rain plus the amount of rainfall on that day. The effect of the rainfall about 20–60 days prior to the storm under consideration is generally taken care of in calculating the API depending on the desired accuracy at the area of application. The value of k ranges between 0.85 and 0.93. The period of the year in which the storm occurs introduces the average interception and evapotranspiration characteristics of the basin during that period which when combined with API provides an index of antecedent soil conditions.

After the volume of runoff has been ascertained for a given storm situation, it is then necessary to determine the distributions this water with respect to time at the forecast point. In this task the unit hydrograph is a simple and generally a very reliable tool. For dealing with uneven distribution of runoff in time unit hydrographs for short period, i.e., 6–12 hours duration are used. The increment of runoff is estimated for each time period with the contribution from each interval superimposed on the previous ones. After the determination of flood hydrograph at a point in a river the estimation of the flood hydrograph at subsequent places downstream for forecasting purpose is done by any of the routing techniques mentioned earlier.

7.6   METHOD OF FORECASTING STAGES

Herein the methods of forecasting the stage for large streams is mentioned. It is assumed that there are no major tributaries or no local inflows from small streams between the upstream station (stage is known) and the downstream station where the stage is to be computed or predicted. A simple method of river forecasting is the method of predicting, the stage at a downstream station, in relation with the stage in the upstream direction. From earlier stream flow records one can plot the stage of the downstream station against the stage of the upstream station and obtain a family of curves. In these sets of curves (Fig. 7.2) a third parameter, namely the time elapsed between the occurrence of the stage in days can be introduced to indicate the time of travel of the flood hydrograph across the reach of the river.

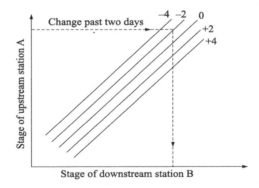

Figure 7.2   Stage at downstream station vs. stage at upstream station.

7.6.1   The Relationship for the Peak Travel Time

The time required for a peak to travel from one station to another is measured by the travel time of the reach of the river. This travel time is of great importance in prediction of the stage. It is not usually constant but will vary with the value of the stages and the channel conditions. Lower river stages are usually shorter as the stage approaches the top of the banks, at which stage the travel time is normally the shortest. As the river becomes high enough to create overbank stages by flooding adjacent areas, the time of travel may begin to increase again, due to the relatively rougher surfaces lying in the overbank stages. Figure 7.3 represents the nature of variation of the travel time with the stage in the river.

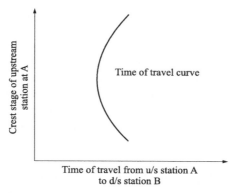

Figure 7.3 Variation of time of travel with stage in the river.

Observations have shown that the peak travels at a ratio of 1.4–2.0 times the mean velocity of water. Further, it is possible to develop a relationship in terms of the characteristics of the river. If the roughness and the hydraulic radii of the streams are similar, the time of travel can be derived as follows.

From Chezy's equation:

$$V_1 = C \sqrt{m_1 s_1}; \; V_2 = C \sqrt{m_2 s_2}, \tag{7.2}$$

where

V_1 and V_2 are the respective velocities of the reaches,

C is the Chezy's constant,

m_1 and m_2 are the respective hydraulic mean radii,

s_1 and s_2 are the respective slopes of the reaches.

Thus, the time of travel,

$$t_1 = \frac{L_1}{V_1} \text{ and } t_2 = \frac{L_2}{V_2}$$

or

$$t_2 = \frac{L_2}{V_2} \; t_1 \; \sqrt{s_1/s_2}, \tag{7.3}$$

where

t_1 and t_2 is the time of travel of the first reach and second reach respectively

s_1 and s_2 are the respective slopes of the reaches,

L_1 and L_2 are the respective lengths of the reaches.

Storage routing is the usual approach in flood routing problems, which is usually done by the methods outlined earlier in Chapter 3. In river routing usually the Muskingum method is adopted.

Apart from the short-term forecasting which has been discussed above in detail, the problem of long-term forecasting is of great significance for the proper utilization of water resources, like any other natural resources. However, its solution depends to a great extent, on solving the problem of long-term forecasting of precipitation. The reliable forecasting of rainfall is of particular importance in regions like India with a tropical and monsoon climate where the main source of runoff is rainfall. The current knowledge of the meteorological science has not yet reached a stage that would permit reliable long-term forecasting of rainfall. This limits the possibilities of the success of long-term runoff forecasting.

7.6.2   Example on Forecasts Reporting

A simple procedure required to prepare a flood forecast for a small drainage basin is illustrated in the following pages. A hypothetical river basin is shown in Fig. 7.4 for which forecasting is to be done at stations A and B. Unit hydrographs are used for the distribution of runoff and the Muskingum routing technique for the flood routing from stations A and B.

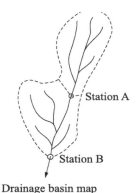

Station A

Station B

Drainage basin map
Figure 7.4   Drainage basin.

Let us assume that a storm began over the given basin (Fig. 7.4) at 7.00 p.m. on May, 17 and that a forecast is to be prepared on the basis of rainfall reported up to 7.00 a.m. on May, 19. The computations of runoff are shown in Table 7.1. The antecedent precipitation index (API) used is the last value prior to the storm. The week of the year determined by the date of beginning of the storm, i.e., May, 17 is the 20th week. The average rainfall amounts above station A and between stations A and B for 12 hour increment are entered in lines 5 and 14.

In Fig. 7.1, the dashed lines indicate the computed values of runoff for the area above station A for 7.00 a.m. on May, 18. Enter the relative value of API (49), move, left to the week of the year (20), down to storm duration (12), to the right storm precipitation (26), and obtain storm runoff which is say 9 cm. The process is repeated at the end of each 12-hour period using accumulated values or precipitation up to that time. The 12-hour increment of runoff (Table 7.1), lines 9 and 18 are determined by subtracting each total storm runoff values from the previous total and are entered in lines 1 and 12 of the forecast sheet in Table 7.2.

The 12-hour runoff increments are converted to discharges using the 12-hour unit hydrograph for the station A (Fig. 7.5). Each 12-hour ordinate of the unit hydrograph is multiplied by the first runoff increment (9/10) and entered in line 2 with the first value in the same column as the runoff increment (this being the ending time of the 12-hour period when the runoff occurred). This process is repeated on lines 3 and 4 for the same increments of runoff and the total for each time entered in line 6. Base flow given by line 7 includes all flow events preceding the storm (Table 7.2).

The computed forecast, the sum of the total runoff (line 6) and base flow (line 7) is entered on line 8 and plotted on the hydrograph (Fig. 7.6). The computed forecast is the adjusted result of the forecast procedures, and the forecaster must then draw an adjusted forecast reconciling the computed forecast with the available observed data. The adjusted forecast is shown as a dash line in the forecast period. The adjusted values are entered in line 9 for routing to station B.

Table 7.1 Computation of Storm Runoff.

Month	Year		Date	16		17		18		19	
May			Hour	7 a.m.	7 p.m.	7 a.m.	7 p.m.	7 a.m.	7 p.m.	7 a.m.	7 p.m.
Drainge	1.	0.9 of previous days API	—	54	—	49	—				—
(area above	2.	Precipitation in past 24 hours	—	0	—	0	—				—
station A)	3.	API for to-day	—	54	—	49	—				—
	4.	Week of the year	—	—	—	20	—				—
	5.	12-hour precipitation increments (mm)	—	—	—	—	—	26	17	48	—
	6.	Total storm precipitation (mm)	—	—	—	—	—	26	43	91	—
	7.	Duration (hours)	—	—	—	—	—	12	24	36	—
	8.	Total storm runoff (mm)	—	—	—	—	—	9	19	48	—
	9.	12-hour runoff increments (mm)	—	—	—	—	—	9	10	29	—
Drainage	10.	0.9 of previous day API	—	60	—	54	—				—
between	11.	Precipitation in past 24 hours	—	0	—	0	—				—
stations	12.	API for to-day	—	60	—	54	—				—
A and B	13.	Week of the year	—	—	—	20	—				—
	14.	12-hour precipitation increments (mm)	—	—	—	—	—	23	18	54	—
	15.	Total storm precipitation increments (mm)	—	—	—	—	—	23	41	95	—
	16.	Duration (hours)	—	—	—	—	—	12	24	36	—
	17.	Total storm runoff (mm)	—	—	—	—	—	8	18	47	—
	18.	12-hour runoff increments (mm)	—	—	—	—	—	8	10	29	—

Table 7.2 Forecast Computation Sheet.

Month: May	Date	17		18		19		20		21		22		23	
	hour	7 a.m.	7 p.m.	7 a.m.	7 p.m.	7 a.m.	7 p.m.	7 a.m.	7 p.m.	7 a.m.	7 p.m.	7 a.m.	7 p.m.	7 a.m.	7 p.m.
Station A															
1. Forecast runoff (mm)	12	–	–	9	10	29	–	–	–	–	–	–	–	–	–
2. Distribution of runoff		–	–	20	40	49	36	22	12	6	2	1	1	–	–
3. -do-		–	–	–	22	45	55	40	24	13	7	2	6	3	–
4. -do-		–	–	–	–	64	131	161	116	70	38	20	6	–	–
5. -do-		–	–	–	–	–	–	–	–	–	–	–	–	–	–
6. Total runoff		–	–	20	62	158	222	222	152	89	47	23	7	3	–
7. Base runoff		35	33	31	30	25	28	28	28	28	28	28	28	28	–
8. Computed forecost (m³/s)		35	33	51	92	183	250	250	180	117	75	51	35	31	–
9. Adjusted forecast (m³/s)		35	33	60	100	205	260	260	190	125	80	55	38	–	–
Routing from Section A to B															
10. $(I_1 + I_2)$ for routing purpose		–	68	93	160	305	455	510	450	315	205	135	93	–	–
11. A routed to B		35	40	43	62	107	170	215	220	145	111	111	79	–	–
Station B															
12. Forecast runoff (mm)	12	–	–	8	10	29	–	–	–	–	–	–	–	–	–
13. Distribution of runoff		–	–	36	61	59	35	15	6	1	1	–	–	–	–
14. -do-		–	–	–	45	76	74	41	19	7	1	–	–	–	–
15. -do-		–	–	–	–	131	221	216	119	55	20	3	–	–	–
16. -do-		–	–	–	–	–	–	–	–	–	–	–	–	–	–
17. Total runoff		–	–	36	106	266	330	272	144	63	21	3	–	–	–
18. Base flow		28	27	35	24	23	22	22	22	22	22	22	22	–	–
19. Computed forecast (m³/s)		63	67	114	192	396	522	509	386	230	154	136	101	–	–
20. Adjusted forecast (m³/s)		63	66	90	175	370	495	490	370	265	180	130	98	–	–

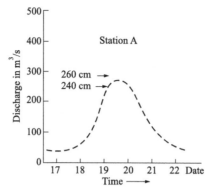

Figure 7.5   Unit hydrograph of 12-hour duration for stations A and B.

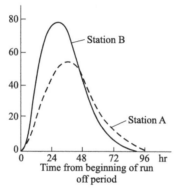

Figure 7.6   Discharge hydrograph at station A.

The final step in preparing the forecast is the conversion of forecast discharge to stage using the stage discharge relationship as shown in curve (Fig. 7.7). The forecast for station A could be given as 'crest of 250 cm at 1 a.m. on May, 20' or as 'crest of 240 to 260 cm early on May, 20'. The adjusted flows for station A (line 9) are routed to station B using Muskingum routing diagram in Fig. 7.8. Successive pairs of inflow values are added to obtain $(I_1 + I_2)$ values (line 10). The computation of the routed value for 7 p.m. on May 19, 170 m³/s is indicated by dashed lines in the routing diagram.

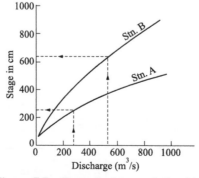

Figure 7.7   Stage discharge relationship.

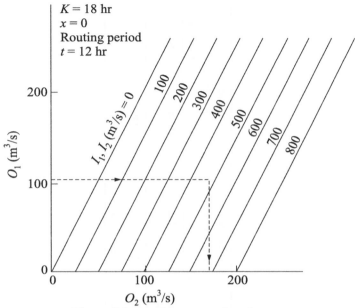

Figure 7.8 Muskingum routing diagram.

The forecast of flow from the local area (Table 7.1) is made in the same manner as for station A. The computed forecast is the sum of the routed values (line 11), the total runoff (line 17) and the base flow (line 18). These values are plotted on the hydrograph and adjusted on the basis of the observed data (Fig. 7.9). The forecast for station B might be given as 'crest of 640 cm at 2 a.m. on May, 20 or as 'crest of 630 cm to 650 cm early on May, 10. It is a good practice to maintain a record of these forecasts on a tabulation sheet such as Table 7.3 in order to minimise the chances of errors in transmitting the forecast to the users.

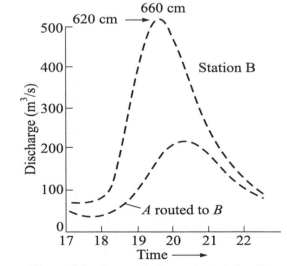

Figure 7.9 Forecast hydrograph at station B.

Table 7.3 Forecast Record Sheet.

Forecast for	Forecast crest (cm)	Time forecast issued			Latest stage available when forecast was prepared			Based on precipitation upto		
		Hour	Date	By	Stage (cm)	Hour	Date	Hour	Remarks Date
Station A	250	1 a.m.	20	Mr. X 9 a.m. 19 Mr. X	210	7 a.m.	19	7 a.m.	19
Station B	640	2 a.m.	20	Mr. X 9 a.m. 19 Mr. X	490	7 a.m.	19	7 a.m.	19

7.7 FLOOD WARNING

The flood warning authority, usually the engineer in charge, divides the vulnerable areas broadly into two categories viz., (i) protected and (ii) unprotected areas. Areas in flood plains which have been protected from inundation by construction of embankments along river banks have been termed as protected areas. The rest of the areas in flood plain which are liable to be flooded are unprotected areas. The entire vulnerable areas are usually grouped against the flood warning stations, i.e., for each flood warning station a zone usually selected where warnings are issued whenever the river reaches the warning stage at the particular station.

Usually flood warnings are given through different types of signals:

(i) *Yellow signal*: This indicates in advance the approach of flood in the river in the reach concerned.

(ii) *Amber signal*: This signal indicates in advance the approach of a high flood and is meant to alert the departmental staff for increasing their vigilance and taking necessary measures. This signal is not meant for unprotected plain areas.

(iii) *Red signal*: This would indicate the approach of high flood of such an order as may cause damage to the embankment in cases of protected areas, to cause inundation to a depth of 1.0 m or more over the general level of the agricultural land in case of unprotected areas.

The yellow warning is not to be conveyed to the public but is meant for the departmental staff and the authorities of vital installations such as General Administration, District Magistrate, Sub-divisional Officers, Block Development Officers, Police, Health, Civil Defence, Food and Supply, Public Works Department, Food Corporation of India, Army, Air Force, Railways, Posts and Telegraphs, Electricity Boards, Water Supply, Oil Storage and Installations.

The red warning is not only conveyed to all the officers and authorities to whom the yellow warning is conveyed but is also conveyed by the quickest possible means to the general public of the area. When the protected areas in respect of which the red warning has been issued is a notified Civil Defence Area and has the Civil Defence sirens, the red warning is conveyed to the public by sounding the siren for a period of three minutes with alternate steady and wobbling notes. When the area in respect of which the red warning is issued is not a civil defence area, the red warning is conveyed to the public of the area by beating of drums and through relays of special messengers who may be equipped with microphones or loudspeakers. On receipt of the red warning the administration brings the rescue parties, the other emergency units to a stage of instantaneous readiness. The cancellation of or withdrawal of red warning is signalled to the public by sounding the sirens on all clear signals for a period of two minutes or by other suitable methods outside the Civil Defence Area.

7.8 ENGINEERING MEASURES FOR FLOOD FIGHTING

Apart from administrative measures as outlined above a number of emergency construction activities are necessary for relief against the fury of flood. In most cases havoc due to flood are caused as a result of failure or breaching of the protective embankment. The causes of bank recession and the protective measures needed have already been discussed in Chapter 6. Herein the emergency engineering measures that may be necessary to take an extraordinary high flood which is likely to cause serious breaching problem or actually breached an embankment are discussed briefly.

(i) Protection against sliding

At flood time excessive seepage through the body of the embankment may lead to piping underneath the foundation. This phenomenon is known as sand boiling. This actually occurs at the toe of the embankment. To prevent that a ring of sand-filled bags should be placed around the boil. The resultant rise in water level in the ring will reduce the critical hydraulic gradient thereby controlling seepage and loss of material and ultimately failure of the embankment.

(ii) Protection against sliding of the landside level slope

Usually slope drains coupled with a horizontal channel at the toe of the levee is provided to tackle seepage water from a levee. When it is excessive as may be during a high flood, the situation may become very bad and in such a case there is inherent danger of failure of the levee by sliding. In order to prevent this a wooden mattress is laid over the slope which is then loaded by gunny bags filled with sand or gravel, the maximum load being placed near the berm. The above measure will permit easy drainage as well as prevent failure of the slope.

(iii) Protection against overtopping

Depending on the flood forecast if it is anticipated that at a particular reach the embankment is likely to overtop, immediate steps should be taken to increase the height of the embankment as an emergency measure, for if once the flood water overtops the levee it will wash away the soil and the failure of the levee will occur in no time. For heightening the embankment various methods can be adopted depending on the availability of material. The easiest one is of course to put up gunny bags filled with sand or spoilt cement bags when the increase in height required is not much. For greater heights box-type structures made of timber or bamboos with earth-filled capping is generally adopted.

(iv) Protection against scouring of the riverside slope

During high floods the protective arrangement of the riverside levee slope may not be sufficient. This may result in scouring of the slope and by the time scouring face appears above the water surface a good deal of the levee is lost. As an emergency measure cribs filled with stones or sand bags are placed over the scoured areas. This will prevent further erosion and stabilise the embankment.

(v) Closure of breaches in the embankment

In case the levee has actually breached at a certain place, steps should be taken to repair the breach as early as possible so as to prevent the breach from spreading as well as to shorten the period of inundation of the flooded areas. Generally speaking, the repair of the breach is carried out in parts. First of all a ring around the breached portion is made by sinking timber trestles filled with sand bags. This will prevent free flow of water. After that a second ring is placed at its back with earth-filled mud box to make the structure watertight (Fig. 7.10).

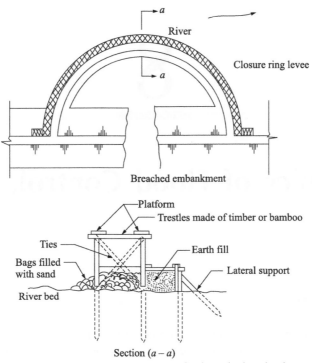

Figure 7.10   Closure arrangement of a breached embankment.

8

Economics of Flood Control, Project

8.1 GENERAL

The importance of scientific assessment of flood damage aims at a realistic assessment of the benefit-cost ratio of a flood control scheme. The word 'control' in real terms means protection. Assessment of the worth of any flood protection scheme involves a study of applied economics in the sense that annual savings due to a flood control measure in terms of severity, protection of life and property must be evaluated against the cost of construction and maintenance of the engineering structures needed for it. Generally speaking, the condition of economic analysis requires that the system needs to be, developed to the level at which the marginal benefit of producing output equals to the marginal cost of producing it. Marginal cost and marginal benefit are respectively the first derivative of the cost and the first derivative of the benefit with respect to the output. For a multi-purpose project for optimization the above condition for a particular purpose may be written as:

$$\frac{MB_1}{MC_1} = \frac{MB_2}{MC_2} = --- = \frac{MB_n}{MC_n} = 1, \tag{8.1}$$

where MB_n is the marginal gross benefit and MC_n is marginal cost.
Or

$$\frac{\partial B_1}{\partial C_1} = \frac{\partial B_2}{\partial C_2} = --- = \frac{\partial B_n}{\partial C_n} = 1, \tag{8.2}$$

In analysing a project for the above-mentioned economic conditions, several types of economic and technical relationship should be considered such as input-cost function, the production function, the annual-cost and benefit function. The input-cost function is the multi-dimensional relationship between the total cost of the project and the magnitudes of the project variables.

The production function is an input-output function which shows the relationship between the magnitudes of the project variables and their feasible combinations of outputs. From the input-cost function and production function, the annual cost function can be derived to show the relationship between the annual cost and output of the project variables. Finally, the annual-benefit function is the relationship between the average annual gross benefits and the outputs of the project variables.

The above principles may be illustrated by a simple example such as the construction of a flood embankment for flood control purpose. Here the input-cost function can be represented by the relationship between the total cost in rupees to be invested in the construction and the height of the embankment or the storage provided. The production function is the relationship between embankment height and the project output. The output in this case can be measured by the peak flood flows in terms of storage for which complete protection is provided. In order to do so let us consider how assessment of flood damages and benefits as a result of flood control are generally arrived at.

8.2   ESTIMATING FLOOD DAMAGES

Flood damage can be direct or indirect. These terms are applied in a physical classification of damage. Direct damage results due to physical contact with flood waters whereas the indirect damage results to property or services not touched by flood water but harmed as a result of interrupted trade or diversion of rail or road traffic or other effects of the flood.

The following may be considered as examples of direct and indirect damages. Suppose a man who is the owner of a land in the path of flowing flood water has grown, stored or constructed something on that land which can be harmed by water, he may suffer damage during the flood. Inundation by flood water and sediment deposition on buildings, equipment or even on the land itself may constitute direct flood damage. The damage suffered is usually obtained by the cost of rehabilitation, replacement or in terms of reduction of the original value. The monetary loss suffered by a man who is put out of work because of the factory where he is employed has been closed due to a flood, or a truck driver who fails to cross a bridge by which he normally reaches the market and has to travel many miles as a detour, can be considered to have suffered indirect flood damage. Damages caused due to interruption of the normal course of everyday affairs may be considered as indirect. Such damages although important are, however, difficult to objectively estimate.

Generally speaking, damage to the agricultural products forms the major portion of the total damage caused by floods, amounting to, say 70–80 per cent. Then come, houses and buildings, which constituting about 20–25 per cent and the remaining is accounted for by public utility services. The main purpose of the survey of the damages caused by flood is to relate it to the flood stage. It is common knowledge that damage is a function of the area of overflow. However, the depth of overflow, duration of inundation and the time of the year when flood occurs have all important bearing on flood damage values. With the data thus obtained it is possible to derive functional relationship between stage, damage and duration of inundation for a particular river basin.

In connection with flood control projects protecting highly industrialised areas it is customary to relate damage to the magnitude and frequency of floods. For this purpose

(a) the graphical plot of the flood stage vs. flood discharge at the gauging station corres-ponding to the property is combined with the plot of the flood discharge vs probability, (b) the curve of flood stage vs probability is combined with flood stage vs property damage described previously to obtain the curve of flood damage vs probability for the property. The latter curve shows the probabi-lity of occurrence of flood damage equal to or greater than various values as shown in Fig. 8.1. The total area under the damage probability curve is a measure of the average annual damage caused by all floods that can be anticipated for a long period of years.

Usually to ascertain net flood loss, a statement of flood damages and relief measures given to the people affected is prepared by appropriate authority in the Table 8.1 [28].

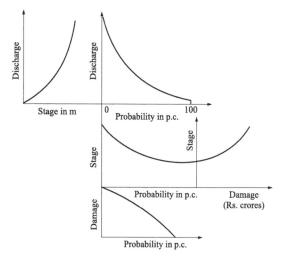

Figure 8.1 Relationship between stage discharge probability and damages.

Table 8.1 Flood loss estimation

Name of the river basin: **Name of scheme:**

		Nature of flood damages				
Year	*Area Affected (hectares)*	*Frequency of floods*	*Probable depth of inun- dation (m)*	*Duration of inun- dation (days)*	*Damage to crops*	
					inun- dation area (hectares)	*Value of crops (in ₹ crores)*
..	..	..	..	..	..	..
	Damages to houses			Cattle lost		Damages to public utilities (in ₹ crores)	Total damages (in ₹ crores)
No.	Values (in ₹ crores)	Huamn lives lost	No.		Value (in ₹ crores)		
..	..	..	..		..	..	..
			Relief measures			
Gratuitous relief (in ₹ crores)	Agricultural and other loans (in ₹ crores)	Remission of land revenues (in ₹ crores)	Total relief expenditure (in ₹ crores)	Total cost or relief and damages (in ₹ crores)		
..	..	..	..	..	..	
	Extent of beneficial value of fertilising slit (in ₹ crores)		Net flood loss (in ₹ crores)			
	..					

8.3 ESTIMATES OF BENEFIT OF FLOOD CONTROL

The benefits from flood control are of two kinds—(a) those arising from prevention of flood damage, and (b) those resulting from more intensive use of protected land. The benefit arising from the prevention of flood damage is the difference in expected damage throughout the life of the project with and without flood control, and these benefits include—(i) cost of replacement or repairing of damaged property, (ii) cost of evacuation, relief rehabilitation of victims and emergency flood protection measures, and (iii) losses due to disruption of business and losses due to crops or cost of replanting crops. Land protected from flood may be utilised for more productive purposes when it is not subject to flood hazard. The method for determining the benefit gained by improved land use is to estimate the difference in net revenue from the property with and without flood control. Firstly, benefits from flood control may arise in activities which results from use or processing of products and services directly affected by floods. As for example, if a factory manufacturing steel or cement is closed as a result of flooding other industries depending on these may have to be closed down.

Secondly, benefits are difficult to assess and generally are not included in estimates of flood control benefits. Intangible benefits of flood control includes saving of lives, reduction in diseases resulting from flood conditions. Although it is difficult to put a monetary value to these intangibles it may be remarked that the death benefits payable under workman's compensation law may be taken as the economic value of human lives. Average annual benefits may be computed by multiplying the benefits expected by prevention of flooding at a given stage by the probability of occurrence of that stage in any year. This is shown in Table 8.2 for a typical case.

Table 8.2 Cost benefit analysis

Peak stage (m)	Total damage below indicated stage	Increment of loss	Return period (years)	Annual benefits for protection of increment	Total benefits with protection below given stage	Project cost	Ratio of benefit to cost
6.0	0	40	10	4.00	0.00	4.0	0
6.5	40	50	20	2.50	4.00	6.00	0.67
7.0	90	60	25	2.40	6.50	7.0	0.93
7.5	150	70	30	2.33	8.90	8.9	1.00
8.0	220	80	50	1.60	11.23	9.5	
8.5	300	90	100	0.90	12.83	10.0	
9.0	390	110	150	0.73	13.73	11.0	
9.5	500				14.46	12.0	

All costs in crores of ₹.

8.4 COST-BENEFIT ANALYSIS FOR A FLOOD CONTROL PROJECT

The table shows the increment of prevention of loss in 50 cm intervals of stage divided by the expected return period of a flood at the mid-height of the interval. The summation of the above increments indicates the total benefit with protection against floods of various sizes. In the above table, the cost of various degrees of protection by constructing a flood bank below the given stage is indicated. It ignores possible intangible benefits that may warrant the selection of another design stage. Usually any flood control project should show economic feasibilities before it is constructed and only those portions of a comprehensive project which are in themselves beneficial should be constructed. Thus flood banks and channel improvements for the protection of an agricultural land alone might not be justified by the expected returns. However, if a proposed reservoir will have some benefit for the agricultural area, those savings should be included in considering the justification of the building up of the reservoir.

The benefits obtained from a typical flood control project shown in Table 8.2 is measured by the reduction in flood damage as a result of construction of flood embankment. The flood damage as has been mentioned has been evaluated by proper flood survey. Occurrence of flood being stochastic in value, its analysis is usually based on probabilistic concept. From the Table 8.2, the peak stage for the given recurrence interval or probability, i.e., $(P = 1/T)$ can be obtained. It is, therefore, possible to represent graphically (Fig. 8.2), the total benefits against peak stage for given probabilities. The marginal benefit can be represented by the slope of the tangent to the benefit curve. Similarly, the marginal cost is represented by the slope of the tangent to the project cost curve. At the optimal scale of development the two slopes should be equal or the two tangents should be parallel which is indicated in Fig. 8.2. According to Eq. (8.1)

$$MB = MC$$

or

$$\frac{d\ (benefit)}{d\ (output)} = \frac{d\ (cost)}{d\ (output)}$$

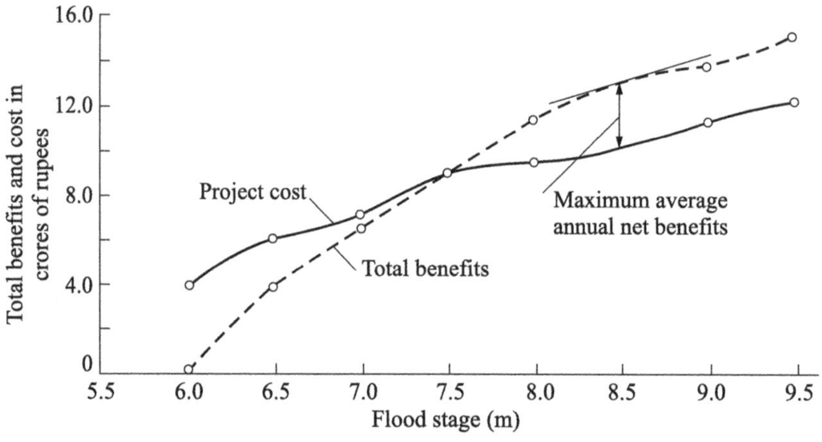

Figure 8.2 Benefit-cost analysis of a flood control project.

Mathematically, the above condition may lead to either a maximum or minimum. Hence, to assure the optimization being a maximum, the following secondary condition must hold valid.

$$\frac{d^2\ (benefit)}{d\ (output)^2} < \frac{d^2\ (cost)}{d\ (output)^2}.$$

In making economic division by theoretical analysis, the result cannot always be dependable as a good deal of uncertainty is involved in the hydrologic and economic data. It is, therefore, necessary for such a decision to be modified by experienced judgement in accordance with intangible factors involved.

8.5   FLOOD CONTROL PLANNING THROUGH REMOTE SENSING

8.5.1   General

Flood plains and highly susceptible flood-prone areas are being gradually occupied as centres of population and increased economic activities. Also in many instances there is a gradual depletion of vegetal cover and change in landuse practices in the catchment areas. These also result in increased runoff and change in river regimes in the downstream areas. National Water Policy of 1987 recommends that while physical flood protection works like embankments and dykes will continue to be necessary the emphasis should be on non-structural measures for the minimisation of losses such as flood forecasting and warning and flood plain zoning so as to reduce the recurring expenditure on flood relief. Further, with the emphasis now on comprehensive river basinwise regional planning it is necessary to obtain current informations on the flood plain and its response to floods. In the above context remote sensing techniques can play an important role for making necessary assessment for planning over large areas.

8.5.2   Remote Sensing Techniques

Remote sensing techniques involve collection of terrain features by employing sensors in aircrafts and satellites. The sensors act to the reflected and emitted electromagnetic radiation from the earth's surface to produce images. The images are then analysed to yield information of importance for planning. Generally, for the following types of flood problems remote sensing techniques can provide necessary inputs.

(a) Flood inundation mapping and its areawise estimate.
(b) Flood plain landuse mapping and its areawise estimate.
(c) Mapping flood susceptibilities indicators in the flood plain.
(d) Mapping river channel—migration and study of river behavioural changes.
(e) Flood warning.

(a) Flood inundation mapping

Near-infrared spectral bands provide an unique recognition characteristics for mapping water surface. As such, an extent of the area inundated by flood can be obtained easily from satellite observation. Both optical and digital data processing techniques are of help for such delineation. The above information is helpful for the flood control planner to take decisions as under— (i) extent of flood damage, (ii) focussing relief effort where it is needed, (iii) identifying areas requiring post-flood alleviative measures and (iv) suggestions regarding strengthening measures or for taking up additional works.

(b) Flood plain land use information

An important input to flood-prone area planning is the landuse information of flood plain of the flood-prone river. They provide important inputs for immediate flood damage assessment and in the long run such information helps to develop (i) flood hazard zoning with respect to varying degrees of flood hazard and (ii) control on activities of man on the flood plain and finally drawing up a long-term basinwise flood control measures.

(c) Mapping of flood susceptibility indicators

Preparation of flood hazard zoning along with inundated areas and flood plain landuse information are required in the case of a design flood which may be standard project flood. For these purposes certain natural flood susceptibility indicators are helpful. They can be obtained by air-photo interpretation techniques that are available. These are respectively— (i) Upland physiography, (ii) Basin characteristics such as shape, drainage density, etc., (iii) Degree of abandonment of levees, (iv) Occurrence of stratified sand dunes of river terraces, (v) Channel configuration and fluvial geomorphic characteristics, (vi) Backswamp areas, (vii) Soil moisture availability, (viii) Soil differences, (ix) Landuse boundaries, (x) Agricultural development, and (xi) Flood alleviation measure in the flood plain.

(d) River behavioural changes

Most of the river systems in India show migratory behaviour. As for example, the Ganga and the Brahmaputra river systems tributaries exhibit vivid dynamic and migratory changes during flood season. Also the river system undergoes considerable morphological changes in the downstream-side after the construction of dams and barrages. In such a situation satellite remote sensing is of great help for mapping and studying river situation.

(e) Flood warning

An effective flood warning system can be based on data collection systems on board the satellite. The satellite obtains the data from the ground-based remotely located automated data collection platforms which are equipped with transducers. Different organisations and scientific departments transmit the data to the ground receiving station with their own transducers from their computer facilities. The transducers measuring the variables such as water stage, streamflow, precipitation, temperature, wind speed and its direction, humidity, snow depths/density, water equivalent, soil moisture, water quality parameters, etc. can be interfaced with Data Collection Platform.

This ground designed hydrologic data can be relayed via satellite to the flood control management centre within a fraction of a second and many times each day from the farflung and inaccessible areas of a larger river basin. India's multipurpose geostationary Satellite INSAT-1B has the DCS facilities on board which relay 10 ground-based DCP data at present for meteorological monitoring. For the flood warning systems the following informations are required. They are respectively—(i) time and location of breaking of the storm event on the basin, (ii) amount of precipitation, (iii) rate of runoff, (iv) water stage records at selected points of the river reach, and (v) reservoir water levels.

All the above can be monitored in real time through the Satellite Data Collection system. The above data is vital for management decisions for efficient operation of reservoir, issuance of flood warning to the flood-prone areas, and timely action for protective measures.

Problem 8.1

The magnitude of a large flood occurring in a river on which a barrage is being constructed is defined by the distribution.

$$f(x) = Ke^{-kx} \quad \text{for} \quad 0 < x < \infty$$

The damage caused by the flood is given by the function

$$D(x) = B - e^{-K_1(x-A)} \quad \text{for} \quad 0 < x < \infty$$

Determine the average value of damage from a flood.

Solution

The expected value of damage

$$\int_A^\infty \left[B - e^{-K_1(x-A)} \right] K e^{-Kx} dx$$

$$= B \int_A^\infty K e^{-Kx} dx - \int_A^\infty K e^{-(K_1+K)x + AK_1} dx$$

$$= \left[-Be^{-Kx} \right]_A^\infty + \frac{K}{K_1 + K} e^{AK_1} \left[e^{-(K_1+x)} \right]_A^\infty$$

$$= -Be^{-KA} - \frac{K}{K_1 + K} e^{AK_1} \left[e^{-(K_1+x)} A \right]$$

$$= -e^{-KA} \left[B - 1 + \frac{K_1}{K + K_1} \right]$$

Problem 8.2

Consider the construction of a protection wall for a riverside project to be protected from floods. A minimum height of the wall would provide the barely needed protection and there exists every possibility of overtopping the wall by flood waters, which will result damage to construction work. The damage can be reduced by increasing the wall height which will then result in increase the investment cost.

Table 8.3 furnishes the probabilities of floods of different magnitudes above the minimum wall height of increments of 2 m and expected cost resulting from flood damage. The additional cost of constructing the wall in increment of 2 m is also given in the Table 8.3.

Table 8.3

Flood level above min wall height in range (m)	Prob. of occurrence	Flood damage (₹)	Cost to construct wall upto lower value of range (₹)
0-2	0.20	80,000	150000
2-4	0.20	150,000	200000
4-6	0.10	200,000	300000
6-8	0.08	250,000	400000
Over 8	0.00	0	450000

Find out the minimum expected annual cost for the desired wall height. Assume the work is likely to continue upto 3 years and the rate of interest on investment is given as 10 per cent.

Solution

The initial cost of the wall is converted to equivalent annual cost by multiplying it with CRF (Capital Recovery Factor).

This cost for a wall of minimum height will be

$$150,000 \left[\frac{10(1+10)^3}{(1+10)^3 - 1} \right]$$

$$= 150,000 \times 0.402$$

$$= ₹ 60,3000$$

The expected value of flood damage, if wall of minimum height is constructed, will be

$$= 80,000.00 \times 0.2 + 150,000.00 \times 0.2 + 200,000.00 \times 0.1 + 250,000.00 \times 0.08$$

$$= 16,000.00 + 30,000.00 + 20,000.00 + 20,000.00$$

$$= ₹ 86,000.00$$

Total expected cost for minimum wall height = 60,300.00 + 86,000.00

$$= ₹ 146,300.00$$

The values for different height are furnished in Table 8.4.

Table 8.4

Wall hight above min	Flood damage from incremental flood levels above wall hight				Exp. value of flood damage	Value of investment	Total expected cost
(m)	0–2 m	2–4 m	4–6 m	6–8 m	(₹)	(₹)	(₹)
0	80000	150000	200000	250000	86000	50250	146300
2	0	80000	150000	200000	47000	80400	127400
4	0	0	80000	150000	20000	120600	140600
6	0	0	0	80000	64000	160800	167200
6	0	0	0	0	0	180900	180900

The minimum expected annual cost is found to be for a wall height 4 m above minimum height stipulated initially, i.e. 2 m.

Problem 8.3

A project area is subject to occurrence of floods of varying magnitude every year. A severe flood would result in dislocating the construction work and is causing damage to work already done. An insurance company is prepared to provide cover at a cost of Rs. 80,000 every year but would not include all losses caused by flood. It is estimated the true loss resulting from the flood would be at least 100% greater than the physical loss covered by insurance. The insurance company is prepared to reduce the insurance premium to 50%, if the contractor provide same flood control measures at the site. The cost of the measures is estimated to be ₹250,000.00 and the annual maintenance and repairs for these measures would be ₹5000.00. The project is scheduled to be completed in 5 yrs and the capital can be borrowed at an interest rate of 10%.

Should the contractor provide flood control measures?

Solutions

When no flood control measures are adopted and a flood occurs the liability of the contractor will be:

Payment to insurance company annually   = ₹80000.00
Uncompensated loss due to floods   = ₹80000.00
Total   = ₹160000.00

When flood control measures are provided and a flood occurs liability will be as follows:

Capital recovery annual   = ₹250,000.00 × CRF

where
$$CRF = \left[\frac{i(1+i)^n}{(1+i)^n - 1} \right]$$

where
$$i = 10\%$$
$$n = 50 \text{ yrs}$$

So
$$CRF = \left(\frac{10(1+10)^5}{(1+10)^5 - 1} \right) = 0.264$$

So capital recovery annual   = 250,000 × 0.264 = ₹66,000.00
Maintenance and repair annual   = ₹5000.00
Payment to Insurance company   = ₹40000.00
Uncompensated loss annual   = ₹40000.00
Total   = ₹151000.00

So it would seem providing flood control measures is advantages though benefit is marginal.

9

Design of Subsurface Drainage System

9.1 INTRODUCTION

The term 'drainage' has been used to describe all processes whereby surplus water is removed from lands. This includes both internal drainage of soils and the collection and disposal of surface runoff.

Water-logged land is of little use. However, it can be utilised after providing proper drainage arrangement. Usually in undulating country, the surface slopes are sufficient to carry off this surplus water into the ditches and streams without any engineering construction. Low-lying flat areas are usually invariably near or below the flood level of the river. In order to utilise the area, the river must first be prevented from flooding it, usually by construction of flood embankment. Arrangements have then to be made to collect and dispose of the water otherwise entering the area. The area may be so low that gravity, discharge is not possible all the time. Such conditions occur in the lower reaches of a tidal river. The problem in such cases is to collect and convey the water entering the area to a discharge point and then pass it into the river either by gravity or by pumping.

9.2 NECESSITY FOR DRAINAGE

Soil has the capacity for holding water which enables plants to grow by drawing water and nutrients in solution in the water from the soils through their root systems. The structure of a soil consists of a framework of solid material enclosing a complex system of pores and channels which provide space within the soil for air and water. When all these spaces are filled with water, the soil is termed saturated. A soil can only remain in a saturated condition if it is below water table and cannot drain truly. It may be temporarily saturated during and immediately after irrigation or heavy rainfall. The maximum amount of water or moisture which a soil can hold at saturation depends on the volume of its pore spaces and is known as its saturation capacity.

In order that plant to grow, apart, from availability of water, air is also needed, and hence soils should not be permanently saturated with water. A good soil, therefore, has good internal drainage characteristics which means water must be able to move fairly easily through the soil so that excess water can be removed when required.

9.2.1   Topographic Factor

Good topographic maps are required in the preparation of a drainage plan. The map provides the essential features such as the location and type of outlet, i.e., whether gravity or pumped, surface slope and the degree of land preparations. The topographic survey should indicate the position, alignments and the gradient of existing ditches, streams, culverts and other natural and artificial features that may influence the drainage system. Possible outlets should also be included and where possible the dimension and capacities of the drainage channel and the natural water courses be also determined. The location of pumped wells should also be indicated along with their capacities, drawdowns and the zones of influence. The use of aerial photography including satellite imageries will be of assistance in locating areas of poor drainage.

9.2.2   Drainage Factor

Drainage is greatly hampered by the presence of impermeable substratum. Submergence of lands due to flood also affects growth and yield of crops. Soil drainage affects growth and yield of crops. Soil drainage refers to the rapidity of removal of water added to the soil and also the frequency and duration of periods when the soil is free of saturation. Under drainage factor there are four characteristics—(i) water table, (ii) impermeable substratum, (iii) submergence due to floods, and (iv) soil drainage class.

9.2.2.1   Water table

Presence of high water table affects soil aeration which, in turn, affects root growth microbial activity and nutrient availability. Accumulation of soluble salts results most commonly in consequent of a high water-table formation. Water table should not be within a depth of 2 m from the soil if recharge to the soil profile with salts through capillary rise of groundwater is to be avoided. The critical depth of water table may range from zero in case of rice to about 150 cm in case of many other crops. The National Commission on Agriculture in the year 1976 have indicated the critical depths of water table for the following crops:

Crops	Ground-nut	Maize	Potato, Wheat	Sugarcane, Grams	Onion	Tomato, Cotton	Bajra	Soyabean
Depth of water table (cm)	60	75	50	70	40	100	125	125–150

Measurement of depth to water table in the dug-wells in pre- and post-irrigation soil surveys can help to know the rise of water table due to irrigation schemes, if any, as well as to delineate the water-logged and potentially water-logged areas.

9.2.2.2 Impermeable substratum

Impermeable substratum at a depth within 10 cm of the soil surface is considered unsuitable for irrigation. The following table gives the irrigability rating for depth of impermeable substratum:

Irrigability class	1	2	5	4	3
Impermeable substratum depth from soil surface, (m)	3	1–3	0.5–1	0.1–0.5	0.1

9.2.2.3 Submergence due to floods

Submergence due to flood affects crop growth. Agricultural land rating incorporating flood rating is prescribed in some countries. The land affected by flood once in 6–12 years are classified to be under rare flooding and those affected under 1–5 years are under common flooding. When there is considerable sedimentation or scouring the lands are marked to be under damaging flooding. Depending on the type of flooding different points are allocated as flood numbers and lands are rated accordingly. Under the climatic condition prevailing in monsoonic humid and sub-humid region of India and other countries flash flood is a common feature. Intensity and frequency of flood varies from year to year. An idea of irrigability rating based on submergence due to flood is furnished below:

Irrigability class	1	2	3	4
Submergence due to flood. Probable occurrence out of 10 years,	No flooding nil	Occasional flooding 0.1	Common 0.4	Regular 1

9.2.2.4 Soil drainage class

The procedure as described in *USDA Soil Survey Manual* is followed in making a land into the relative soil drainage class. As per the Manual there are seven soil drainage classes such as (i) very poorly drained, (ii) poorly drained, (iii) imperfectly drained, (iv) moderately well drained, (v) well drained, (vi) somewhat excessively drained, and (vii) excessively drained. Based on above well drained and moderately well drained conditions are taken to be most favourable for a wide variety of crops and are marked under irrigability class 1. The following table provides irrigability rating as per soil drainage class:

Irrigability class	1	2	3	4	5
Soil drainage class	Well to moderately well drained	Imperfectly drained	Poorly and somewhat excessively drained	Very poorly drained	Excessively drained

9.3 REMOVAL OF DRAINAGE WATER—UNDERDRAINS AND THEIR LAYOUT

Subsurface drainage is meant to remove the excess water present below the ground surface. Generally, agricultural lands affected by high water table requires subsurface drainage. The following are the means of accomplishing the subsurface drainage. They are respectively:

(i) Tile drains,
(ii) mole drains,
(iii) deep open drains,
(iv) drainage wells, and
(v) combination of tile and open drains.

The commonest type of underdrains used are unglazed claytile or concrete pipe. Plain pipes without special joints are used and are placed with the ends of adjacent pipes butted together. Water enters the drain through the space between the abutting sections. The space should not be larger so that soil cannot enter and clog the drain. The drains are laid following the contours of the field. The various types of layout are indicated in Figs. 9.1 through 9.5, which are respectively designated as natural, gridiron, herringbone, doublemain and intercepting [31].

The natural system is used in natural topography where drainage is necessary in valleys. For draining an entire area gridiron layout is usually economical. The gridiron system may be used when the land is practically level or where the land slopes away from the submain on one side. The herringbone pattern is employed when the submain is laid in a depression, the laterals joining it from each side alternately. If the bottom of the depression is wide a double main system is usually employed. This reduces the length of the laterals and eliminates the break in the slope of the laterals at the edge of depression. For draining excess water from hill lands an intercepting drain along the toe of the slope is required for protection of the bottom land.

9.4 DESIGN OF CLOSED UNDERDRAINS

9.4.1 Depth and Spacing of Drains

In order to estimate the flow of groundwater to drain, it is necessary to determine the position of the water table at equilibrium with the rainfall or irrigation water. Essentially, this means, how high will the water table rise for a given rainfall, permeability of soil, depth of drain, their spacings, for a given depth of barrier layer which restricts the downward flow. If drains are installed the water table will rise until the flow into the drains just equals the amount of rain or irrigation water infiltrating through the soil surface. At this time the water table is said to be in equilibrium with the rainfall or irrigation water. The position of the water table will primarily depend on the following:

(i) The rainfall rate or the rate at which irrigation water is applied,
(ii) The hydraulic conductivity,
(iii) The depth and spacing of the drains, and
(iv) The depth to an impermeable layer.

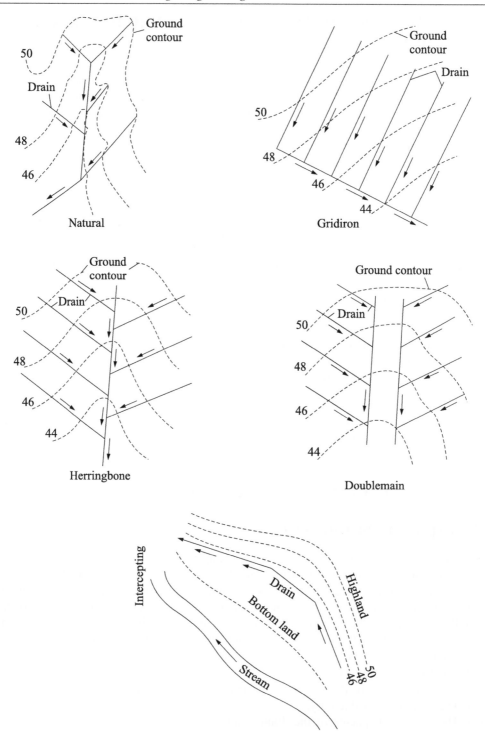

Figures 9.1–9.5 Different subsurface drainage layouts.

The analysis of the problem presented in Fig. 9.6 by Hooghoudt* is (vide ref. 29, pp. 151–153) is based on the following assumptions:

(i) The soil is homogeneous and of hydraulic conductivity k,

(ii) The drains are spaced a distance S apart,

(iii) The hydraulic gradient at any point is equal to the slope of the water table above the point, i.e. (dy/dx),

(iv) Darcy's law is valid for flow of water through soils,

(v) An impermeable layer underlies the drain at a depth d,

(vi) Rain is falling or irrigation water is applied at a rate q_0, and

(vii) The origin of coordinates is taken on the impermeable layer below the centre of the drains.

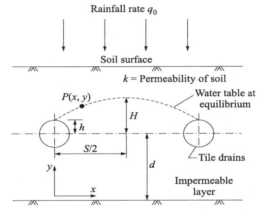

Figure 9.6 Analysis of flow in an underdrain.

Consider flow through a vertical plane drawn from the point P on the water table to the impermeable layer. It is evident from the figure that all the water entering the soil to the right of the plane upto the division plane which is at the centre of the two drains must pass through it on its way to the drain. If q is the quantity of water entering a unit area of soil surface then the total quantity of water passing through the plane will be

$$q = \left(\frac{S}{2} - x\right)q_0 \tag{9.1}$$

Similarly, by applying Darcy's law to the flow through the plane, we can obtain the second equation, i.e.

$$q = ky\frac{dy}{dx} \tag{9.2}$$

since the $\left(\dfrac{dy}{dx}\right)$ is the slope of the water table above the point.

Equating Eqs. (9.1) and (9.2), we have

$$\left(\frac{S}{2} - x\right)q_0 = ky\,\frac{dy}{dx}$$

---

*Original publication in: Versi, Land bouwk und 46: pp. 515–540, 1940, Netherlands.

or
$$q_0 \frac{S}{2} \, dx - q_0 \, x \, dx = ky \, dy$$

Integrating
$$\left(\frac{q_0 S}{2}\right)x - \frac{q_0 x^2}{2} = \frac{ky^2}{2}$$

the limits of integrations being $x = 0$, $y = h + d$, and $x = \dfrac{S}{2}$, $y = (H + d)$.

On substitution, we have
$$S^2 = \frac{4k(H^2 - h^2 + 2Hd - 2hd)}{q_0} \tag{9.3}$$

For practical purpose the drain is considered empty and hence the Eq. (9.3) reduces to
$$S^2 = \frac{4kH(2d + H)}{q_0} \tag{9.4}$$

where $h = 0$

The spacing of the drains must be such that the water table on its highest point between drains does not interfere with plant growth. The water table should be below the root zone to permit aeration of the soil. A high water content also reduces soil temperature and retards plant growth. The necessary depth to the water table depends on the crop, the soil type, the source of the water and the salinity of the water. Generally, the water table should be lowered more in heavy clay soils than in light sandy soils. It is not desirable to lower the water table far below the recommended minimum depth as they deprive the plants of capillary moisture needed during dry season. The variation in water table elevation in the drained land should not preferably be more than 30 cm which means the drains should be placed about 30 cm below the desired maximum ground water level.

The diameter of the tile drains usually ranges from 15 cm to 45 cm and they are laid in short pieces of length of about 50 cm with open jointed ends. The joints are enclosed by suitable gravel filter so as to prevent fine sand and clayey material entering the drains. The drains are usually laid to a slope so that these can carry the discharge and usually the slope is calculated from the drain outlet end. In case the natural ground slope is insufficient the drain ends into a sump from where the water is pumped out for maintaining free drainage condition in the well. For proper maintenance manholes are usually provided at suitable interval, say about 300 m depending on the type of subsoil and to effect change in the size of the drain. The bottom of the manhole is placed below the invert of the drain pipe.

9.4.2 Hooghoudt's Equation for Layered Soil

For layered soil consider that the hydraulic conductivity above the drain line as k_a and that below it by k_b, the Hooghoudt's formula then becomes

$$S^2 = \frac{4}{v}(k_a H^2) + \frac{8}{v}(k_b d'H) \tag{9.5}$$

where v, rainfall rate, d' is the equivalent depth, which is to be obtained from Hooghoudt's curves (Fig. 9.7) given for drain radius, $r = 0.7$ ft. Suppose the layer above the drain line consists of three layers of conductivity, k_1, k_2, k_3 and having thicknesses l_1, l_2, and l_3, then

$$k_a = \frac{k_1 l_1 + k_2 l_2 + k_3 l_3}{l_1 + l_2 + l_3} \qquad (9.6)$$

The values of equivalent depths obtained by Hooghoudt's curve are to be substituted in the equation for calculating the spacing. For practical purpose if the underlying layer has one-tenth of the permeability of the upper layer then it can be considered as impermeable.

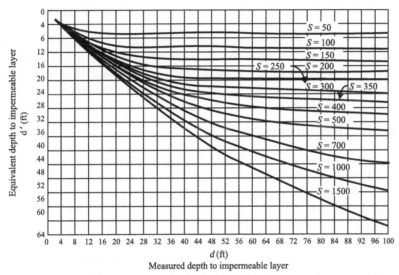

Figure 9.7 Relationship between d and d' where $r = 0.7$ ft and S is the spacing between the drains in feet. Curves are based on Hooghoudt's correction (after Bureau of Reclamation).

9.4.3 Determination of Irrigation or Rainfall Rate

This is actually determined by the infiltration rate of the soil, although in the Hooghoudt's equation, the value of v is based on the rainfall rate or the irrigation rate, although in reality it is dependent on many factors as rainfall rate is seldom constant. Really v will be the rate at which water replenishes the soil beneath the water table. Factors which influence the replenishment rates are, soil moisture at the time of rain storm, interception loss, deep seepage, artesian seepage, surface runoff, evapotranspiration. Therefore, the best way to obtain a value of v is to measure the value of outflow of existing drains. Surface runoff must be excluded from these measurements.

9.4.4 Hooghoudt's Equation in Humid Areas

The Dutch practice is to take drainage rates, i.e., v as equal to 5 to 7 mm/day, and a surface runoff of 15 mm/day is assumed. The height to which the water table will rise is plotted as a function of the rainfall rate. The frequency of rainfall rate is indicated on the diagram. Say, the height of the water table is plotted for rainfalls which are known to occur in two days, three days, etc. From a study of the rainfall distribution throughout the year and a consideration of the tolerance of the plant to high water table conditions, seriousness of the drainage problem is ascertained. The time of the year is taken into account by the temperature, which is to be taken as an indicator of the likely damage to the plant from high water table conditions. The number of day degrees that the temperature is above zero centigrade is directly related to the damage that will result to the plant, although plants like tomato do not suffer from high water table conditions until the temperature is about 6 to 7 degrees above zero. The equation to be used is

$$t \text{ (in days) } T° \text{(temp in degrees above zero)} = 30 \tag{9.7}$$

In case it exceeds the above value the plants will suffer serious damage thereby necessitating additional drainage. In irrigated areas the problem of determining υ is similar. Extensive application of Hooghoudt's equation has been made for drainage of irrigated areas. In this connection it is important to satisfy the leaching requirement to keep the water at a safe level.

9.4.5 Leaching Requirement

The leaching requirement is defined as the fraction of irrigation water that must be leached through the root zone of the plants to prevent the soil salinity from exceeding a specified level. It depends on the salinity of the applied irrigation water as well as the maximum concentration of salts that can be permitted in the rootzone of the plants. The concept is very useful in computing the amount of drainage water which has to be removed from a large area and it presupposes that there is adequate control of applied irrigation water which should be just equal to the crop needs plus the leaching requirement. The maximum concentration of salts will be at the bottom of rootzone excluding the surface crusts formed due to evaporation. This concentration will be the same as that of the concentration of the salts in the drainage water provided there is no excess leaching and irrigation water is applied uniformly. The increase in concentration of salts in the drainage water over that of the concentration in irrigation water is due consumptive use of crops. The crop will extract the water from the soil leaving most of the salts behind.

The assumptions inherent in application of leaching requirement are Uniform application of irrigation water, no rainfall, no removal of salts in harvested crops, no precipitation of salts in the soil. The calculation is based on the total equivalent depths of water over a period of time, ignoring the moisture and salt storage in the soil, cation exchange reactions, depth of rootzone and crop use of salt.

The leaching requirement is equal to the ratio of equivalent depth of drainage water to the depth of irrigation water, D_{dw}/D_{iw} expressed in percentage or ratio, alternatively it may be expressed in terms of the electrical conducting of the drainage water compared to the irrigation water. So the leaching requirement becomes

$$LR = \frac{D_{dw}}{D_{iw}} = \frac{EC_{iw}}{EC_{dw}} \tag{9.8}$$

Here it may be mentioned that winter precipitation may be adequate to leach the soil. All the water that passes through the rootzone of the plant must be considered in the use of the equation. Conductivity of the irrigation water should be the weighted average of the conductivities of the rain water EC_{rw} and the irrigation water EC_{iw}, that is

$$EC \, (rw + iw) = \frac{D_{rw} EC_{rw} + D_{iw} EC_{iw}}{D_{rw} + D_{iw}} \tag{9.9}$$

where D_{wr} and D_{iw} are the depths of the rainwater and irrigation water that enter the soil. The amount of irrigation water will be equal to the sum of consumptive use plus the drainage water

$$D_{iw} = D_{cw} + D_{dw}, \tag{9.10}$$

Eliminating D_{dw} [vide Eq. (9.8)], the depth of irrigation water in terms of consumptive use and leaching requirement

or

$$D_{iw} = \frac{D_{cw}}{1 - LR}$$

$$D_{iw} = \left(\frac{EC_{dw}}{EC_{dw} - EC_{iw}}\right) D_{cw} \tag{9.11}$$

In the above equation, EC_{dw} represents the salt tolerance of the crop to be grown.

The EC_{iw} can be taken to be about 1 m mho/cm and EC_{dw} can be taken as about 8 m mho/cm. Considering consumptive use as 1 cm/day, the leaching requirement can be estimated as follows:

$$D_{iw} = \left[\frac{8}{(8 - 1)}\right] 1 = 1.14 \text{ cm/day and } D_{dw} = \upsilon = \left(\frac{1}{8}\right) 1.14 = 0.14 \text{ cm/day.}$$

9.4.6 Krikham's Formula [44]

Kirkham has given the following relationship for calculating the drain spacing by using mathematical procedures

$$H_d = \left(\frac{2SR}{k}\right) F\left(\frac{2r}{2S}, \frac{h}{2S}\right) \tag{9.12}$$

where

H_d = maximum height of the water table above the drains,
R = rate of rainfall = υ
k = hydraulic conductivity
h = distance from impermeable layer to water table immediately over the drains,
$2S$ = spacing of drains
r = radius of drains

and

$$F = \frac{1}{\pi} \left\{ \ln \frac{2S}{\pi r} + \sum_{m=1}^{\infty} \left[\frac{1}{m} (\cos \frac{m\pi r}{s} - \cos m\pi)(\cot h \frac{m\pi K}{S} - 1) \right] \right\} \tag{9.13}$$

For the solution of the Eqs (9.12 and 9.13), graphs have been prepared and are shown in Fig. 9.8.

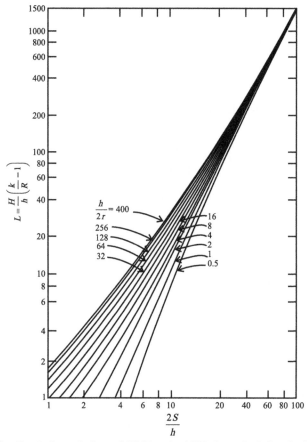

Figure 9.8 Graph for solution of Kirkham's 1958 formula (after Sadik Toksoz).

9.4.7 Bureau of Reclamation (USBR) Formula

The Kirkham's formulae have been developed considering the water table in equilibrium with the rainfall or irrigation water. In actual situation it is in transient state as the hydraulic head varies with time. The equation governing the transient case of the moving water table can be expressed as

$$\frac{KD}{s}\left(\frac{dy}{dt}\right) = \frac{d^2 y}{dx^2} \tag{9.14}$$

where

k = hydraulic conductivity,

D = average depth of ground water stream,

s = specific yield,

y = elevation of water table above a datum

A number of solutions of the equation for the drainage of land by parallel sub-surface drains were obtained by scientists of the USBR, with the initial water table flat and also with other shapes. The solution which is recommended is the one having initial water table shape corresponding to a fourth order polynomial, that is at $t = 0$, the water table shape is as given by

$$y = \frac{8H}{L^4}(L^3x - 3L^2x^2 + 4Lx^3 - 2x^4) \tag{9.15}$$

and at the two drains the water table is taken to be at the same elevation as the drains or $y = 0$, $t = 0$, $x = 0$ and $y = 0$, $t = 0$, $x = L$ where L is spacing between the drains. The solution of interest is in the height of the water table at the mid-point between the drains, i.e., for y at $x = x = L/2$, and which can be expressed as

$$H = \frac{192}{\pi^3} \sum_{n=1,3,5}^{\infty} (-1)^{(n-1)/2} \frac{n^2 - 8/\pi^2}{n^5} \exp\left(-\frac{\pi^2 n^2 \alpha t}{L^2}\right) \tag{9.16}$$

Taking only the first term of the series an approximate solution can be obtained. A rough approximation of D is given as follows, curves (Fig. 9.9) show the relationship between

$$\frac{y}{y_0} \quad \text{and} \quad \frac{KDt}{sL_0^2} \tag{9.17}$$

A rough approximation of D is given as follows

$$D = d + \frac{y}{2},$$

where d is equivalent depth from the drain to the barrier as obtained from Hooghoudt's graphs.

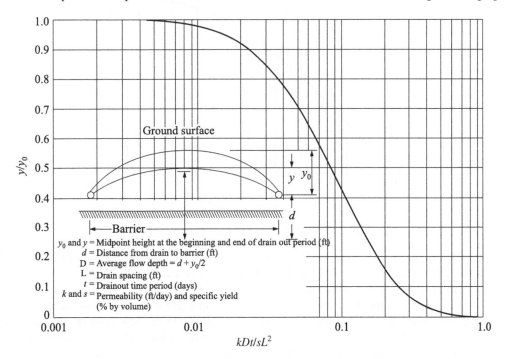

Figure 9.9 Curve showing relationship between y/y_0 and kDt/sL^2 at midpoint between drains (where drain is not on barrier).

Ground water hydrographs analysis will indicate the variation of water table during the irrigation season, being the lowest at the beginning of irrigation and highest at the end. If the fall of the water table during lean season does not equal the net rise during the irrigation season, the water table will rise progressively over the years, and for prevention of this rise, the soil has to be drained. So the main emphasis on the analysis is to arrive at a drain spacing which will result in a dynamic equilibrium with a specific water table height under specific soil, irrigation, crop and climate characteristics of the area. It is, therefore, necessary to know the consumptive requirements, the cropping practices and the irrigation schedules and requirements. In this connection the concept of specific yield is very important, where all the soil above the water table is assumed to be drained to the same moisture content. The specific yield is a measure of the pore size distribution of soil and so a correlation between the pore size distribution and soil hydraulic conductivity is expected. The relationship between permeability and specific yield is shown in Fig. 9.10. Since specific yield pertains only to the soil layer which is being drained, so permeability has to be measured in this layer.

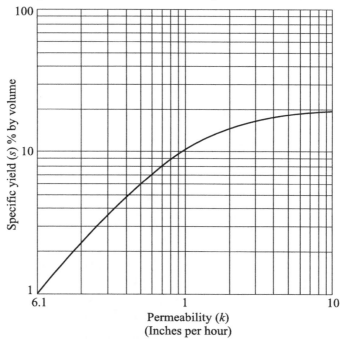

Figure 9.10 Specific yield vs. permeability.

Calculations are made after the last irrigation, then the average flow depth D for the maximum height that the water table will reach needs to be calculated. Assume a value for L, then the values of k, t and s which have been measured or calculated are used to follow the water table fluctuations for the assumed spacing. By using graphs the drop in the water table during the time interval t can be calculated. After each recharge the new position of water table is calculated and then after recharge, the drop is calculated. At the end of the season the water table should be in the same position or lower than at the start of the calculations. If the water table is higher then the drain spacing be altered and calculations repeated till proper spacing is reached.

The drain discharge formula for a drain which is located above a barrier can be expressed as

$$q = 2\pi k y D/L \qquad (9.18)$$

where

q = drain discharge in cubic feet per linear foot of drain per day,

k = the permeability in cubic feet per square feet per day and y, D and L as defined earlier and the discharge for parallel drains on a barrier is

$$q = \frac{4k\,y_0^2}{L} \qquad (9.19)$$

9.4.8   Discharge Capacity of Drains

The carrying caracity of drains have been investigated by Yarnell and Woodword [29]. They have provided an excellent chart for use in the design of a subsurface drainage system. The discharge in the lines is based on the Mannings formula having a roughness coefficient of 0.0108. In order to use the chart, it is necessary to know the drainage coefficient. The drainage coefficient is the quantity of water that must be removed within a period of 24 hours. In humid area, the drainage coefficient varies between 3 mm and 25 mm per day. It depends on the rainfall rate and the amount of surface drainage water that is admitted to the drainage system. The following tentative design criteria for determining the amount of water to be drained from an irrigated area can be adopted.

Area in acres	*Required drain capacity*
0–40	0.4 cfs
41–80	0.7 cfs
81–900	0.2 cfs for each additional 40 acres over 80 acres
1000–3000	0.1 cfs for each additional 40 acres over 1000 acres

9.4.9   Slopes of Drain Lines

Subsurface drains can be laid on a variety of slopes. However, the slope that will give a self-cleaning velocity is adopted. A velocity of 20–25 cm/s is considered to be adequate for removal of silt in European country. In USA, however, a velocity of the range of 30–45 cm/s is required. The velocities assume the drain line is running full. They may be taken from the Yarnell and Woodward chart.

9.4.10   Gravel Filter

Criteria for the design of a gravel filter which will prevent the movement of fine sand and silt into the drain have been developed. The design is based on the ratio between certain size ranges in the soil and certain size ranges in the filter material. The following criteria for the design of gravel filter is proposed by USBR.

Uniform material—D_{50} filter/D_{50} base = 5–10. Graded material—D_{50} filter/D_{50} base = 12–58. Apart from quick drainage the placement of gravel materials beneath the drain improves the bedding of the tile.

9.5 DESIGN OF OPEN UNDERDRAIN

Sometimes subsurface drains are excavated as an open drain. This will induce seepage flow into the drain and help ultimately to lower the water table. The depth of the seepage drains depends on soil stratification but usually it is more than 1.5 m. Sometime the top soil is clayey which is underlaid by a pervious strata under artesian pressure. In such a case the open drain should be cut below the clayey layer so as to provide effective drainage.

Apart from above excavation below water table have to be made for engineering construction. In such cases it is necessary to arrange for drainage of the excavation. The usual methods adopted are—(i) drainage from open sump, (ii) multiple stage pumping, (iii) pumping from deep wells, or (iv) relief well for draining from confined aquifer.

Under method (i), the seepage water which emerges mostly from the toe of the excavation is guided through drains into one or more sumps from which it is pumped out and it is normally restricted to soils with high permeability coefficient $k > 0.1$ cm/s. In method (ii), pipes approximately 6 cm diameter with 1 m strainer at the bottom are jetted into the ground roughly at a spacing of 1.5 m. The upper ends of the wells are connected to a horizontal pipe which is connected to self-priming pumps at suitable intervals. For excavation greater than 6 m below water level a multiple stage setup is necessary. Here additional rows of wells are provided at successive levels each about 4.5 m lower than the first, the last being at the toe of the excavation (Fig. 9.11). When the excavation is greater than 20 m, the water is pumped out by a submersible pump which can be lowered into the well casing. The seepage water in between the deep wells may be collected by a single row of well points at the toe. This method is effective for soils of medium permeability k lying between 10^{-3} cm/s and 10^{-1} cm/s. To

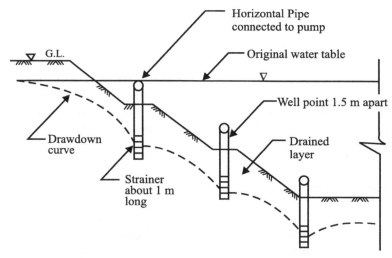

Figure 9.11 Drainage by multiple well points.

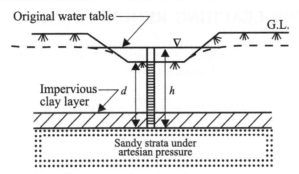

Figure 9.12 Drainage through open cut drain.

understand the application of method (iv) consider a situation as depicted in Fig. 9.12 where a thin impervious layer exists at some depth below the bottom of the excavation. The water level in the excavated area is lowered to the bottom of the pit. However, below the impervious layer there is no change in the piezometric head. As such, if the downward pressure due to unit weight of saturated soil of depth d is not balanced by the upward pressure of the piezometric head 'h', it is likely that the soil will blow out. To prevent this, a series of pressure relief wells are needed.

Problem 9.1

Calculate the spacing required for the water table to drop from the soil surface to a depth of 30 cm in a two-day period over an area of 200 ha from the following data.

　　The hydraulic conductivity, k = 4 cm/hr. Tile drains are to be placed 105 cm below the soil surface. The impermeable layer is 195 cm below the soil surface.

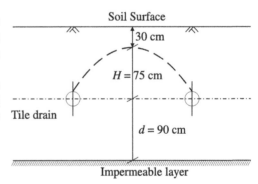

Using the formula:

$$S^2 = \frac{4kH(2d + H)}{q_0}$$

where

　　q_0 = 15 cm/day,

　　k = 4 × 24 = 96 cm/day,

　　H = 75 cm, d = 90 cm, and substituting these values in above formula, we have

$$S^2 = \frac{4 \times 96 \times 75\,(180 + 75)}{15}$$

$$= 489600 \text{ cm}^2$$

$$S = \sqrt{489600} \simeq 7\text{m}.$$

or

9.6 DESIGN FOR LEACHING REQUIREMENT

The leaching water requirement to leach out undesirable salts, etc. can be estimated from the following relationship proposed by Physusnin [34]:

$$M = P - m + \left(\frac{H - h}{\gamma}\right) 10000 \qquad (9.20)$$

where

M = leaching rate (m³/ha)

P = field limit water capacity of the given layer of soil (m³/ha)

m = existing water content in the given layer of soil prior to leaching (m³/ha)

H = ground water level prior to leaching

h = permissible level of groundwater immediately after leaching

$\gamma = 100/(A - K)$ where A is overall space of soil above ground water table in volumetric percentage, and K is water content in the soil layer in volumetric percentage.

10

Design of Surface Drainage System

10.1 NECESSITY OF SURFACE DRAINAGE

Surface 'drainage' problems occur in nearly flat areas, uneven land surfaces with depression or ridges preventing natural runoff and in areas without outlet. Soils with low infiltration rates are susceptible to surface drainage problems. It is also important in both rainfed and vegetated areas. Surface drainage is intended for safe removal of excess water from the land surface through land shaping and channel construction. It uses the head difference that exists due to land elevation to provide necessary hydraulic gradient for the movement of water.

Functionally the system may be considered as—(i) Collection system, (ii) Conveying system, or (iii) Outlets. Water from the individual field is collected and is then removed through the conveying system to the outlet. Four types of surface drainage system are used in flat areas having a slope less than two per cent for drainage of agricultural lands. They are—(i) random drain system, (ii) parallel field drain system, (iii) parallel open ditch system, and (iv) bedding system. In sloping areas with a slope greater than two per cent the system consists of—(i) bench terracing, (ii) graded bunding, and (iii) cross-slope ditch system or intercepting system. Generally open ditches are commonly adopted. Further surface drainage is required for:

(a) the removal of storm rainfall where the subsurface drainage is not economically feasible,

(b) the collection and disposal of surface irrigation runoff, and

(c) the collection and disposal of drainage in deltaic areas.

The first situation is commonly found in heavy soil situation in the tropical area. Generally, wet season rains of high intensity fall on soils with low infiltration rates and surface drainage is required to remove the excess. Therefore, estimation of excess storm runoff is necessary in order to design a field surface drain which will shed it without erosion and to design disposal by the channel having adequate capacity. The third situation arises because the land elevation of the lower estuarine region is nearly the same or sometimes lower than the tidal high water level of the adjacent estuary or tidal creeks.

The river water levels frequently rise above the levels of the land surface which slope away from the river margins and there is the liability to frequent and excessive submergence with corresponding risk of serious damage to occupied areas and cultivation by the intrusion of saline water. Great difficulties are experienced in disposing of and drainage of this water and also of the rain falling in the area. In earlier days, these difficulties were not accentuated because only the elevated grounds were occupied and the great extent of unoccupied low-lying lands in depressions gave storage capacity into which drainage was disposed off without inconvenience. Since the land in this area is very fertile, cultivation extended and the lands become more valuable, blocks lying at lower and lower levels were occupied, drainage difficulty appeared and became more and more accentuated. Attempts have always been made to save the areas from inundation by constructing embankments on the tidal creeks and estuaries and by excavating a network of drainage channels. The outflow of these drainage channels is obstructed by the continuous variation of tidal levels at the outfall. Therefore, lands up to a certain elevation can only be drained by gravity flow with suitable sluice at the outfall. Below this level, the gravity drainage through such channels becomes more and more difficult and sometimes, it becomes necessary to install large pumps near the outfall. At this lower level, it is not economically feasible to secure adequate drainage, and cultivation of the land must be attended with risk of great losses of crop from submergence after any abnormally heavy rainfall.

In the upper region of the estuary, the drainage problem is not so acute as the ground elevation is in general higher than the tidal high water. Drainage channels in this region may not be sluiced at the outfall and the tide water is allowed to play freely in such a channel.

Thus the selection of the type of drainage channel in deltaic areas is mainly governed by the following factors:

(i) Relative elevation of land surface to be drained with respect to tidal levels at the outfall.

(ii) Volume of water to be drained.

10.1.1 Surface Drainage System for Agricultural Land

For agricultural land where the land slope is less than two per cent the following types are adopted.

(i) Random drain system

It is used where small scattered depressions to be drained occur over the areas. Here, drains are designed to connect one depression to another and water is conveyed to an outlet. The drains connecting the depressions could be surface drains or underground pipe lines.

(ii) Parallel field drain system

It is the most effective method of surface drainage and is well suited both for irrigated and rain-fed areas. Here individual fields are properly graded such that they discharge into the field laterals bordering the fields and the laterals in turn lead into the mains. Laterals and mains should be deeper than the field ditches to provide free outfall.

(iii) Parallel open ditch system

This system is applicable in soils that require both surface and subsurface drainage. It is similar to the parallel drain system except that in this system the drains are replaced by open ditches which are comparatively deeper and have steeper side slopes than the field drains. The open ditches cannot be crossed by farm machinery. The spacing of the ditches depends on the soil and water table conditions and may vary from 60 to 200 m. This system is also known as diversion ditch system.

(iv) Bedding system

The bedding system is essentially a land-forming process. The land is ploughed into beds, separated dead furrows which run in the direction of prevailing slope. Ploughing is to be done parallel to the furrows. All other furrowing operations can be done either across the beds or parallel to the furrow. Bedding is suitable for poorly drained soils and on flat lands and lands with slope of 1.5 per cent. The bed width depends on the land use slope of the field, soil permeability and farming operations. Length of the bed depends on field conditions and may vary from 100 to 300 m. The maximum bed height is 20–40 cm. Water from the dead furrows discharges into a field drain constructed at the lower end of the field and normal to, dead furrows. The field drains discharge into field laterals and ultimately into the main drains. The bedding system has some disadvantages such as—(i) it does not operate satisfactorily if the crops are grown parallel to the furrows and on ridges; (ii) the furrow require regular maintenance; (iii) due to movement of topsoil some reduction in yields nearer to the furrows could occur; and (iv) the slope of the furrows may not be enough for drainage.

For sloping areas greater than two per cent cross slope ditch system are sometimes used for drainage of sloping lands. It consists of a series of shallow open drains constructed across the slope with a mild grade. This system is also known as interception system. Open ditches are commonly adopted for surface drainage because of their convenience of construction. The rate at which the open drains have to remove water depends on (i) rainfall, (ii) size of drainage area, and (iii) drainage characteristic and nature of crops including degree of protection needed from flooding. The design discharge of the open ditches for drainage may be estimated from Cypress creek formula which can be expressed as

$$Q = CA^p,$$

where

Q = design discharge in cfs,

C = coefficient depending on degree of protection,

p = a coefficient (general value 5 to 6),

A = area in square miles.

The formula is empirical and is derived from a number of observations in USA.

10.2   SURFACE DRAINAGE CHANNELS DESIGN CONSIDERATIONS

For the hydraulic design of drainage channels, it is necessary to have an estimate of total storm runoff and peak flow rates. Since agricultural catchments are generally not gauged it is common practice to estimate runoff from agricultural lands directly from rainfall.

Generally, when rain falls on a dry catchment a part is intercepted by the vegetable cover, part infiltrates into the soil and part fills the depressions in the ground before overland flow begins. For a single period of rain and a given type of cover interception can be taken as a fixed depth of water but infiltration proceeds at a rate which falls with time in a manner dependent on the soil and its internal drainage. The proportion of rainfall which becomes surface runoff, therefore, varies with the intensity and duration of precipitation and also with the initial moisture content of the soil.

Frequency, intensity and duration of storms are related by statistical analysis. Frequency is the average number of times an event occurs in a given number of years. Recurrence interval is the average number of years between occurrences of that event. Design storm for drainage works is that for which the recurrence interval is acceptable. That means, if an interval of 10 years is chosen it can be expected that on averaging over many years the following will be observed:

(i)   The drainage system will run at a capacity or overflow once every 10 years.

(ii) Simple frequency/intensity curves can be plotted, provided there are rainfall data for a period of at least 20 years.

(iii) Critical duration of rainfall is taken as the time of concentration of the catchment.

Time of concentration of a catchment is the time taken after the onset of rainfall for all parts of the catchment to contribute to streamflow at the point of measurement and is equal to the time of flow from the most remote point of the catchment to the gauging point. It is found that average storm intensity declines with duration and it is, therefore, assumed that the storm giving the greatest flow from a catchment is that with duration equal to the time of concentration t_c.

10.2.1   Characteristics of a Storm

The intensity of the rainfall is the rate at which it falls at any one time. It is usually expressed in cm/hr. The rate at which rains fall changes continuously during a storm. It may rain 1 cm during one hour, giving an average rainfall rate of 1 cm/hr. However, during that hour there will be many times when the rainfall rate greatly exceeds 1 cm/ hour, on the other hand, it might be considerably lower at other moments. Figure 10.1 shows an idealised storm. Initially, the intensity increases more or less linearly to a peak. Then, it decreases at

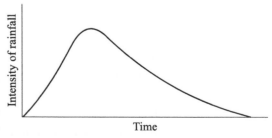

Figure 10.1   Intensity duration curve for a design storm.

a slower rate to the end of the storm. Figure 10.2 shows the total cumulative amount of water plotted against time. The curve rises steeply at the start of the storm and falls off towards the end of storm.

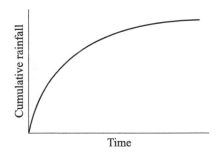

Figure 10.2 Cumulative rainfall vs. time.

10.2.2 The Design of Storm

In surface drainage, the structures and facilities are designed for a storm of a certain intensity and duration. The capacity of the drains must be adequate to handle the anticipated runoff for a given storm. The problem is to decide on the storm to be used in the design. The first step is to tabulate all available data on rainfall. From these data, the most intense storms are selected. Data for each severe storm is then tabulated to indicate the most severe rain that fell during a particular time interval. In other words, the 5-minute duration in which the most rain fell during the storm. Similarly, 10-minute duration is the maximum amount of rain that fell during any 10-minute period of the storm. Next is to tabulate the data for the severe storm according to their severity. As for example, see Table 10.1.

It is now necessary to consider one of these composite storms for the purpose of design. The storm that is selected is called the design storm. The design storm selected is based on the premise that one such storm might be expected to occur in a given number of years. If the design storm is based on the assumption that such a storm or one more severe will occur in a set period of years

$$\frac{1}{T} = \frac{M}{N}$$

where

N = number of years of the rainfall records,

$\frac{1}{T} = f$ = design storm frequency,

M = ranking of the severe storms.

As an example suppose that rainfall records have been kept for a period of 40 years. It is desired to pick the storm that will be expected to appear once in 10 years, so that $T = 10$ years. Since $1/T = (M/N)$, $M = 40/10 = 4.0$ or ranking of the storm is 4.0. Hence, select the fourth-most severe composite storm for the design. There will be more severe storms but this would be expected to occur more rarely. The data from the composite storm which is to be used for design purpose are then plotted. The intensity duration is plotted (Fig. 10.3) as a function of time.

Table 10.1 Tabulated Data for the Severe Storm according to their Severity.

	5 min.			10 min.			15 min.			30 min.			60 min.			90 min.			120 min.		
											Duration											
Year	pptn	intensity		Year	pptn	intensity	Year	pptn	intensity	Year	pptn	intensity	Year	pptn	intensity	Year	pptn	intensity	Year	pptn	intensity
	cm	i = cm/hr			cm	i = cm/hr		cm	i = cm/hr		cm	i = cm/hr		cm	i = cm/hr		cm	i = cm/hr		cm	i = cm/hr
1918	0.85	10.2		1918	1.2	7.2	1918	1.74	6.96	1925	1.74	3.48	1918	2.15	2.15	1918	2.46	1.64	1918	2.97	1.48
1914	0.76	9.12		1914	1.04	6.24	1925	1.55	6.2	1918	1.55	3.10	1925	1.92	19.2	1925	2.38	1.59	1931	2.63	1.32
1925	0.73	8.76		1925	0.93	5.58	1914	1.36	5.44	1914	1.36	2.72	1931	1.70	1.70	1931	2.14	1.43	1925	2.34	1.17
1931	0.72	8.64		1931	0.88	5.28	1931	1.22	4.88	1931	1.22	2.44	1914	1.45	1.45	1914	1.81	1.21	1944*	2.12	1.06
1936	0.66	7.92		1936	0.84	5.04	1936	1.18	4.72	1936	1.18	2.36	1936	1.40	1.40	1936	1.65	1.1	1939*	1.83	0.92
1941	0.62	7.44		1924	0.80	4.80	1924	1.15	4.60	1927	1.10	2.20	1944	1.33	1.33	1944	1.50	1.0	1936	1.64	0.82
1944	0.51	6.12		1941	0.78	4.68	1941	1.05	4.20	1924	1.05	2.10	1939	1.25	1.25	1939	1.40	0.93	1941	1.55	0.78
1939	0.45	5.40		1944	0.68	4.08	1927	1.01	4.04	1941	1.01	2.02	1941	1.20	1.20	1941	1.36	0.91	1914	1.51	0.76
1921	0.36	4.32		1939	0.52	3.12	1944	0.95	3.80	1944	0.95	1.90	1921	1.14	1.14	1921	1.34	0.89	1927	1.41	0.73
1927	0.28	3.36		1921	0.51	3.06	1939	0.83	3.32	1939	0.83	1.66	1927	1.11	1.11	1927	1.27	0.85	1924	1.41	0.71
1924	0.21	2.52		1927	0.39	2.34	1922	0.71	2.84	1921	0.79	1.58	1924	1.09	1.09	1924	1.23	0.82	1921	1.34	0.67

Note: 5-minute, 10-minute or 15-minute rainfall is not taken from only one storm but represents the maximum for all storms, if we take each of the maxima the result is called a composite storm since it consists of data several storms.

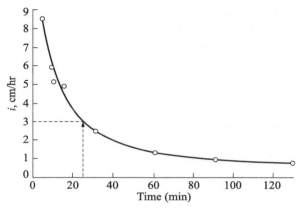

Figure 10.3 Rainfall intensity of an ideal storm.

10.2.3 Storm Runoff

Runoff is the portion of precipitation that makes its way towards streams, channels, lakes or oceans as surface or subsurface flow. Common usage refers to the surface flow alone. The design of drainage channels, bridges, culverts and other engineering structures depends on a knowledge of the amount of runoff that will occur in a given area. It is desirable to know that peak rate of runoff, the total volume of runoff and also the distribution of runoff rate throughout the year.

For small watershed, the rational method is widely used for estimates of peak runoff rate:

$$Q = 0.277 \ CiA \qquad (10.1)$$

where

  i = rainfall intensity in mm/hr for the design recurrence interval and for the duration equal to the time of concentration for the watershed,

 A = area in km^2,

 C = runoff coefficient.

The time of concentration t_c (Fig. 10.4) being the time required for water to travel from O to A.

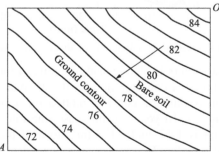

Figure 10.4 Concept of time of concentration.

There are a number of different individual times of flow that influence the time of concentration. For example, the time required for water to enter a ditch or drainage channel, and the time required for the water in the ditch to reach the pond being considered. Another factor is the time of overland flow, i.e., *TOF*.

The time of overland flow varies with the slope, the type of ground, the length of the flow path, and a number of complicating factors. In Fig. 10.5, some values are given for times of flow for various condition of coverage, slope, and length. If the drainage area consists of several different types of surfaces, the time of overland flow must be determined by adding together the respective times. Computation for flow over lengths of different surfaces along the path from the most remote point to the inlet for estimation of *TOF* is done first. For the water shed considered if the distance from *O* and *A* is 150 m and the slope of ground is 0.5 per cent then the time of overland flow will be 25 minutes.

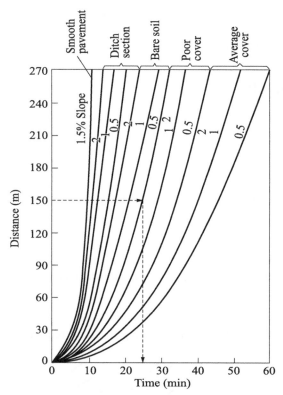

Figure 10.5   Time of concentration as a function of distance for various ground conditions [29].

In the rational formula *i* refers to the intensity of rainfall index. It is determined by using the design storm curve and time of concentration for the drainage area given. It may be recalled that in Fig. 10.3, the intensity duration curve is plotted for a composite design storm having a recurrence interval of once in 10 years. In this case, the intensity of rainfall is 1.45 cm/hr, that means the design storm is 1.45 cm/hr. For this storm if the concentration is 25 minutes, then the intensity *i* will be 3 cm/hr. The discharge is dependent on the estimation of *C* known as runoff coefficient. The value of *C* depends on many factors such as soil type, vegetation cover, etc. Since $(1-C)$ represents the loss that takes place and as it is very difficult to assess the loss accurately, there is some arbitrariness involved in the estimation of *C*. The value of *C* has been estimated by a Committee of ASCE and these values are used for the estimation of runoff and they are applicable for storms of 5–10 years frequency. The recommended values are given in Table 10.2.

Table 10.2 Values of Runoff Coefficient *C*. (ASCE)

Sl. no.	Type of drainage area	Runoff coefficient
	Lawns	
1.	Sandy soil flat 2 per cent	0.05–0.10
	Sandy soil average 2–7 per cent	0.10–0.15
	Sandy soil steep 7 per cent	0.15–0.20
	Heavy soil flat 2 per cent	0.13–0.17
	Heavy soil average 2–7 per cent	0.18–0.22
	Heavy soil steep 7 per cent	0.25–0.35
2.	Business	
	Down town area	0.70–0.95
	Neighbourhood area	0.50–0.70
3.	Residential	
	Single family areas	0.30–0.50
	Multi-unit detached	0.40–0.60
	Multi-unit attached	0.60–0.75
	Suburban	0.25–0.40
	Apartment dwelling houses	0.50–0.70
4.	Industrial	
	Light areas	0.50–0.60
	Heavy areas	0.60–0.90
	Park, cemeteries	0.10–0.25
	Playground	0.20–0.35
	Railroad yards	0.20–0.40
	The improved area	0.10–0.30
5.	Streets	
	Asphaltic	0.70–0.80
	Concrete	0.80–0.95
	Brick	0.75–0.85
	Drives and Mall	0.75–0.85
	Roofs	0.75–0.95

After having obtained maximum flow for a given storm over the given area, the main drainage should be designed to carry this flow to the outfall. The level outside the outfall is that of the river and hence for design of the internal drainage system, the hydrograph of the river at the outfall is also required. It should show the worst river conditions likely to occur with a certain frequency. For gravity flow drainage the 10-year probability is the most suitable. The probability of worst rainfall in 10 years coinciding with the worst river condition in the same period offers ample protection.

10.2.4   Design of Drainage Channel

Two most common equations of flow are the Chezy's and Manning's equations:

Chezy's equation

$$V = C\sqrt{RS},$$ (10.2)

where

$V =$ mean velocity of flow at a section,
$R =$ hydraulic mean radius,
$S =$ slope,
$C =$ Chezy's coefficient,

Manning's equation

$$V = \frac{R^{2/3}\,S^{1/2}}{n}$$ (10.3)

where

$n =$ Manning's coefficient.

The equations show that a particular amount of discharge (Q) can be passed through a channel with different combination of width (B), hydraulic mean radius (R) and the slope (S). But experience have shown that only a few of them provide a stability of cross-section. It is a matter of great importance to design and construct earthen channels with such dimensions that they will neither be obstructed by the deposition of silt or injured by the erosion of banks and bed. Design of channel should, therefore, be guided by the laws which specify such a stability. These laws are largely empirical and are obtained by a study of the condition of flow of a number of irrigation channels of varying capacities, velocities and relations between bed width and depth which led some to a permanent regime, i.e., they were neither silting nor scouring. This does not imply absence of periodic seasonal deposits of scour but that on the whole year's flow and counting from year to year, there is no change in the silt deposits in the channel beds.

A number of equations known as Lacey's regime equations, provide the facility to obtain the dimension of such a channel if the discharge (Q) and the type of soil over which the channel is supposed to run is known. In the irrigation system of canals where the discharge is controlled and sufficient latitude is available in selecting the bed slope, it is possible to obtain a cross-section geometry by satisfying the regime laws. But in the case of drainage channels such advantages are in most cases, not available and, therefore, design of regime drainage channels pose great difficulty. This is the reason why the drainage channels are not stable and require more frequent maintenance than an irrigation canal. However, the ground condition permitting, it is advisable to construct a channel geometry as near to the regime geometry as possible by satisfying relationships developed by Lacey [30].

Lacey's regime equations

$$S = \frac{f^{5/3}}{3340\,Q^{1/6}}$$ (10.4)

$$V = \left(\frac{Q\,f^2}{140}\right)^{1/6}\ \text{m/s}$$ (10.5)

$$R = \frac{2.50\,V^2}{f}$$ (10.6)

$$P = 4.75\,\sqrt{Q}$$ (10.7)

where

$P =$ wetted perimeter in m,
$Q =$ discharge, in cumec,

V = velocity at which the channel, regarded from point of maintenance, neither silted nor scoured in m/s,

R = hydraulic mean radius in m,

S = slope,

A = cross-sectional area in m^2 and

f = silt factor equal to 1 for standard silt, less than 1 for finer than standard silt and more than 1 for coarser than standard silt. For all practical purpose it ranges between 0.6 and 1.2, alternately $f = 1.76 \sqrt{d_{50}}$, where d_{50} = average particle size in mm.

Table 10.3 gives trapezoidal sections, with 1 : 2 side slopes, and slopes for a range of 'f' that should cover almost all practical cases. Sections cease to be accurate in the region of 0.28 m^3/s as they deviate from trapezoidal and become elliptic.

Table 10.3 Cross-sectional Dimensions as per Regime Equations.

Discharges (m^3/s)	Bed width (m)			Depth (m)			Slope in 10^3		
	$f = 0.6$	$f = 0.9$	$f = 1.2$	$f = 0.6$	$f = 0.9$	$f = 1.2$	$f = 0.6$	$f = 0.9$	$f = 1.2$
283.17	71.6	73.20	73.8	3.97	3.51	3.05	0.051	0.100	0.16
141.5	49.7	51.2	51.85	3.20	2.75	2.44	0.058	0.113	0.18
70.8	35.1	35.99	36.6	2.52	2.2	1.68	0.065	0.136	0.20
28.3	21.35	21.96	22.26	2.98	1.68	1.53	0.075	0.150	0.24
14.2	14.34	14.9	15.55	1.59	1.34	1.22	0.085	0.166	0.27
7.1	9.76	10.37	10.68	1.22	1.04	0.95	0.095	0.186	0.30
2.8	5.8	6.1	6.4	0.98	0.82	0.73	0.11	0.22	0.35
0.85	2.7	3.0	3.2	0.73	0.58	0.52	0.135	0.27	0.43
0.28	1.1	1.5	1.68	0.64	0.46	0.396	0.16	0.32	0.52

10.3 GENERAL DESIGN CONSIDERATION OF OUTFALL CULVERT [31]

To allow the flow from each watershed across an embankment, outfall in the form of culverts are built at the lowest point of the valleys. In case, the river level rises above the adjacent land level, which land has been protected by ernbankments from flooding, there must be some sort of control at the outfall of the area to prevent the river flowing up the drain and at the same time allow local discharge from the area when conditions are favourable. The hydraulic design of most of the culverts consists in selecting a structure which will pass the design flow without an excessive headwater elevation. The other consideration is the prevention of scour at the culvert outlet. The maximum headwater elevation must provide a reasonable freeboard against flooding of the embankment, also it must be low enough so that there is no damage to property upstream from the highway. In culvert design the limiting headwater elevation will be well above the top of the culvert entrance and flow in the culvert will take place as shown in Fig. 10.6. Under conditions (i) and (ii) of the figure the culvert is said to be operating with outlet control and the headwater elevation required to discharge the design flow is determined by the head loss h_l an the culvert. When the inlet and outlet of the culvert are submerged, flow will be as in condition depicted under (i) regardless of the culvert slope. When the bottom slope is such that the normal depth d_n for the design flow is greater than the culvert height, D, and the inlet is submerged the culvert will flow full although the outlet is not submerged.

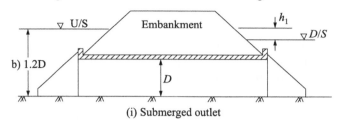

(i) Submerged outlet

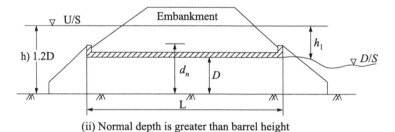

(ii) Normal depth is greater than barrel height

Figure 10.6 Flow in culverts with outlet control.

The head loss h_l consists of the following

$$h_l = h_e + h_f + h_v = K_e \frac{V^2}{2g} + \frac{n^2 V^2 l}{R^{4/3}} + \frac{V^2}{2g}$$

where

h_e = head loss at inlet,

h_f = head loss due to friction.

h_v = velocity head.

or

$$h_l = \left[K_e + 1 + \frac{19.6 n^2 l}{R^{4/3}} \right] \frac{V^2}{2g} \tag{10.8}$$

Since $Q = AV$, where V can be obtained from above, A = culvert area.

The entrance coefficient K_e is about 0.5 for a square edged entrance and 0.05 if the entrance is well rounded.

When the normal depth in the culvert is less than the barrel height with the inlet submerged, the flow conditions from the outlet will be as shown in Fig. 10.7 and in this case the culvert is said to be flowing under inlet control, i.e., the inlet will not admit water fast enough to fill the barrel. The discharge is determined by the entrance condition and is given by

$$Q = C_d A \sqrt{2gh},$$

where

h = head on the centre of the orifice,

C_d = coefficient of discharge

Or

$$h = \frac{1}{C_d^2} \left(\frac{Q^2}{2g \, A^2} \right)$$

(10.9)

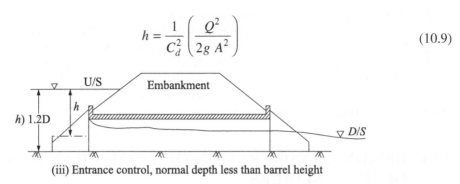

(iii) Entrance control, normal depth less than barrel height

Figure 10.7 Flow in culverts with inlet control.

The value of C_d varies with the entrance condition and hence it is not possible to cite the appropriate value in advance. Using for sharp-edged entrance without suppression of the contraction $C_d = 0.62$ while for a well-rounded entrance $C_d \simeq 1$. Under conditions of inlet control flaring the culvert entrance will increase its capacity and permit it to operate under a lower head for a given discharge. Experimental results have indicated that the best results are obtained by flaring the entrance of a box culvert to about double the barrel area in a distance equal to 4 times the offset from the barrel.

Sometimes the drop culvert may be adopted when there is very little head room at the entrance or where pounding is permissible to reduce peak flow or shorten barrel length (Fig. 10.8). The required barrel size may be estimated from Eq. (10.8) assuming $K_e = 0$ and including a bend loss of $K_b V^2/2g$, where K_b varies from maximum of 1.5 for a square bend

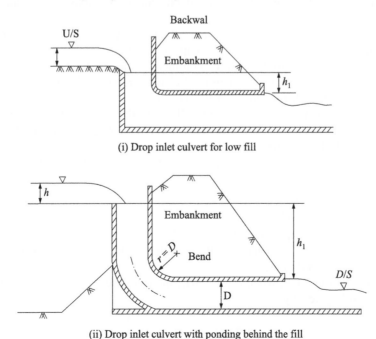

(i) Drop inlet culvert for low fill

(ii) Drop inlet culvert with ponding behind the fill

Figure 10.8 Drop inlet culverts with and without pounding.

to about 0.45 for a circular bend with radius $r = D$. The inlet of a dropinlet culvert functions as a weir as long as it is not submerged and its length must be sufficient to discharge the design flow without an excessive headwater elevation and if necessary the inlet may be flared to obtain adequate crest length. In case the head loss in the barrel is insufficient to keep the riser filled to the top, the C_d for the weir will be approximately 0.47 while the riser filled, negative pressure exists under the nappe as long as vortex action is prevented by a backwall and the weir coefficient will be about 0.73.

10.4 DESIGN CONSIDERATION OF TIDAL CHANNELS AND OUTFALL SLUICES

When the tidal channel is not sluiced at the outfall, tides enter freely and propagate upstream. During the rainy season when the channel is fed at its landward end with runoff from the drainage area, there is an interplay of fresh and tidal water. Depending on the relative intensities either of them dominate the flow condition of the channel. During the dry months of the year the tidal flow dominates. Normally, significant runoff from the drainage area is generated only for a small number of days in a year. For the major period, the tides control the flow conditions of the entire length of the channel. Therefore, the design of such a channel should be based on tidal flow modified to accommodate the upland discharge during the rains.

In a sluiced channel, there is no oscillation of flow and the movement of water is unidirectional but unsteady. Such a channel can be designed adopting the criterion governing the unidirectional flow, but considerable caution is necessary in fixing the sill level of the outfall sluice, its capacity and, therefore, the design depth and dimensions of the drainage channel leading to the sluice.

A channel with a sluice at its outfall, is not influenced by the tidal variation and the design of such a channel is mainly governed by the capacity of the pumps.

The present discussion will be mainly confined to the hydraulic design of a drainage channel with a sluice at its outfall. An open tidal channel requires the use of tidal computation technique for its, design and, therefore, an involved process.

A drainage channel with a sluice at the outfall where the tides are to play has to bear some relationship with the dimension of the sluice. Therefore, neither the channel nor the sluice can be designed independent of each other. The sluice with rectangular vents are generally considered as broad-crested weir and the following types of condition may appear while water flows over the weir.

(i) Free flow condition,

(ii) Submerged flow condition.

When the downstream depth over the weir is less than 2/3 of the upstream depth over the weir, then downstream water level has no effect on the discharge over the weir. This condition is known as 'free flow condition'.

With reference to Fig. 10.9, when $H_d < 2/3 H_u$, the free flow condition exists. The discharge Q at this condition is given by

$$Q = 1.72\, C_w\, B \left(H_u + \frac{V_u^2}{2g} \right)^{3/2}$$

(10.10)

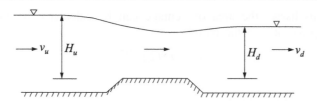

Figure 10.9 Flow over sill.

Taking the coefficient $C_w = 0.94$ and neglecting $V_u^2/2g$ being small compared to H_u

$$Q = 1.61\, BH_u^{3/2}, \qquad (10.11)$$

where

H_u = upstream depth of the sluice,
B = width of the openings,
V_u = approach velocity.

When the downstream depth over the weir is greater than 2/3 of the upstream depth, the condition is known as 'submerged condition' and the performance of the opening is no longer unaffected by downstream depth (H_d). The discharge over the weir under this condition is given by:

$$Q = 4.19\, BH_o\, (H_u - H_d)^{1/2} \qquad (10.12)$$

neglecting the contribution of approach and exit velocities, where H_0 = sluice opening.

At a condition when downstream depth is equal to the 2/3 of upstream depth, any one of the above equations be used for both, the equations give same value.

The present discussion will be mainly confined to the hydraulic design of a drainage channel with a sluice at its outfall, first following a simplified procedure and then followed by a rigorous procedure.

10.4.1   Computation of Discharges through the Sluice

10.4.1.1   Simplified approach

The discharge through the sluice at the outfall can be estimated from the following relationship once, the period of tidal blockage is determined.

$$Q = 27MI \left(\frac{24}{24 - P} \right)$$

where

Q = discharge through the sluice in cusecs
M = drainage area in square miles
I = drainage index in inches per day
P = period of tidal blockage in 24 hours.

Knowing the discharge, the area of ventage can be calculated based on the following velocity relationship for flow in the barrel

$$V = 0.8(2gh)^{0.5},$$

where

h = depth of flow in the sluice barrel.

10.4.1.2 Rigorous approach

10.4.1.2.1 Computation of discharge through the sluice

With the aid of the discharge formulae, it is now possible to find out the discharges through a sluice where both the upstream and downstream water levels vary. The variation of downstream water levels may be due to tidal fluctuation at the outfall and the variation of upstream level may be due to variation of basin level on account of rainfall or outflow through the sluice.

Let it be considered that the tide levels as shown in Fig. 10.10 are operative at the outfall and for the simplicity of case consider that this tidal fluctuating remains invariant due to outflow from the sluice. Let the basin water level remain invariant at +1.83 m GTS. In the actual case, this level will also vary because of the imbalance of inflow and outflow from the basin. Consider that the clear width of the sluice is B m and the sill is placed at −0.92 m GTS.

Referring to Fig. 10.10, it is seen that water from the basin cannot flow out till the tide level falls below +1.83 m GTS. As soon as the tide level falls below this level, the outflow commences and the flow would be of submerged type till the tide water reaches a level of +0.92 m GTS. As the tide level continues to fall further, the free flow condition will be attained and this condition will persist till the tide level again rises to +0.92 m GTS during flood tide. The cycle will continue.

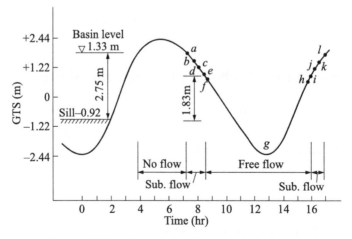

Figure 10.10 Tidal fluctuations at outlet channel.

Let us now calculate the discharges through the sluice at different instants of time a, b, c, d, e ... l as shown in Fig. 10.10. The discharges are shown in Table 10.4.

Table 10.4 Discharges through the Sluice at Different Tide Levels.

Time instants Ref. Fig. 10.11	U/S W.L. of sluice GTS (m)	D/S W.L. of sluice GTS (m)	U/S depth of sluice (H_u) (m)	D/S depth of sluice (H_d) (m)	Type of flow	Discharge (Q) (m³ s)
a	+ 1.83	+ 1.83	2.75	2.75	No flow	0
b	+ 1.83	+ 1.53	2.75	2.44	Submerged	5.69 B
c	+ 1.83	+ 1.22	2.75	2.14	Submerged	6.00 B
d	+ 1.83	+ 1.07	2.75	1.98	Submerged	6.48 B
e	+ 1.83	+ 0.92	2.75	1.83	Free	7.34 B
f	+ 1.83	+ 0.61	2.75	1.53	Free	7.34 B
g	+ 1.83	– 2.44	2.75	—	Free	7.34 B
h	+ 1.83	+ 0.76	2.75	1.68	Free	7.34 B
i	+ 1.83	+ 0.92	2.75	1.83	Free	7.34 B
j	+ 1.83	+ 1.28	2.75	2.19	Submerged	6.87 B
k	+ 1.83	+ 1.53	2.75	2.44	Submerged	5.69 B
l	+ 1.83	+ 1.83	2.75	2.75	No flow	0

Figure 10.11 shows the discharges plotted against time. The area under the curve *a*, *b*, *c*, *d* ... *k*, *l* gives the volume of water that a sluice with a clear width '*B*', sill level at –0.92 m GTS and constant basin level at +1.83 m GTS will discharge between two consecutive high water under the specified tidal variation at the downstream of the sluice.

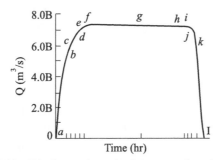

Figure 10.11 Discharge through sluice as a function of time.

10.4.2 Fixation of Sill Level, Clear Width of the Sluice and Channel Dimensions

The above computations show the quantum of water that a particular sluice is capable of discharging under specified conditions. In case of design of a new channel and sluice the sill level and the clear width are to be ascertained for discharging a specified quantity derived from the drainage area upstream of the sluice.

The problem, therefore, reduces to:

Given:

  (i) Basin level to be maintained
  (ii) Volumes of water to be drained
  (iii) Tide levels at the location of outflow.

To find out:

(i) Sill level of the sluice

(ii) Clear width of the sluice

(iii) Dimension of the drainage channel leading to the sluice.

In order to do this design, it would be necessary to obtain the quantum of outflow through the sluice in terms of clear width (B) at different sill levels. Table 10.5 shows the volume of outflow between two consecutive high water at different sill levels and basin levels.

Table 10.5 Outflow through the Sluice at Different Sill and Basin Levels.

Basin level (m) GTS	Sill level (m) GTS	H_u (m)	Max. discharge (m³/s)	Volume of outflow (m³)
1	2	3	4	5
+ 1.83	− 2.44	4.27	14.2 B	$0.43 \times 10^6\ B$
+ 1.83	− 1.83	3.66	11.27 B	$0.35 \times 10^6\ B$
+ 1.83	− 1.22	3.05	8.58 B	$0.27 \times 10^6\ B$
+ 1.83	− 0.92	2.75	7.34 B	$0.23 \times 10^6\ B$
+ 1.83	− 0.61	2.44	6.14 B	$0.198 \times 10^6\ B$
+ 1.83	− 0.00	1.83	3.99 B	$0.130 \times 10^6\ B$
+ 1.83	+ 0.61	1.22	2.17 B	$0.078 \times 10^6\ B$
+ 1.83	+ 1.22	0.61	0.77 B	$0.025 \times 10^6\ B$
+ 1.83	+ 1.83	0.00	0.00	0.00
+ 1.22	− 2.44	3.66	11.27 B	$0.288 \times 10^6\ B$
+ 1.22	− 1.83	0.05	8.58 B	$0.223 \times 10^6\ B$
+ 1.22	− 1.22	2.44	6.14 B	$0.167 \times 10^6\ B$
+ 1.22	− 0.61	1.83	3.99 B	$0.112 \times 10^6\ B$
+ 1.22	0.00	1.22	2.17 B	$0.061 \times 10^6\ B$
+ 1.22	+ 0.61	0.61	0.77 B	$0.021 \times 10^6\ B$
+ 1.22	+ 1.22	0.00	0.00	0.00
+ 0.16	− 2.44	3.05	8.58 B	$0.187 \times 10^6\ B$
+ 0.16	− 1.83	2.44	6.14 B	$0.139 \times 10^6\ B$
+ 0.16	− 1.22	1.83	3.99 B	$0.094 \times 10^6\ B$
+ 0.16	− 0.61	1.22	2.17 B	$0.051 \times 10^6\ B$
+ 0.16	0.00	0.61	0.77 B	$0.018 \times 10^6\ B$
+ 0.16	+ 0.61	0.00	0.00	0.00
+ 0.00	− 2.44	2.44	6.14 B	$0.111 \times 10^6\ B$
+ 0.00	− 1.83	1.83	3.99 B	$0.074 \times 10^6\ B$
+ 0.00	− 1.22	1.22	2.17 B	$0.042 \times 10^6\ B$
+ 0.00	− 0.61	0.61	0.77 B	$0.014 \times 10^6\ B$
+ 0.00	0.00	0.00	0.00	0.00
− 0.61	− 2.44	1.83	3.99 B	$0.056 \times 10^6\ B$
− 0.61	− 1.83	1.22	2.17 B	$0.033 \times 10^6\ B$
− 0.61	− 1.22	0.61	0.77 B	$0.011 \times 10^6\ B$
− 0.61	+ 0.61	0.00	0.00	0.00

Figure 10.12 shows the graphical representation of the informations as tabulated above.

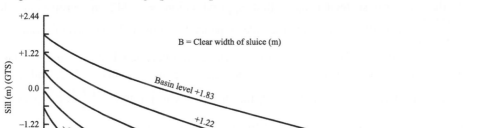

Figure 10.12   Outflow volume as a function of sill and basin level.

CASE STUDY 1

Design the surface drainage system for a flat area with ground level at 3.05 m of 404.69 ha in extent with a main drain 609.6 m long fed by subsidiary drains at regular 152.40 m intervals (Fig. 10.13). The hydrograph for the worst condition in the river based on a 10-year probability is given in Fig. 10.14. The maximum anticipated discharge form the catchment area for the worst possible anticipated storm equals 4.726 m³/s. Consider the duration of design storm to be 1 hour.

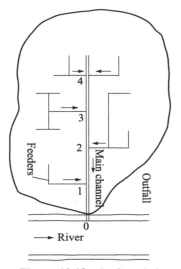

Figure 10.13   Surface drainage layout.

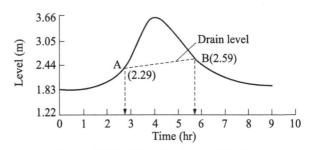

Figure 10.14   Flood hydrograph in the river.

Solution

Steps:

    (i) Assume gravity flow and a channel of uniform grade.

(ii) Normal water level at outfall = 1.834 m above datum. Assume *w/l* at location at 4, the furthermost feeder as a first approximation = 2.442 m permissible slope

$$= \frac{0.610}{609.6} = 1 \text{ in } 1000;$$ Assuming the side slope as 2:1 (dictated by typed of ground), the

discharge at 0–1, 4.726 m³/s, and for the whole area between 1–2, 2–3 and 3–4 taking proportionately, i.e., total discharge is brought by 5 inlet points 1, 2, 3, and at 4 on

either side. Hence, between 1 and 2, the discharges is $\left(4.726 - \dfrac{4.726}{5} \right) = 3.78$ m³/s.

(iii) It will be observed that the channel illustrated in Fig. 10.15 will carry the discharges to these conditions.

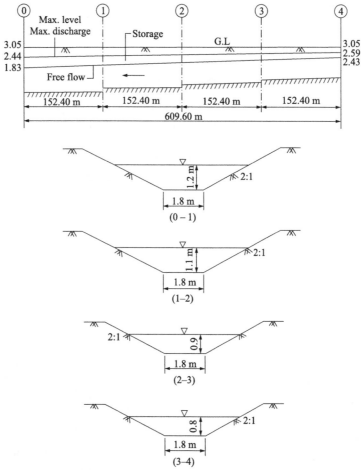

Figure 10.15 Channel section for drainage during normal flow in river.

(iv) When a storm in the area coincides with a flood in the river, the whole of the runoff from the storm will have to be stored in the area.

The storage required will be 4.726 × 3600 = 17013.6 m³. Computation of storage available in the main drain, subsidiaries and the field ditches.

(v) *Storage capacity of feeders and ditches*

The bulk of this storage is only available when the water level in the drain approaches the maximum. The amount of storage can be estimated from survey. In the present problem, it can be taken as per 40.47 ha of land, more than 0.76 m above drain level zero; per 40.47 ha of land between 0.76 and 0.608 m above drain level: 283–566 m^3; per 40.47 ha of land between 0.304 m and 0.608 m above drain level (566–1132 m^3). The whole of the area is between 0.304 m and 0.608 m of the water surface at maximum conditions, therefore, maximum storage available in feeders and ditches

$$= 1132 \times \frac{404.69}{40.47} = 11320.0 \text{ m}^3.$$

(vi) *Storage in the main drain*

Allow the water level to rise to 2.440 m above datum at the outfall and calculate the surface slope with the maximum discharge. First approximation—assume the section of flow 0 to 1 as at the outfall and calculate the necessary surface slope with this section of flow. This will determine the water level at 1 which determines the section of flow 1 to 2 and so on up the drain. It will be found that the water level at 4 is 2.50 m above datum. The storage available in the main drain is the volume between the free flow surface and this surface, which represents the maximum water level conditions. The storage is 1839.5 m^3. Hence, the total storage is 13159.5 m^3. The storage lacking is 3854.1 m^3. This storage must be provided by enlarging the main drain, near the outfall, no purpose would be served by deepening the drain. The free flow surface is controlled by the river level. At the upper end, however, the free flow level is near maximum, i.e., 2.43 m and little storage is available. Assume another section larger than before with a uniform bed grade and no steps. Determine the free flow profile and the maximum level profile and redetermine the storage available. It should be noted that as the drains section is now so much larger the surface slope will be smaller, therefore, the water level at the outfall can rise higher than before and still satisfy the condition that the water level is everywhere 0.456 m below the land. It will be found that a channel section shown in Fig. 10.16 with a slope of 1:6500 will satisfy all the requirements.

The area will not, therefore, be flooded by storms of short duration. Consider continuous rain over the area for longer periods, say 24 hours. The corresponding runoff has been observed to be 1.132 m^3/s which the drain can obviously carry to the outfall. Determine the water surface profile in the drain when the river is at its normal level, 1.834 m (Fig. 10.14). If the maximum rate of rise of the drain exceeds the rate of rise of the river, the level inside the outfall will be that of the river. When the rate of rise of the river exceeds that of the drain the gate will close and there will be no further discharge until the river level falls below that of the drain. Plot the rise in the drain with an inflow of 1.132 m^3/s against time. The level at which the slope of the tangent to this curve is equal to the slope of the tangent to the hydrograph is the level at which the gates will be closed or by calculation shown in Table 10.6, the gates close at 2.29 m. The point of closure is A in Fig. 10.14. Continue to plot the rise time curve for the drain from point A on the hydrograph. Discharge will commence at B when this curve cuts the hydrograph again. If B is below the maximum permissible level there is sufficient storage on the drain, if above extra storage must be provided. Time for the drain to rise from 2.29 m to 2.59 m is 3 hours. So 3 hours after the gates are closed the river has fallen to below 2.59 m, therefore, the discharge will have recommenced.

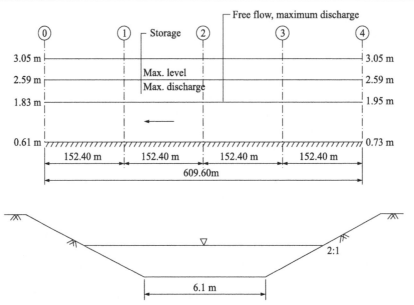

Figure 10.16 Channel section for drainage during flood in the river.

Table 10.6 Gate Closure Calculations.

Level (m)	Storage between levels (m³)	Time to fill storage at 1.132 m³/s (s)	Rate of rise (cm/hr)	Rate of rise of river from hydrograph (cm/hr)	Remarks
1.83					
	1047.10	925	58.38		
1.98					
	1103.70	975	55.38	22.88	
2.13					
	573.08	506	56.92	38.1	
2.21					
	587.23	519	55.49	48.01	
2.29					
	2016.38	1781	14.15	51.56	Gates closed.
2.36					
	3445.53	3044	9.46	63.75	
2.44					
	3459.68	3056	8.25		
2.51					
	3473.83	3069	9.38		
2.59					

Pumping

Provision for pumping the water from the area into the river is made when

    (i) It is not possible for the water to flow by gravity.

    (ii) The size of the channel is such that its cost and maintenance exceed the cost of installing, maintaining and running the pumping installation required for equal benefit.

The cost of the gravity scheme will be made up of (a) excavation of drains, (b) construction of ancillary works such as bridges, outfall, etc. (c) Maintenance. The cost of the pumping scheme is made up of (a) excavation of channel or drain, (b) ancillary works to the drains, (c) pumping station structure which may or may not incorporate the outfall, (d) pumps, and engines, (e) maintenance, running costs and attendance.

A. For the gravity flow, the pumping scheme is arrived at by a system of trial and error. The cost of pumping installation depends on:

(1) Maximum demand—The station must be capable of pumping the maximum demand and thus determining the size of the pump and engines,

(2) Number of hours of pumping per year, and

(3) Efficiency of pumping.

It is essential, therefore, to discharge as much as possible by gravity to reduce the maximum demand by making use of all the available storage, to balance the drain system to the pumping capacity. The drains must bring the water to the pumps, the pumps cannot draw it down the drain. The drains must, however, be as small as is compatible with this condition.

Example

In the example that has been given earlier consider the area sloping uniformly from 3.350–3.05 m at the upper end of the main drain. The maximum runoff is 4.726 m^3/s and the drain required to carry this discharge to the outfall under gravity flow condition is that given in Fig. 10.16. If a storm occurs when the river is high the pumps must operate. If the pumps are used to pump 4.726 m^3/s, the level at the outfall will remain at 1.834 m. If the highest river level is 3.668 m, the water horsepower required would be $\dfrac{1000 \times 1.830 \times 4.726}{75} = 115.3$. Such a capacity is unnecessary. A portion only could be pumped into the river and the remainder would then raise the level in the drain. It has been shown that if the level at the outfall rises 2.440 m, ,there is no flooding in the area. The total runoff during the storm has been calculated as 17,013.60 m^3. The storage in the drain is 1839.5 m and in the ditches, etc. 2830 m^3.

$$\text{Total quantity stored} = 4669.5 \text{ m}^3$$

$$\text{Amount to be pumped} = 12344.1 \text{ m}^3$$

$$\text{Average rate of pumping} = 3.424 \text{ m}^3/\text{s}$$

As the water in the drain rises the head to be pumped decreases. The power required, therefore, is less than that to pump 4.726 m^3/s through 1.219 m, i.e., 76.82 mhp. To determine the horsepower required, to estimate the storage in the drain between each 0.15 m rise in level, assume a power less than 76.81 and for each 0.15 m rise interval determine the discharge of the pump. The rate of storage for this interval is 4.726 m^3/s less than the pump discharge. Calculate the time to fill the storage. Add the storage time for each interval, this will give the time to reach the maximum permissible level. If the time exceeds the duration of the storm the pump is too big, if less than the duration of the storm it is too small. In this case it will be found that 63 mhp will suffice.

The pump suppliers will require to know the discharge at the lowest drain level and the discharge at the highest drain level. Allowing for losses at inlet in the pipes and at outlet they will determine the diameter and speed of the pump and the mhp of the engine.

CASE STUDY 2

Let us consider a specific case of a basin 259 km² area. The ground levels of the basin vary between −1.22 and + 4.27 m GTS. About 20 per cent of the area is below + 1.83 GTS and the distance from this low pocket to the possible location of outfall is 6.442 km. It is desired to maintain a water level of + 1.83 m GTS in the basin by constructing a suitable drainage channel and sluice.

Steps

To find out the volume of outflow consider 2.54 cm runoff/day from the entire 259 km² area. Therefore, total volume of water that accumulates at the low pocket is 6.57×10^6 m³/day. Considering the time interval between two consecutive high water to be $12^h \ 30^m$. The volume of water to be discharged by the channel and the sluice is $6.57 \times (12.5/24) \times 10^6 = 3.4 \times 10^6$ m³/tide. To start with assume that the channel water level upstream of the sluice is the same as that of the basin level to be maintained, i.e., + 1.83 m GTS.

Case   (i) take sill level = −2.44 m

volume of outflow $= 0.43 \times 10^6 \ B$ (from Table 10.5)

or $B = 7.95$ m

Max. discharge = 14.2 B (from Table 10.5)

$\qquad = 14.2 \times 7.95 = 112.89$ m³/s

The wetted perimeter $P = 4.75 \sqrt{Q}$ (from regime equations) = 50.47 m

Depth of water upstream of sluice

$\qquad = + 1.83 - (-2.44)$

$\qquad = 4.27$ m

Consider that the channel has the same depth, i.e., 4.27 m with side slope as $1\frac{1}{2} : 1$, the bed width should be 35.07 m for maintaining the wetted perimeter of 50.47 m cross-sectional area = 177.1 m²

$$\text{Average velocity, } V = \frac{112.89}{177.1}$$

$$= 0.637 \text{ m/s}$$

Hydraulic mean radius, $R = 3.51$ m

Slope $S = \dfrac{V^2}{C^2 R}$ (from Chezy's equation).

Considering $C = 57$ m$^{1/2}$/s

$$S = \frac{0.637 \times 0.637}{57.00^2 \times 3.57} = 0.356 \times 10^{-4}$$

The following dimensions are obtained:

Sluice	*Canal*
Sill at, − 2.44 m	Depth = 4.27 m
Clear width = 7.95 m	Bed width = 35.07 m
	Volocity = 0.637 m/s
	Slope = 0.356×10^{-4}

Table 10.7 Dimension of the Sluice and Canal at Different Sill Levels.

Case no.	Basin level or U/S W.L. of sluice (m) GTS	Sill level of sluice (m) GTS	Vol. of outflow in terms of 'B' (m³)	Clear width 'B' (m)	Max. Q (m³/s)	Wetted perimeter 'P' (m)	Canal depth U/S of sluice (m)	C.S. area of the canal (m²)	Velocity 'V' (m/s)	Bed width (m)	Slope of canal in 10^{-4}
Case (i)	+ 1.83	– 2.44	0.43×10^6	7.95	112.89	50.47	4.27	176.1	0.640	34.84	0.356
Case (ii)	+ 1.83	– 1.83	0.35×10^6	9.77	110.10	49.84	3.66	154.3	0.71	36.66	0.500
Case (iii)	+ 1.83	– 1.22	0.27×10^6	12.67	108.71	49.52	3.05	130.9	0.83	38.36	0.803
Case (iv)	+ 1.22	– 1.83	0.22×10^6	15.34	133.3	54.84	3.05	147.7	0.90	43.86	0.926
Case (v)	+ 1.22	– 1.53	0.195×10^6	17.54	128.2	53.78	3.05	144.49	0.89	42.80	0.900

Comparing the dimensions of the canal thus obtained with that given in Table 10.3, it is seen that the canal is narrower and deeper than the regime canal. So, this combination is not acceptable. Similar trials are made with other sill levels and the results are given in Table 10.7.

Let us accept the dimensions of case (iii) tentatively. The slope at maximum discharge is 0.803×10^{-4}. The length of the canal is 6.44 km and if the basin level is maintained at 1.83 m GTS then the water level upstream of the sluice would be $6.44 \times 10^{-4} \times 0.803 \times 10^{3} = 0.52$ m lower. So the water level upstream of the sluice would no longer be + 1.83 m but it would be $1.83 - 0.52 = + 1.31$ m GTS. So in this case (iii) is not acceptable and the design should be re-examined. Taking a water level of + 1.22 m GTS at maximum discharge case (iv) has, therefore, been evaluated.

For case (iv), the drop of water level between the basin end and sluice end of the canal would be $0.926 \times 6.44 \times 10^{-1} = 0.600$ m. Therefore, the water level at the sluice end of the canal would be at + 1.23 m GTS during the maximum discharge condition when the basin level is maintained at + 1.83 m GTS. For placing the sill of the sluice a little higher than the adjacent bed level of the canal case (v) has been worked out for a sluice sill 0.305 m higher than the canal bed upstream. For automatic operation of the sluice gates, flap shutters are generally used. The shutters should be such that they become operative at a small difference of head. Normal size of flap shutters are 1.22 m wide and 1.83 m high, hinged at the top. The thickness of pier in between the gates is normally 0.61 m. The required 17.54 m of clear width is, therefore, provided by 15 vents and accordingly the width of the sluice between the abutments is $15 \times 1.22 + 14 \times 0.61 = 26.81$ m. The dimensions of the sluice and the canal thus become as follows

Sluice	*Canal*
Width between abutments = 26.81 m	Length = 6.44 km
Number of vents = 15	Canal bed near basin = – 1.22 m
Size of vent = 1.22 × 1.83 m	Canal bed near sluice = – 1.83 GTS
Type of gate = Flap shutter	Bed width = 42.8 m
Sill level = –1.53 m	Slope = 0.9×10^{-4}
Upstream WL at max. discharge condition = + 1.22 m	Side slope = $1\frac{1}{2}$, 1
Max. discharge = 128.20 m³/s	Depth at max. discharge = 3.05 m

Due to, fixation of specified type of flap shutters, the restriction of the depth of flow to a maximum 1.83 m over the sill does not impose restriction to maximum discharge under free flow condition, because this height is equal to two-thirds of the depth over sill immediately upstream of the vents.

It is now necessary to check the discharge capacity of the design of the sluice and the canal. It is to be noted that the basin level cannot remain constant at +1.83 m GTS due to runoff and, therefore, the canal level upstream of the sluice would fluctuate. Consider a possible case when the entire quantum of runoff has taken place in the basin during the time of no outflow through the sluice, i.e., between 4th and 7th hour of the tide curve (Fig. 10.10). The rainfall is 2.54 cm over the 259 km² area and this water accumulates in the basin having 20 per cent of the area, i.e., 51.8 km². The rise in the basin level would be 0.13 m. So the basin level would be $1.83 + 0.13 = 1.96$ m GTS and the canal level upstream of the sluice would also be

+1.56 m GTS at the 7th hour since no outflow has taken place. The sluice starts discharging as soon as the downstream water level falls below +1.96 m GTS.

Time instant = 0630 hr

W.L. *U/S* of sluice = +1.96 m GTS

W.L. *D/S* of sluice = +1.96 m GTS. Hence, outflow $(Q) = 0$

Time instant = 0700 hours.

W.L. *U/S* of sluice = +1.96 m GTS

W.L. *D/S* of sluice = 1.83 m GTS

$$H_u = 3.49 \text{ m (because sill is at } -1.53 \text{ m)}$$

$$H_d = 3.36 \text{ m}$$

Outflow $Q = 4.19 \times 33.49(3.49 - 3.36)^{1/2} = 50.59 \text{ m}^3/\text{s}$

Canal depth at the sluice = 3.79 m (because canal bed is at -1.83)

Canal depth at basin = 3.18 m (because canal bed is at -1.22)

Average canal depth = 3.485 m

Wetted perimeter = 56.76 m

C.S. area = 173.54 m^2

Hyd. mean radius $R = \dfrac{173.54}{56.76} = 3.06$ m

Velocity $V = \dfrac{50.59}{173.54} = 0.292$ m/s

Slope $S = \dfrac{0.292^2}{57^2 \times 3.06} = 0.085 \times 10^{-4} = 8.5 \times 10^{-6}$

Drop of water in the 6.44 km canal = $6.44 \times 10^3 \times 8.5 \times 10^{-6} = 0.055$ m.

Canal W.L. *U/S* of sluice = $1.96 - 0.055 = +1.905$ m GTS

But at the beginning of calculations the water level upstream of the sluice has been taken as +1.96 GTS. The calculation is repeated with an assumed water level of +1.93 m GTS.

W.L. *U/S* of sluice = + 1.93 m

W.L. *D/S* of sluice = 1.83 m

$$H_u = 3.46 \text{ m}$$

$$H_d = 3.36 \text{ m}$$

Outflow $(Q) = 4.19 \times 3.49(3.46 - 3.36)^{1/2} = 44.37 \text{ m}^3/\text{s}$

Canal depth at sluice = 3.76 m

Canal depth at basin = 3.18 m

Average canal depth = 3.47 m

Wetted perimeter = 56.71 m

C.S. area = 172.3 m^2

$$R = \dfrac{172.3}{56.71} = 3.04 \text{ m}$$

$$S = \dfrac{0.066}{57^2 \times 3.04} = 6.71 \times 10^{-6}$$

Drop of water level in the 6.44 km canal = 0.043 m

Canal level *U/S* of the sluice = 1.920 m GTS (calculation started with +1.93 m GTS level)
Volume of water that has passed through the sluice between 0630 and 0700 hours

$$= \frac{0 + 44.37}{2} \times 30 \times 60 = 3993.3\,\text{m}^3$$

Surface area of 51.8 km^2 basin = 51.8 × 10^6 m^2

Therefore, drop in the basin level during this time $\dfrac{3993}{(51.8 \times 10^6)}$ = 0.00077 m.

Similar calculations for all other instant show that during the period in between two tidal high water, the basin level falls from +1.960 m to +1.900 m GTS the canal level upstream of the sluice fluctuates between +1.960 m and 1.340 m GTS.

The tide curve against which the design has been performed is within the spring range. It would be necessary to check the design against a standard neap tide at the outfall.

Effect of sluice outflow on the outfall channel

In the design, it has been assumed that the tide levels at the outfall channel is not altered by the sluice discharges. In reality, the tide levels are affected if the sluice discharge is significant as compared to the discharge capacity of the outfall channel. If the maximum sluice discharge is within 10 per cent of the maximum flood or ebb discharges of the outfall channel, the alteration of tide levels may be ignored. But in case of increased percentage, it is necessary to evaluate the modified tide levels due to outflow. In general, the effect is to elevate the low water and high water levels resulting in a reduced sluice outflow period. If drainage of other adjacent areas is contemplated using the same channel, the combined effect should be taken into account.

In evaluating the changes of tide levels, the problem has to be dealt in a similar way as that of an open tidal channel where tides operate at the downstream end and fresh water is fed at the upstream end or at intermediate points. Computational technique exists for solving such problem. A discussion on the technique is beyond the scope of this book. Interested readers may refer to appropriate books on tidal hydraulics [32].

Problem 10.1

A catchment area is to be drained with the help of pipe drains the layout of which is shown in Fig. 10.17. Find out the required pipe diameters using the rational method based on the following assumptions and data given in Table 10.8 and 10.9.

Table 10.8

Rainfall duration (minutes)	Rainfall intensity (mm/hr)
2.0	122
2.5	108
3.0	97
3.5	89
4.0	82
4.5	77
6.0	70
7.5	65
9.0	60
9.5	55

Table 10.9

Data	Manhole levels from datum level (m)
A	101
B	100.5
C	99.5
D	100.6
Outfall	98.6

Pipe no	Pipe length (m)	Catchment area (ha)
1.0	60	0.2
1.1	70	0.19
1.2	75	0.18
2.0	75	0.21

Assume flow in all pipes can be modelled by Darcy-Weisbach formula using $f = 0.02$ and the time of entry to the manholes be 3 minutes. Table 10.10 provides the solution.

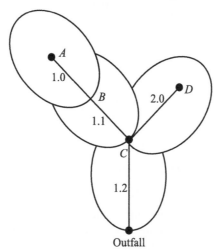

Figure 10.17   Schematic of catchment

Solution

Table 10.10

Pipe no.	Level difff. (m)	Pipe length (m)	Grad. (1 in)	Pipe			Time of flow t_f (mins.)	Catchment			Q_p (l/s)
				Trial pipe dis. (mm)	V (ms⁻¹)	Q_c (l/s)		Time of conc. T_c (mins.)	Rate of rain i (mm/h)	Cum. imp. Area A (ha)	
1.0	0.5	60	120	150	1.11	19.6	0.9	3.9	82	0.2	45.6
				250	1.43	70	0.7	3.7	90.8	0.2	50.0
1.1	1.0	70	70	250	1.87	91.7	0.62	4.32	79	0.39	85.6
2.0	1.1	75	68	250	1.89	93	0.66	3.66	87.5	0.21	51
1.2	0.9	75	83	350	2.03	557	0.62	8.64	56	0.78	122

Calculations

Pipe No. 1 Manhole A = 101 m B = 100.5 m

So drop = 0.5 m

An initial pipe diameter of 150 mm or 15 cm is assumed

Pipe gradient = $\dfrac{0.5}{60} = \dfrac{1}{120}$ using Darcy formula

$$h = \frac{flv^2}{2gd} \quad \text{or} \quad 0.5 = \frac{0.02 \times 60 \times v^2}{2 \times 9.81 \times 0.15} \text{ or } v = 1.11 \text{ m/s}$$

Flow capacity of the pipe

$$Q = \frac{\pi \times (0.15)^2}{2gd} \times 1.11 = 0.0196 \text{ m}^3/\text{s or } 19.6 \text{ lps.}$$

Flow time though the pipe length of 60 m = $\dfrac{60}{1.11}$ sec or 0.9 mins.

So time of concentration = 3.9 mins.

Corresponds to that rainfall intensity = 82 mm/hr

So peak storm runoff

$$Q_p = \frac{kiA}{3.6} \text{ m}^3/\text{s}$$

where i = mm/hr, A in km^2, k = 1 for paved surface.

or

$$Q_p = \frac{82 \times 0.2 \times 10^4}{3.6 \times 10^6} \text{ m}^3/\text{s or } 45.6 \text{ lps.}$$

Since the storm runoff exceeds the carrying capacity of the pipe, the pipe diameter has to be changed. From the table it may be noticed that for pipe 1.2, pipe diameter 350 mm is too large, so 300 mm diameter may be adopted.

11

Water-logging and Salinity

11.1 INTRODUCTION

Water-logging, high salinity levels and poor drainage in agricultural lands are believed to be the primary reasons for the collapse of many earlier civilizations. Indiscriminate and excessive use of water results in bringing the water table to the surface. This also brings salts present in the soil to the surface, thereby creating saline and alkaline conditions. It is believed that about 8×10^6 hectares of land is affected by salinity and alkalinity conditions in India including the coastal regions which are subject to periodical inundation. Of these, about 5 million hectares constitute noncoastal regions and the balance 3 million hectares are in the coastal regions. The statewise breakdown of the two components are roughly as follows:

Table 11.1 Salinity affected regions of India

State	Noncoastal area (10^6 ha)	Coastal area (10^6 ha)
West Bengal	—	0.82
Gujarat	0.50	0.714
Orissa	—	0.40
Andhra Pradesh	0.042	0.276
Tamil Nadu	—	0.100
Karnataka	0.404	0.086
Maharashtra	0.534	0.063
Kerala	—	0.026
Goa	—	0.418
Puduchery	—	0.001
Andaman and Nicobar Islands	—	0.015
Uttar Pradesh	1.295	—
Madhya Pradesh	0.224	—
Rajasthan	0.728	—
Haryana	0.526	—
Punjab	0.688	—
Salt-affected area under mangrove cultivation	—	0.573

11.2 CAUSES OF WATER-LOGGING PROBLEM

Development of modern civilization, resulting in a network of railways, embankments and canals running across the country and expansion of inhabitation with population growth has very seriously affected the natural runoff conditions of ground over vast areas of the country. The principal factors that contribute to water-logging in noncoastal areas are continuous rainfall, seepage, flooding of rivers, over-irrigation, faulty crop rotation, and blocking of natural drainage systems. In the case of coastal regions the problem areas are concentrated mainly at the confluence of rivers with seas which are essentially lowlying areas. They are subject to inundation by tidal waters and the inland drainage water almost simultaneously and regularly.

Generally speaking, the deltas of the eastern coast are low-lying with marshy lands of the Sunderbans in the north and natural lakes like Chilka, Kolleru and Pulikat along the coast. In the case of the western side the coast of Kerala and Valsad and Surat districts of Gujarat are low-lying. The causes of water-logging, the different aspects of its management, the various degrees of relief possible and the reclamation problems associated with it are all briefly dealt with in Fig. 11.1.

11.3 CAUSES OF THE SALINITY PROBLEM

The principal factors that contribute to soil salinity are—(i) high water table, (ii) inadequate drainage, and (iii) hot and dry climate. In the first case, due to irrigation the water table rises near the surface level which causes significant evaporation from soil water leaving salt accumulated in the root zone. In the second case, the irrigated river plains are often low-lying and flat. As a result, gravity drainage systems are difficult due to intrusion of saline sea water.

Salt infestation problem is very acute in the arid and semiarid tracts of the Indo-Gangetic alluvial plains and the major problem there is due to soil alkalinity. In Hissar and Rohtak districts of Haryana, Agra and Mathura district of Uttar Pradesh, Ferozepur, Faridkot and Bhatinda districts of Punjab, the problem is mainly due to salinity. In the coastal regions the soil regime is classified according to the distance of the land from the sea. There soil may be water-logged, non-saline to less saline in the wet season and dry and strongly saline in the dry season. In these areas, soils are situated in deltas, estuaries or coastal fringes in a strip ranging from few kilometres to about 50 km from the coast.

As for example, in the State of West Bengal, the 0.82 m.ha of area lying within the agricultural districts of south and north 24 Parganas, Howrah and Midnapore is salinity affected. Of these, the Sundarban area, i.e., in the north-south 24-Parganas constitute sixty per cent of the area. The regime poses peculiar problems of salinity, water-logging and irrigation. The tidal nature of rivers like Rasulpur, Matla, Herobhanga, Gosaba, Raimangal which get barely any upland discharge and acts mostly as estuaries and helps to carry the tidal saline water much further inland. The hydrology of Bhagirathi-Hooghli system and deposits of sediment specially during high tides also contribute to the formation of saline soil in the Sundarban delta.

The extent and intensity of soil salinisation depends on distance from the sea, dyke position and presence or absence of tidal creeks inside the cultivated area. The occurrence of tides and their levels, have a direct bearing on the formation of coastal saline soils and their development. The tidal flow repeatedly inundates the soils and impregnates them with salts. The high tides from the 12th to the fullmoon and from 1st to 5th day of the lunar fortnight inundate areas in the coastal plains.

CAUSES
1. Rainfall (as a cause)
2. Seepage from rivers and canals (as a cause)
3. Area under irrigaion (as a cause)
4. Other geological formation (as a cause)
5. River floods
6. Ingress of sea water
7. Relative importance of different causes.

MANAGEMENT
1. Wet side of water-site plant relations.
2. Plant breeding.
3. Cropping pattern.
4. Associated problem of water-quality, salinity and alkalinity.
5. Engineering systems.
   (i) System for water table control.
   (ii) System for microdrainage.
   (iii) Drainage return system.
   (iv) Land grading.
   (v) Machinery for water-logged lands.
   (vi) System for sea water barrages.
6. Optimum level of communication on drainage and other inputs.

Problems of water-logging and drainage

RECLAMATION
1. Planning and design of reclamation system
   (i) Surface drainage
   (ii) Sub-surface drainage
   (iii) Pumped drainage.
2. Construction of reclamation system.
3. Maintenance and operation of reclamation system.
4. Ensuring permanent benefits from the installed reclamation systems.

DEGREE OF RELIEF
1. Priority between water-logged areas.
2. Shallow versus deep drainage
   (i) Degree of relief necessary
   (ii) Cost benefit ratios in economics.
3. Drainage coefficients for surface and sub-surface drainage.

Figure 11.1 Water-logging phenomena.

11.4   SALT WATER INTRUSION IN COASTAL AQUIFERS

11.4.1   General

One defines coast as where land, water and air interact. Coastal areas in general comprise lands with very small slope bordering sea, estuaries or lower reaches of rivers, coastal marshes and lagoons. In view of its economic significance and large human habitat within a 100-km strip of coast, environmental management of coastal waters, lands and echo systems are vital for its sustenance. Coastal communities rely on ground water for their water supply. In the process, the hydrological equilibrium/balance of the coastal aquifer which might have been established over a long period of time, is highly vulnerable. Lowering water table as a result of pumping may contaminate the coastal aquifer with salt water. The migration of salt water into fresh water aquifers under the influence of ground water development is termed as salt-water intrusion. (Freeze et al. [45]). There is likelihood of salt water intrusion into surface water bodies as well as fresh water aquifer in low lying coastal areas. Saline water originates mainly from sea into open estuaries. Penetration of sea water into rivers is induced by the density difference between fresh and saline water and also head differences during low water flow. Normally, the denser sea water forms a deep wedge that is separated from fresh water by a transition zone. Under undisturbed conditions the saline water body remains stationary its position being defined by the fresh water potential and hydraulic gradient. When, however, the aquifer is disturbed by activities like pumping of fresh water or changing recharge conditions, the saline water body may gradually advance until a new equilibrium is reached, Problem arises when saline water from the deeper saline wedge enters the wells thereby affecting the water quality. Freshwater aquifers under coastal areas may become saline due to overdraft of fresh water pockets, reclamation of low lying areas, tidal effects, sea level changes and aquacultural (fisheries activities).

11.4.2   Salt Water Intrusion in Coastal Aquifers

Seawater intrusion in coastal aquifers occurs when permeable formations outcrop into a body of seawater and when there is a land word hydraulic gradient. Sea water can be prevented from intruding by maintaining a head of fresh water above it. According to Ghyben-Herzborg principle the interface will occur at a depth h_s below mean sea level, (Fig. 11.2) where

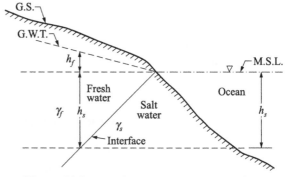

Figure 11.2   Fresh water-salt water interface.

$$h_s = \frac{1}{(S-1)} h_f \qquad (11.1)$$

and S is the specific gravity of water. Considering $S = 1.025$, $h_s = 40\, h_f$ which means for a rise of ground water table by 1 m will induce a fall of 40 m or a fall of 1 m will induce a rise of 40 m in the underlying salt water level even though the response will be very much delayed.

11.4.2.1 Slope of the interface

If the water slope is given by '*i*' then by Darcy law, one can express, (Fig. 11.3)

$$\sin i = \frac{dh}{ds} = \frac{h}{s} = \frac{v}{K} \tag{11.2}$$

where

   v = the velocity of fresh water flow and

   K = the permeability of the media or strata.

Along the slope, the water table elevation decreases in the direction of flow and hence fresh water–salt water interface should rise and its slope is given by

$$\sin \phi = \frac{\rho_f}{\rho_s - \rho_f} = \frac{v}{K} \tag{11.3}$$

where ρ_s and ρ_f are the densities of salt and fresh water.

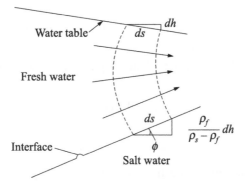

Figure 11.3 Relation between slopes of the water table and the fresh water-salt water interface.

   Since the boundaries converge, the velocity of fresh water flow increases with distance and the magnitudes of the slopes increase accordingly, resulting thereby a parabolic interface, (Figs 11.4 and 11.5. The parabolic is very similar to Dupuit's parabola and can be expressed as, (Rumer et al. [51]).

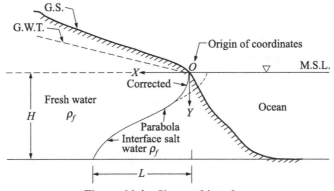

Figure 11.4 Shape of interface.

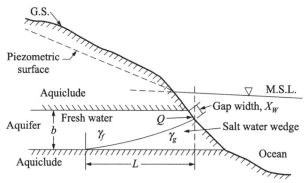

Figure 11.5   Salt water wedge in a confined aquifer.

$$y = \sqrt{2\left(\frac{q}{K'}\right)x + 0.55\left(\frac{q}{K'}\right)^2} \cdot K' = K(S-1) = 0.025\, K \qquad (11.4)$$

where

$\quad q$ = sea ward fresh water flow per unit width of ocean front,

$\quad K'$ = permeability of the aquifer, and

$\quad x, y$ = co-ordinates of the interface with the origin at the contact of mean sea level with land, (Fig. 11.4).

When $x = 0$, $\qquad\qquad\qquad\qquad y_0 = 0.741\,\dfrac{q}{K'}$

and when $y = 0$,

$$x_0 = -0.275\,\frac{q}{K'}$$

The total length of intrusion measured from $x = 0$, is given by

$$L = \frac{K'H^2}{2q}, \text{ when } L > H \qquad (11.5)$$

where H = thickness of the aquifer.

The time t in days required for the toe of the wedge to move the length L is given by

$$t = \frac{\ln K' H^3}{q_{ul}^2} \qquad (11.6)$$

where

$\quad I$ = dimensionless factor,

$\quad n$ = porosity of the aquifer material, and

$\quad q_{ul}$ = ultimate fresh water flow per unit width.

As an example, consider a typical problem designated as number one, Courtsey Raghunath [46].

Problem 11.1

From an examination of hydrological, geological and geo-chemical data of a coastal aquifer, the following information is gathered. Width of aquifer equals, 2.8 km, thickness of aquifer equals, 30 m, porosity of the aquifer material is 10%, permeability of the aquifer is 48.9 m/day.

From the conductivity measurements in two observation wells located at 150 m and 225 m from the shore in the land ward side the 1500 ppm line was found to be located at 15 m and 22.5 m respectively below the top of the aquifer. The problem is to determine the freshwater–sea water interface.

Solution

From Darcy's law,

$$Q = KAi$$

$$q = K' \, (1 \cdot y) \, \frac{dy}{dx}$$

$$q \int_{x_1}^{x_2} dx = K' \int_{h_1}^{h_2} y \, dy$$

$$q = \frac{K' \, (h_2^2 - h_1^2)}{2 \, (x_2 - x_1)} = \frac{48.9 \, (0.03) \, (22.5^2 - 15^2)}{2 \, (225 - 150)}$$

$$= 2.76 \text{ m}^3/\text{day/m width of coast line}$$

Length of intrusion,

$$L = \frac{K' H^2}{2q}$$

$$= \frac{48.9 \, (0.03) \, 30^2}{2 \, (2.76)}$$

$$= 239 \text{ m}$$

The solution of the problem yield $q = 2.76$ m/day/m, width of the coast line and the length of intrusion $L = 239$ m landward from the shore at the time of investigation. Since L is inversely proportional to q if further ground water exploitation is proposed say at the rates of $Q/4$, $Q/3$, $Q/2$ and $3Q/4$, where $Q = 2800 \, q$, the fresh water flows will be $3Q/4$, $2Q/3$, $Q/2$ and $Q/4$ respectively and the corresponding lengths of intrusion will be $4L/3$, $3L/2$, $2L$ and $4L$ respectively. The time when the toe will advance and reach its final length of intrusion or any other length can be calculated from the equations already furnished. This will indicate the rate of landward sea intrusion and is very useful for planning ground water exploitation in coastal areas, (Fig. 11.6 and Fig. 11.7).

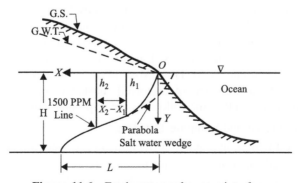

Figure 11.6   Fresh water–salt water interface.

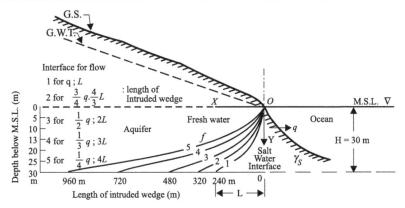

Figure 11.7 Advancement of salt water wedge with increased ground water exploitation (Courtsey Raghunath [46]).

A continuous programme for the collection and interpretation of hydrologic, geologic and water quality data has to be initiated to halt and abate sea water intrusion in all basins known to be affected at present. Investigations carried out by UNDP 46 along the Madras coast at the rates of ground water extraction in the year 1966 indicated that the movement of sea water interface is of the order of 120 m per year and it recommended the limit of extraction to halt and abate this intrusion. During 1970, due to over extraction to satisfy the increasing demand, GWT of the Madras coastal aquifers had dropped by more that 25 to 30 m causing a reversal of hydraulic gradient and movement of interface towards land at the rate of 120 to 150 m per year. The ground water is falling at a gradient of 1 m per km towards the sea and the interface was located at 60 m depth below MSL at 3 km inland. Relevant calculation with appropriate K values indicates that approximately, 750 cubic meter of fresh water per kilometer of coast line is lost to the sea.

11.4.3 Methods of Control

Various control/remedial methods were proposed to maintain sustainable development of coastal fresh water aquifer, Todd (47). It is, however, required to assess the spatial and temporal extent of extension phenomena and also the transient nature of the coastal aquifer system in order to initiate preventive measures effectively. Monitoring wells can be installed to get information on salt water–fresh water interface and the rate at which salinity level changes. Prognostic analysis of the problem is carried out based on field observation data and hydro geological information of the aquifer system and remedial measures are evaluated. Mathematical modelling of flow and solute transport in coastal aquifer can assist greatly to devise methods of control of salinity. One such numerical code is named SUTRA, Voss [48]. SUTRA is a computer code which simulates fluid movement and transport of dissolved substances in a subsurface environment. Out of many capabilities of SUTRA, the following are important with regard to coastal system.

I. Assess well performance and pumping test data.

II. Analyse aquifer restoration, hydraulic barriers, liners and water quality protection systems.

III. Model cross-sectional salt water intrusion in aquifers at near well or regional scales with relatively sharp transition zones between fresh water and salt water.

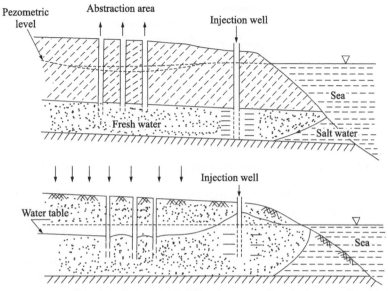

Figures 11.8 and 11.9 Sea water intrusion control by recharge through injection well.

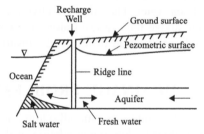

Figure 11.10 Control of sea water intrusion by pressure ridge parallaling the coast.

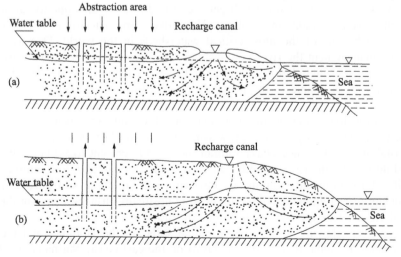

Figures 11.11(a) and (b) Sea water intrusion control by artificial recharge canal.

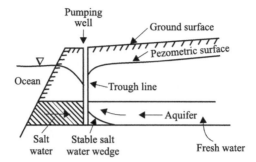

Figure 11.12 Control of sea water intrusion by a pumping through parallaling the coast.

Sea water can be controlled by artificially recharging the intruded aquifer by surface spreading methods and injection well methods. Artificial recharge of ground water produces a hydraulic barrier to the sea water intrusion. Figure 11.8 shows a case of confined aquifer where the recharge through injection wells keeps on check the advancement of sea water interface towards land. Figure 11.9 shows a phreatic aquifer where the mound developed in water tables due to recharge through injection wells control the sea water intrusion. Figure 11.10 shows artificial recharge through injection wells to control the sea water intrusion. Figure 11.11 shows artificial recharge through recharge canals close to the shore. The pressure built up due to recharge controls the ingress of sea water. The other effective methods for controlling intrusion are creation of fresh water ridge above the sea level along the coast, development of pumping trough adjacent to the coast and construction of artificial subsurface barriers, (Fig. 11.12).

11.5   CAUSES OF THE DRAINAGE PROBLEM

In the case of plain areas such as the Indo-Gangetic alluvium and other areas in India drainage congestion during flood time is primarily due to aggradation of the river bed due to silting. The major river systems when flowing through the plains of India, which are responsible for draining the area, carry huge amounts of silt which gets deposited in the plains due to various hydraulic reasons causing severe aggradation problems. These rivers originate from the Himalayas which are considered to be one of the most critical watersheds in the world. Due to deforestation and faulty land management the river systems carry huge amount of sediment during the rainy season. Because of the aggradation of the river bed the capacity of the river decreases year to year and finally it is unable to discharge the yearly flood safely. Due to higher river bed, the flood water elevation also rises and lower areas along the river are not able to drain their runoff. Even during low water, some areas may not drain into the river and hence becomes marshy. These marshy areas gradually grow while inundation lasts longer.

The coastal areas experience heavy rainfall, nearly 80 per cent to 90 per cent of which is concentrated during the monsoon months. In view of flat topography, low infiltration rate and lack of well defined drainage systems the low-lying coastal saline soil have serious surface drainage problems. Over and above this, runoff from upstream surfaces also contributes towards flooding of these areas as the rivers and drains overflow.

The major rivers flowing to the coastal saline soils are the Hooghly and mouths of the Ganges. Mahanadi, Krishna, Kaveri and Godavari on the east coast, Narmada and Tapti on the west coast. The main rivers on their confluences with the sea divide into a number of diversary branches enclosing and intersecting the delta. The rivers of the east coast flow from west to east and confluence with the Bay of Bengal and of the west coast from east to west and confluence with the Arabian sea.

The flood flows in the regions are caused due to the flat nature of the topography in the delta and long and tortuous courses the rivers take before finally emptying into the sea. This causes silting of rivers and drain and create outlet problems for the local drainage. Further, the sand dunes that get formed at the mouths of these rivers near the coast due to littoral drifts of sand also cause drainage congestion. During rainy seasons, the rivers carry huge sediment loads from the hills to the plain coastal reaches. The fine particles like clay and silt remain in suspension and get transported to the sea. At the confluence of river and sea due to reduction of velocity of flow as well as physico-chemical action, the fine particles settle down, resulting in the formation of mud flats, which in turn, leads to flooding with varying degrees of regularity.

The surface drainage from such areas is generally poor as most of the natural drains are choked with silt and also due to tidal blockage. In most of the cases, there is no inland network to provide speedy disposal of excess drainage water of various seasons.

11.6   REMEDIAL MEASURES TO COMBAT WATER-LOGGING AND SALINITY

From a perusal of the above, it can be seen that India faces an enormous task in fighting the triple menace of water-logging, salinity and drainage to improve upon its agricultural productivity and for the reclamation of new areas, specially since increased pressure on land will continue due to increasing population and the consequent demand for greater food production. There occurs at present colossal loss in agricultural productivity due to water-logging, soil salinity and poor drainage. To cite an example, Punjab suffered an annual loss of more than fifty crores of rupees of agricultural productivity due to water-logging effects in the sixties. If one takes into account other losses suffered by the community due to diseases, loss of soil fertility, damages to buildings, roads, the cost will be colossal indeed. It is thus imperative that effective remedial measures to fight the triple menace are devised consistent with benefit–cost ratio through application of new engineering technologies and improved genetically developed seeds.

The measures to combat water-logging may be classified as preventive measures, curative measures and lastly measures for appropriate river training. Under preventive measures included all attempts for economic use of water, soil investigation, canal lining, construction of water courses and field channels, and proper discharge of canal water, prevention of sea water intrusion by constructing dykes, etc. The curative measures constitute the drainage system with appropriate means for removal of excess water through artificial surface or sub-surface drainage or pumping of water from the sub-soil water table.

Under river training the following measures are generally adopted to prevent flooding and consequent water-logging—(i) embankment or levees, (ii) diversion of excess water, (iii) overflow weir in the embankment, (iv) detention storage, (v) improvement of river regime by cutoff, spurs or groynes, and (vi) removal of sand bars near the river mouth by dredging.

Preventive measures for combating salinity include the following—Artificial discharge of ground water with fresh water in areas where ground water withdrawal is higher than the natural discharge from the streams, by storage of fresh water maintaining the head for as long a period as possible to increase ground infiltration. Various techniques such as building of check dams, discharging tanks, spreading channels and discharge wells can be adopted for achieving the above.

Regarding curative measures it appears the only practical means of reducing soil salinity is by leaching saline soil water from the root zone with water of lower soil concentration. The quantity of water required for leaching depends on initial soil salinity, the quantity of leaching water, the method of leaching, the depth of the profile and the soil properties. Salt accumulation is caused by the preponderance of evaporation over drainage and if the water balance can be controlled then the spread and growth of salinity can be prevented which means adoption of an effective drainage system.

11.7   IMPORTANT DRAINAGE PROJECTS AND A CASE STUDY

Some of the important drainage schemes of different states are respectively—(i) drainage schemes in the upper and lower reaches of the Kaveri delta in Andhra Pradesh, (ii) drainage schemes in the Sundarbans and Ghatal in the district of Midnapore of West Bengal, (iii) Kuttanad development scheme and Kole lands development in the Trichur district of Kerala. Essentially, the general features of surface drainage systems are as follows:

- (i) Construction of peripheral bunds to clearly demarcate the catchment and also the various fields so that inflow of excess water from outside the sea into the catchment and the flow from one zone to the other is regulated.
- (ii) Channelisation of the catchment to directly route the excess rainwater from different zones to the outlet.
- (iii) Opening of tidal sluice gates more frequently and for longer periods to maintain the deadwater level in the catchment.

12

Application of Remote Sensing Technology for Flood Control

12.1 INTRODUCTION

Remote sensing can be defined as a technique for the collection of terrain features from measurements made at a distance without coming into physical contact with the objects under investigation. Characteristics electromagnetic radiation that is reflected or emitted by the earth's surface are detected. Sensors can be installed either in aircrafts or satellites. They are intended to react with the reflected and emitted electromagnetic radiation from the earth's surface to produce images. The images are then analysed to yield information of importance for flood control and management.

12.1.1 Airborne Sensing

Sensors which are of 'imaging' type, like cameras, are categorised as airborne sensors. They are of two types—(i) instantaneous imaging sensors, and (ii) line scanning image sensors. In the former case, the area under the field of view is imaged instantaneously. The image is then recorded on a photographic film or electronically. This type of sensors are television cameras or photographic cameras.

In the latter case, the sensor records one picture element at a time. A line is generated by mechanically or electronically scanning the instantaneous field of view in one direction. Successive scanlines are produced through movement of the platform. It then generates a two-dimensional image. This type of sensors are designated as linear image self-scanning (LISS) and multi-spectral scanners.

Panchromatic black and white aerial photographs have provided imagery for topographic mapping, soil mapping and engineering planning. The coverage can be made so as to obtain detailed information. Further, they have good stereoviewing capability. Nowadays black and white infrared, false-colour infrared and true colour aerial photographs are very much in use for flood-related studies. Also advanced airborne missions can employ multispectral scanners. This system obtains terrain data in continuous digital scanning mode in the visible, near infrared and thermal infrared of the electromagnetic specturm. In India, National Remote Sensing Agency (NRSA) undertakes airborne survey missions for the user agencies which are normally Government or autonomous organisations.

12.1.2 Satellite Sensing

Sensors can also be installed in satellites orbiting around the earth to cover large areas and for extracting useful informations for weather forecasting and land water resource surveys. A number of such satellites are in orbit since 1972. In this case, a large area synoptic view can be obtained by the satellite sensors with repetitive imaging capabilities. Tables 12.1 and 12.2 provide a list of satellites which are of interest in flood control planning. The tables also list the data that are readily available and their approximate cost price.

Table 12.1 Land Resources Satellites of Interest to Flood Control Planning.

Satellite/sensor	Resolution			Ground coverage per scene (km^2)	Data products useful for flood controt planning		NDC unit price (₹)*
	Spectral band (µm)	Ground resolu- tion (m)	Revisit period (day)		product type	scale	
Landsat-5 TM	0.45–0.52 0.52–0.60 0.63–0.69 0.76–0.96 1.55–1.75 2.08–2.35 10.40–12.5	30 120	16	185 × 185	– Standard FCC of Band 2, 3 and 4 enlargement paper print – Computer Campatible Tape (CCT) per quadrant of scene	1:50,000	6,000.00 11,000.00
IRS-IA LISS I LISS II	0.45–0.52 0.52–0.59 0.62–0.68 0.77–0.86	73 and 36.5	22	148 × 148	– Standard FCC of Band 2, 3 and 4 enlargement paper print, geocoded product of LISS II – CCT of LISS II	1:50,000	2,250,00 2,600.00
SPOT -1 HRV PLV MLA	0.50–0.50 0.50–0.60 0.61–0.71 0.80–0.91	10 20	26	60 × 60	– Black and white enlargement paper print of PLA, geocoded 28 km ×	1:25,000	5,000.00

28 km		
– FCC	1:50,000	7,500.00
enlargement		
paper print,		
geocoded		
28 km ×		
28 km		
– Black and white	1:25,000	10,000.00
PLA stereo-		
image		
60 km ×		
60 km		
– PLA stereo-		20,000.00
image CCT		

NDC: NRSA Data Centre: *Prices are effective till March 31. 1990.
Information extraction costs are not included here since such costs vary depending on the study problem in question.
Courtesy: Sri A.K. Chakraborty, Head Water Resources Division, Indian Institute of Remote Sensing, Dehradun.

Table 12.2   Meteorological Satellite Coverage over India Available for Flood Control Planning.

Satellite/sensor	Resolution			Flood control planning applications
	Spectral band width (μm)	Ground resolution (km)	Frequency of data availability	
NOAA-10 & 11 AVHRR (Polar Orbiting)	0.50–0.90 0.725–1.10 3.55–3.93 10.50–11.50 11.60–12.50	1.1	3 times daily	1. Cloud index mapping. Weather forecasting. 2. Snow over area mapping—For seasonal snowmelt runoff forecasting.
INSAT-1B VHRR (Geostationary) DCS	0.55–0.75 10.50–12.50 INSAT-1B:	2.75 11.50 8 Test & Evaluation DCPs. 3 DCPs for cyclone warning purpose.	3 hourly interval	3. Reservoir monitoring The Meteorological Payload on-board INSAT consists of a data collection transponder for collection and transmission of meteorological, hydrological and oceanographic data from the ground and sea based data collection platforms (DCP).
	INSAT-1C:	It is planned to instal 100 land based and 10 ocean-based DCPs to work with INSAT-1C series of satellites.		

Courtesy: Sri A.K. Charkraborty, Head Water Resources Division, Indian Institute of Remote Sensing, Dehradun.

12.2   APPLICATION FOR PLANNING FLOOD CONTROL MEASURES

For planning flood control, measures, it is necessary to have the following inputs which can be obtained by using remote sensing techniques. These are:

12.2.1 Flood Inundation Mapping

Surface water bodies can be mapped by adopting the unique recognition characteristics of near infrared spectral bands. Accordingly, the extent of the area inundated by flood can be obtained relatively easily from satellite-based observation. In this case one can adopt both digital and optical data processing techniques as they are helpful in delineating the flooded areas. Based on the information, the planning and decision-making authorities of flood control can take decisions with a fair degree of reliability on such aspects as—(i) extent of flood damage, (ii) structural measures, (iii) areas requiring postflood alleviative measures, and (iv) providing relief to the affected people.

12.2.2 Information regarding Flood Plain Landuse

The landuse information of the flood-prone rivers is very important, which is required for planning measures for flood alleviation. This also helps in the assessment of flood damage. On a long-term basis such information is also of help in the development of (i) necessary measures to control man's encroachment on the flood plain, and (ii) flood hazard zoning giving due weightage on the varying degrees of flood hazard.

12.2.3 Indicators of Flood Susceptibility

The natural flood susceptibility indicators are as below:

(i) characteristics of drainage basin, i.e., drainage density, shape, etc., (ii) channel configuration and geomorphological characteristics, (iii) soil moisture availability and differences in soil type, (iv) upland physiography and agricultural development, (v) landuse boundaries, and (vi) flood alleviation measures and degree of abandonment of levees. The flood plan indicators can be obtained through application of available air-photo interpretation techniques. These parameters are helpful for flood hazard zoning and estimation of inundated areas.

12.3 FLOOD WARNING

The satellite can be fitted with Data Collection System (DCS). These systems gather data from ground-based remotely located automated Data Collection Platforms (DCP). The DCPs are equipped with transducers fitted by the concerned user organisation or department. These data are then transmitted to the ground receiving stations or central computing facilities of the department.

The transducers measure the variables such as stage, flow, precipitation, windspeed and direction, temperature, humidity, snow depth, its density and water equivalent, soil moisture, etc. They can be interfaced with DC platforms at ground level. These hydrologic data can be relayed via satellite to the flood control room in a very short time. The transmission of data can be carried out many times each day from far-flung and inaccessible areas of a large river basin. The multipurpose geostationary satellite INSAT-1B has a DCP on board and relays to ground-based DCP, data for meteorological monitoring. Apart from that the system can be used for operation of flood warning during critical flood periods. For this purpose real time monitoring of the following information are required, i.e., time and location of the bursting of the storm over the basin, amount of rainfall, rate of runoff, stage records and selected locations on the river reach and the reservoir level. This information is useful for decisions regarding issuance of flood warning to the people in flood-prone areas, efficient operation of reservoirs and timely action for flood protective measures.

12.4   CONCLUDING REMARKS

It may be mentioned here that there are some problems with regard to satellite data acquisition for flood control and management purposes. These are—first, due to the difficulty of getting cloud-free coverage of satellite scene during the flood season, i.e., roughly from June to September. Secondly, the revisit period of the scene by the satellite may not be adequate to map the situation corresponding to peak flood. To overcome the first problem microwave sensores like Synthetic Aperture Radar have the capability of all-weather imaging. With regard to the second, the difficulty can be overcome indirectly. It is known that the receding flood waters have identifiable marks specially on soil moisture, vegetation, etc. in the flood plain. They can be identified even days after the flood crest has passed.

13

Flood Plain Delineation and Flood Hazard Assessment

13.1 GENERAL

Flood plain delineation is the process of determining inundation extent and depth by comparing river levels with ground surface elevations. In brief, the process involves making the observed water level points on a topographic map and extending water levels over the map until contours of higher elevation is encountered. The water level intersection points are then connected together by following the contours to get the extent of inundation. The traditional method, therefore, consists of the following steps.

   I. Observe river stages or simulate water levels from surface water models at different locations along the river.

   II. Prepare or collect a topo map.

   III. Mark water levels on topo map.

   IV. Extend water level until impeded by higher elevation.

   V. Trace the contour lines to delineate the flood plain.

   VI. Manually prepare the flood extent map.

Nowadays estimation of flooded area is generally done utilizing the remote sensing (RS) technique with the available Geographical Information System (GIS) data. This will enable preparation of a flood hazard map and a land development priority map. This will help to design operate flood control infrastructure and to provide aid and relief operations for high risk areas during future floods. GIS plays a major role in flood control techniques and the integration of these data in a spatial data base is crucial.

13.2 FLOOD PLAIN DELINEATION FROM REMOTE SENSING

It is necessary to have satellite imaging during the passage of a flood event of suitable quality with low cloud amount and fairly close to avoid geometric distortion from a distorted image. It is necessary to establish a relationship between the image co-ordinate system and the geographic co-ordinate system. For this purpose use of Affine transfer motion equations and positions of ground control positions have to be chosen on the satellite maps. Ground control points have to be chosen so that they can be easily identified on the satellite maps and the topographic maps.

Generally each of the GIS and RS images yields 118.736 pixels on a display monitor and each pixel covers 1.1 by 1.1 km on the ground surface after geo-coding. Finally, all digital GIS data have to be incorporated with geometrically corrected satellite imagery data within a GIS approach.

13.3 ESTIMATION OF FLOODED AREA

From the image data, it is necessary to differentiate between water and non-water which is generally done following an iterative non-hierarchical clustering that uses a minimum spectral distance to assign a cluster for each candidate pixel. Initially, all pixels are classified into several categories and these categories are divided into three classes, water, non-water and cloud. Thereafter interpretation of cloud cover pixels, the three categories are divided into two, i.e., water and non-water. Estimated flood areas can then be obtained after the drainage network map is superimposed onto the flood season images. The flooded area can be estimated after subtracting the normal water area, i.e., river, lake, ponds, etc. from the total inundated area. It can then be converted to a percent of land area, non-water area in the dry season of the whole country.

Flood depth determination from remote sensing imagery is very important, so the flood depth categories of different individual pixels has to be determined adopting the maximum likelihood of supervised classification and can be categorized as shallow, medium and deep. Training areas of shallow, medium and deep floods are selected on each image according to the visual interpretation by differences in coloured gray scales for different categories of depth for supervised classification. These are interpreted after superimposing the images into a digital elevation image. The results of the different categories of flood depth can be examined by the estimated albedo of the pixels with the same flood depth category. Generally deeper water show lower albedo. The ranking for the flood depth can be categorized as no flooding, shallow, medium and deep flooding. So several flood depth images can be constructed for a flood event.

13.4 FLOOD PLAIN DELINEATION FROM DIGITAL TERRAIN MODELS (DTM) [49]

In an automated system of flood plain delineation the topographic map is replaced by a digital terrain model. The observed river stages are either replaced or supplemented by hydraulic model results. The process requires the creation of a water surface using georeferenced water level points. The flood depth map is obtained by subtracting DTM from the water surface. The positive values indicate a flood depth, hence flood plain areas while negative or zero values indicate flood free areas. The automated procedure is more speedy, efficient and focus on flood depths in addition to extent of flooding.

Available tools: With the availability of computer and satellite based surveying technology, GIS has been recognized as a powerful means to integrate and analyze data from various sources. Some of the common methods and tools are:

Flood plain delineation using ARC/INFO, Watershed modeling system (WMS) (EMRL 1998)

Arc Info MIKE-11-GIS (FMM 1994)

Arc View MIKE-ll-GIS (DHI 1997)

HEC-GEORAS (HEC1999)

Coupling of Inundation Mapping capability with FLDWAV and Ensemble Streamflow Prediction (EPS).

13.5 GEO-REFERENCING OF THE WATER LEVEL POINTS

Water levels along the rivers and flood plain are the primary input for automated flood plain delineation. These are observed at gauging sites or simulated in a hydraulic model. Geo-referencing of all points is required to generate a reasonably smooth water surface. Some processes are designed to accept water levels in *X, Y, Z* format.

The DTM is the other principal input to a flood plain delineation system. Traingulated Irregular Networks (TIN) and grids are the two most commonly used data structures for DTM. A TIN can represent a topography using a variable resolution of nodes and triangles. In the grid system speed of analysis is faster.

13.6 CREATING A WATER SURFACE

With geo-referenced water levels and a DTM, creation of a water surface from discrete water level points is the next objective. Using an appropriate hydraulic model water level is simulated at a single point along the cross-section without any regard to the width of the water section. The water level points thus imported in the flood plain delineation provide water levels along the center line of the river network which has to be extended within the river width and flood plains, i.e., WLL (water level lines). The channel boundary lines (CBL) are digitized to define the extent of WLL, as shown in Figs 13.1 and 13.2. A TIN is then created using all available water level points including the vertices along the water level lines to represent a continuous water surface vide Fig. 13.3.

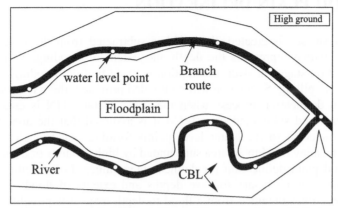

Figure 13.1 Creation of Channel Boundary Lines (CBLs).
(Courtsey Norman et al ASCE [49])

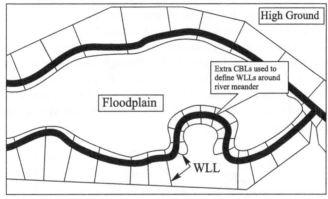

Figure 13.2 Creation of Water Level Lines (WLLs).
(Courtsey Norman et al ASCE [49]).

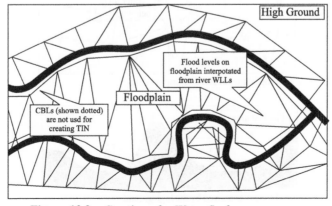

Figure 13.3 Creation of a Water Surface.
(Courtsey Norman et al ASCE [49]).

13.7 FLOOD PLAIN DELINEATION

Once the water surface is created, the DTM is subtracted from the water surface to create another surface which represents the depth and extent of flooding. When a flood plain is delineated based on stages in river, assumption is made that the flooding has occurred due to breaching or over spilling. So an area will be flooded provided, there is hydraulic connectivity. In a flood plain delineation process when the water surface TIN is created from available water level points and subtracted from DTM it is assumed that the area by each triangle is hydraulically connected to the water level points forming that triangle. Ignoring hydraulic connectivity may show flood in an area surrounded by high grounds or a natural barrier. Figure 13.4 shows flood plain delineation for defined stage values. For usability of the flood plain delineation maps it is necessary to have depths and extents of flooding. A classified flood zone map is sometimes obtained from a flood to depth map. It is possible to generate several inundation maps from similar water level time series obtained from a unsteady hydraulic model which are useful to illustrate flood propagation in a plain. Comparing flood maps from two different scenerios the impact of engineering intervention and other changes can be studied. Another application has brought the probabilistic point of view to the flood plain delineation process. The system is tied to a forecasting hydraulic model and generate composite inundation maps that represents the probability of excedence rather than a single event. The inundation mapping software cycles through each probability of excedence in ascending order of, i.e., 25 per cent, 50 per cent and 75 per cent probability of excedence. The result of each probability level can be superimposed to produce a single inundation a map. All the maps can be saved or exported to a common GIS vector or raster formats for further use in flood plain management.

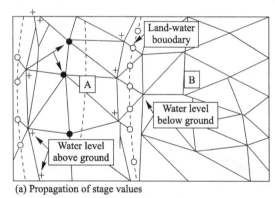

(a) Propagation of stage values

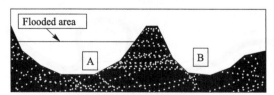

(b) Flood is contained by the high ground

Figure 13.4 Flood plain delineation for defined stage values.
(Courtsey Norman et al ASCE [49]).

A flood plain delineation tool connected to a forecasting model is useful for generating probabilistic flood maps. For this purpose, a flood plain tool must be connected to a hydraulic model in order to be operative in a real time scenerio. The resulting flood maps can be used to determine the flood effected areas and plan a relief programme.

13.8 FLOOD HAZARD ASSESSMENT

It has been established that land cover, physiography and geology is the best combination of GIS data for each of the flood hazard maps created for flood depth and flood affected frequency. GIS are specialized data system that preserve locational identities of the information they record. A computer provides the basis for storage, manipulation and display of large amount of data that have been encoded in digital form. These overlays may depict raw data in numerical form like area crop information or may show thematic information such as soil, land use, geology etc. but, they must have common geographic quantities including geographic co-ordinate system that permit them to be merged into single system. Specialized computer system like ARC-INFO, are available for rapid analysis and display of merged data in a computer screen with large amount of manipulations not possible manually and the final output may be in the form of maps or tables.

Analysis procedure in GIS can be explained as follows:

SOURCE DATA IN THE FORM OF MAPS

I

GEOCODE SOURCE MAPS

II

LAND DATA FILES

III

DEVICE APPLICABLE PARAMETERS FOR THE OBJECTIVE

IV

ANALYSE EACH CELL IN THE UNIT

V

ANALYZED OUTPUT

The object herein is to have flood hazard assessment using land cover physiographic geologic features and drainage network data. Usually flood depth and flood affected frequency are adopted as hydraulic parameters of the flood.

Flood hazard ranks estimate can be based on a weighted score for land cover, physiographical and geological data for each pixel size of land. Islam et al. (2002) proposes hazard rank assessment through land cover, physiography and geology based on a weighted score for each pixel of landuse of Bangladesh. Weighted score was estimated by the following relationship

$$\text{Weighted score} = 0.0 \times A + 1.0 \times B + 3.0 \times C + 5.0 \times D, \tag{13.1}$$

where A, B, C and D represent the occupied area percentage by non-hazard area, low, medium and high flood affected frequency respectively as a hydraulic factor, the coefficients 0, 1, 3 and 5 where adopted to describe the weight for the flooding, when flood affected frequency was considered as a hydraulic factor for each category of GIS components. When flood depth is considered as the hydraulic factor A, B, C and D represent the occupied area in percentage by non-flooded area, shallow, medium and deep flooding respectively. The coefficients of 0.0, 1.0, 3.0, and 5.0 for A, B, C and D in Eq. (13.1) were used to describe the weight for the flood damage for the land cover categories of Bangladesh as shown in Table 13.1. Points for the categories of land cover were estimated on the basis of linear interpolation between 0 and 100, where 0 corresponds to the lowest and 100 to the highest (230.21). Hazard ranks were fixed according to the value of the points, points 0 to 33 corresponds to hazard rank 1, 33 to 66 hazard rank 2, 66 to 100 hazard rank 3.

Table 13.1 Flood Hazard Ranks for Land Cover Categories. (Courtsey Islam et al ASCE [50])

Land cover category	A (%)	B (%)	C (%)	D (%)	Weighted score	Points	HR
(a) For flood-affected frequency ($A + B + C + D = 100.00\%$)							
1. Cultivated land with scattered settlements	36.93	28.49	22.01	12.57	157.38	68.36	3
2. Boro rice field	29.88	19.22	21.73	29.17	230.21	100.00	3
3. Cultivated lowland with scattered settlements	28.29	23.08	26.17	22.46	213.89	92.91	3
4. Dry fallows	55.84	23.14	12.70	8.32	102.82	44.66	2
5. Mixed cropped area with scattered settlements	47.54	26.05	14.58	11.83	128.93	56.01	2
6. Mangrove area	60.93	22.54	13.46	3.07	78.29	34.01	2
7. Highland with mixed forest	77.11	12.59	5.55	4.75	52.99	23.02	1
8. Highland with scattered settlements	42.96	33.66	14.03	9.35	122.52	53.22	2
9. Saline area, cultivated	16.98	30.67	34.84	17.51	222.76	96.76	3
(b) For flood depth of September 18, 1988 ($A + B + C + D = 100.00\%$)							
1. Cultivated land with scattered settlements	53.38	5.17	34.60	6.86	143.24	69.46	3
2. Boro rice field	38.18	8.30	37.19	16.33	201.53	97.73	3
3. Cultivated lowland with scattered settlements	47.81	7.13	37.16	7.90	158.11	76.67	3
4. Dry fallows	70.31	5.63	17.33	6.73	91.27	44.26	2
5. Mixed cropped area with scattered settlements	66.09	3.76	26.54	3.62	101.47	49.21	2
6. Mangrove area	72.42	3.41	19.77	4.40	84.72	41.08	2
7. Highland with mixed forest	78.51	6.03	9.03	6.43	65.27	31.65	1
8. Highland with scattered settlements	60.49	11.07	21.12	7.31	110.99	53.82	2
9. Saline area, cultivated	37.96	3.91	44.18	13.96	206.22	100.00	3

Flood hazard maps were constructed by considering the interactive effect of flood affected frequency and flood depth on the land cover categories and physiographical and geographical divisions. First each hazard map consisted of three ranks (HR 1 to 3) which were developed

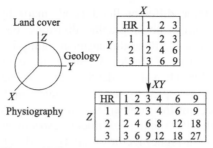

Figure 13.5 Concept of ranking matrix.

Table 13.2 Examination of Results of Flood Hazard Maps Developed by Floodwater Depth. (Courtsey Islam et al ASCE [50])

Year[*]	1	2	3	4	6	8	9	12	18	27	Total
(a) Comparison between hazard maps of 1988 and 1995 (columns and rows represent occupied pixels by hazard rank of 1995 and 1988 hazard maps, respectively)											
1	5,260	85	0	20	0	0	0	0	0	0	5.365
2	4,475	5,640	0	292	0	154	0	0	0	0	10.838
3	0	5,545	7,303	602	5,049	444	0	2,247	0	0	21.190
4	7,522	2,054	2	204	8	28	0	0	0	0	9.818
6	0	5,314	4,784	600	3,652	3	430	189	2	0	14.974
8	487	145	0	2,775	4	0	0	0	0	0	3.411
9	0	0	0	1,085	2,359	0	1,757	25	427	0	5.653
12	0	4,295	3,736	1,463	2,163	0	79	52	4	0	11.792
18	0	0	0	6,589	9,226	274	4,763	140	1,488	255	22.735
27	0	0	0	0	0	1,264	0	3,259	2,532	5,905	12.960
Total	18,021	23,078	15,825	13,630	22,461	2,167	7,029	5,912	4,453	6,160	118.736
(b) Comparison between hazard maps of 1988 and 1998 (columns and rows represent occupied pixels by hazard rank of 1998 and 1988 hazard maps, repectively)											
1	5,067	133	0	165	0	0	0	0	0	0	5.365
2	8,007	2,104	0	378	5	344	0	0	0	0	10.838
3	296	3,679	5,240	1,163	6,043	1,639	0	3,130	0	0	21.190
4	3,380	623	2	5,454	37	202	0	120	0	0	9.818
6	177	2,746	1,600	1,801	6,400	323	433	1.033	461	0	14.974
8	10	54	0	3,318	4	0	0	25	0	0	3.411
9	63	0	0	17	1,727	0	2,263	402	1,181	0	5.653
12	108	657	1,256	1,464	5,201	5	66	2,350	685	0	11.792
18	160	0	0	89	6,335	0	5,533	2,428	6,476	1,714	22.735
27	20	0	0	0	0	0	0	554	2,751	9,635	12.960
Total	17,288	9,996	8,098	13,849	25,752	2,513	8,295	10.042	11,554	11.349	118.736

[*] The top half of the table refers to years 1988 and 1995. The bottom half of the table refers to years 1988 and 1998.

only by land cover categories or physiographic divisions or geologic divisions. Hazard ranks were considered from 1 to 27 after combining the hazard ranks of the land cover categories (HR 1 to 3) physiographic divisions (HR 1 to 3) and geologic divisions (HR 1 to 3) simultaneously using the ranking matrix of the three-dimensional multiplication mode. The concept is shown in Fig. 13.5, for the development of flood hazard map.

Islam et al., developed three different flood depth maps by considering the interactive effect of the land cover categories, physiographic divisions, geologic divisions on the flood depth using the ranking matrix of the three-dimensional multiplication mode. Each hazard map consisted of hazard ranks from 1 to 27. Table 13.2 shows the comparison of the hazard map of 1988 to those of 1995 and 1998 when flood depths were considered independently. The rows represent the number of pixels occupied by the hazard ranks on the hazard maps for the 1988 flood while the columns represent the number of pixels, occupied by the hazard ranks of the hazard maps developed for the 1995 and 1998 floods respectively. It may be mentioned, here that higher hazard ranks means a higher factor of safety for development purposes of the land.

13.9 FLOOD HAZARD ZONE INDEXING—SIMPLIFIED APPROACH

Flood frequency and depth of inundation are important factors for flood mitigation measures planning. High frequency and shallow inundation resulting in extensive flooding of areas may not be very harmful to the people. Generally, the people get used to it as the farmers changes the cropping pattern both species wise and time wise considering the fertility of land into account. Therefore, for gradation of hazard zones it may be assumed that area which are deeply submerged under water for considerable period of time are more hazardous than the areas which are under shallow water for shorter period of time. The first event may be less frequent than the second.

As for example, suppose in an area of a flooded basin, designated as X the depth of inundation is 1.5 metre, average in 20 years while the number of floods that occurred in that period is 20 with average duration of 3 days. In another area of the flooded basin having area equal to Y, the depth of water say is 3 metre, the number of floods being 5, in the same period of 20 years having an average duration of 5 days.

In the first case, it is apparent that loss of life or damage to crop and other properties will be less as the people will be accustomed to such type of occurring every year as they will be well prepared with measure to be adopted for flood mitigation. Due to shallow depth and shorter duration of flooding the chance of crop loss will be less. In the second case, however, due to higher depth of water and longer duration of flood there is greater chance of loss of life and crop. Therefore if we choose a hazard criteria, index based on the following:

$$\frac{\text{Average depth of inundation} \times \text{Number of days of inundation}}{\text{Number of floods occurring within the specified period}}$$

Then in the first case the hazard index will be $(1.5 \times 3)/20 = 0.225$ for the flooded area X in the basin. Similarly, for the flooded area Y will be more hazardous than the area X.

In the above criteria the area of inundation has not been included as they are usually not known. However if it can be ascertained through use of aerial photography or remote sensing it should be included for more rigorous hazard zonation index.

14

Flood Damage Management
Tsunamis and Storm Surges

14.1 INTRODUCTION

By the term "Tsunami" it is meant to denote a set of long period ocean waves generated by any large, abrupt disturbance of the sea surface. The word tsunami is a Japanese word represented by two characters—tsu, meaning "harbour" and nami meaning 'wave'.

Generation—Tsunamis are commonly generated by under sea earthquakes in coastal regions. The earthquakes cause sea bottom movement having a significant vertical component in shallow water. Sea bottom movement results in a large scale disturbance of a mass of ocean water causing displacement of the ocean surface and generation of waves.

Majority of recorded tsunamis have been generated by earthquakes having the focal depth less than 60 km and a magnitude of 6.5 or higher on the Richter scale. The tsunami that caused major devastation in the north eastern coast of Papua New Guinea 1998 was due to an earthquake of magnitude 7.0 in Richter scale, which probably triggered a huge underwater land slide. Three waves of more than 7 m high hit the ten kilometer stretch of coast line within a very short period of ten minutes of the earthquake.

One of the greatest tsunamis due to an earthquake was of 8.5 magnitude with after shock magnitude of 6.5, that occurred at Alaska in the year 1964. The focal length of the earthquake is at depths 20-50 km. The earthquake resulted in land movement over a distance of 800 km in a period of 4 to 5 minutes vide [53]. There occurred rotation of sea bed around a hinge line with underwater uplift and subsidence in excess of 8 m and –2 m. The location of majority of tsunamis is at the active earthquake regions along the rim of the Pacific Ocean covering Aleutian Islands, Japan, New-Zealand and the west coast of South America.

14.2 CHARACTERISTICS OF TSUNAMI WAVES

A tsunami causing disturbance in the ocean will generally consist of a group of waves having periods of 5-60 mins with irregular amplitudes. Figure 14.1 shows the tsunami wave generated by an earthquake. Figure 14.2 shows a record taken at Sanfransisco after 1964 Alaskan earthquake which shows remarkable water level oscillators heights. Figure 14.3 shows the water column height at Rat Island Alaska tsunami of 2003. Generally speaking tsunami waves in the open sea are of the order of 1 m or less. Consider a simple case study—Assume a typical period of 30 minutes and considering the mean ocean depth as 4000 m the wave celerity *C* can be expressed as

$$C = \sqrt{gd} = \sqrt{9.81(4000)} = 62.64 \text{ m/s} \tag{14.1}$$

where

d = ocean depth,

g = acceleration due to gravity

L (wave length) = CT = 62.64 × 1800, with T = Time period in seconds.

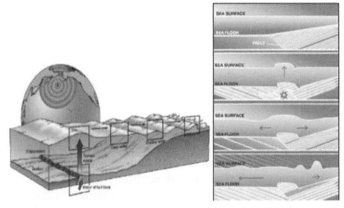

Figure 14.1 A tsunami wave generated by an earthquake.
(*Source*: http://www.tsunami.noaa.gov/tsunami_story.html)

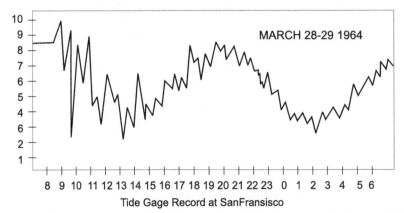

Figure 14.2 Tide Gage record at Sanfrancisco (from Speath & Berkman, 1965).

Now $d/L = 0.0036$ which is less than 0.05, so the use of the above equation is justified. An idea of the changes that occur as the tsunami waves approach shore can be obtained from the analogies given below. Consider the 30-min. wave in a water depth 10 m, in this case

$$C = \sqrt{9.81(10)} \ = 9.9 \text{ m/s}$$

$$L' = 9.9 \times 30 \times 60 = 17820 \text{ m}.$$

Ignoring the effect of friction, refraction, diffraction, etc., if the wave height is 1 m in 4000 m, the height in 10 m of water designated as H_{10} will

$$H_{10} = \sqrt{\frac{L}{L'}} = \sqrt{\frac{112.32}{17.82}} = 2.51 \text{ m}$$

Since $n = 1.0$ throughout, i.e., the ratio of wave group to phase velocity as they are considered as shallow water waves, i.e., $C_g = C$

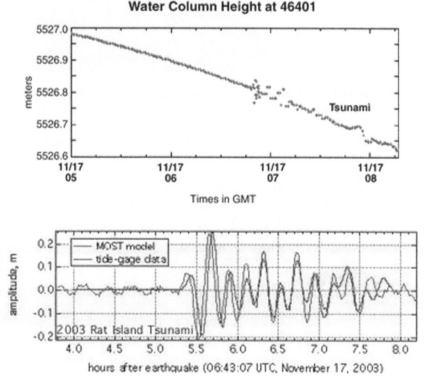

Figure 14.3 Walter Column Height and Tsunami Wave propagation at Rat Island due to Alaska Tsunami of 2003.

(*Source*: http://www.tsunami.noaa.gov/tsunami_story.html)

Now as a result of refraction near shore wave heights will be significantly greater than as indicated above. Further diffraction also causes a spreading of wave energy as they propagate. Tsunami wave groups will be more complex to analysis as they travel further from the source as a result of reflection development of multiple wave trains due to refraction or diffraction and shelf and embankment, resonant oscillation. Figure 14.4 shows changes that take place when the tsunami enters the shallow water. As tsunami approaches the coast and the water become shallow, wave shoaling compresses the wave and its speed decreases below 80 kmph, its wavelength diminishes to less than 20 km and its amplitude grows very high. Except for the very largest tsunamis the approaching wave does not break but appears as a fast moving tidal bore.

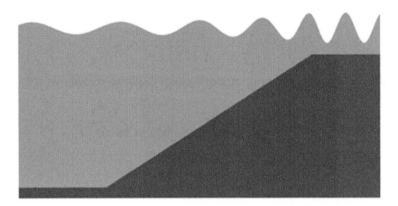

Figure 14.4 Tsunami wave transformation as it enters swallow water.
(*Source*: http://en.wikipedia.org/wiki/Tsunami Generation_mechanisms)

In case the first part of a tsunami to reach land is a trough called a drawback rather than a wave crest the water along the shore line recede dramatically, exposing normally submerged areas. The drawback occurs because the water propagates outwards with the trough of the wave at its front. Drawback begins before the wave arrives at an interval equal to half of wave's period, and drawback can exceed hundreds of meter and people unaware of its danger sometimes remain near the shore (Fig. 14.5).

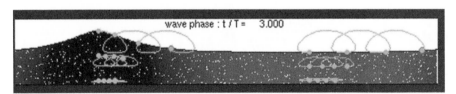

Figure 14.5 Initial drawback effect of a tsunami wave.
(*Source*: http://en.wikipedia.org/wiki/Tsunami Generation_mechanisms)

14.2 TSUNAMI RUN UP AND DAMAGE

Another important aspect of tsunami waves is its low steepness defined as (H_0^1 / T^2) where H_0^1 represents the hypothetical refracted wave height and as a result the relative run up can be quite high. The relationship governing tsunami vertical run up elevation above, sea water level, i.e, SWL can be expressed in empirical form depending on slope, incident wave height and length at the toe of the slope.

The tsunami run up is generally high and is very irregular along the same coastline for a particular tsunami. Prediction of tsunami run up is required for planning and design purpose of tsunami barriers. In the case of 1964 Alaska tsunami the wave run up is of the order of 6.4 m above MLLW [54].

Tsunamis cause damage by two mechanisms the destructive force of a wall of water travelling at high speed and the destructive power of a large volume of water draining off the land. Flooding is also due to very high velocity in the run up surge and to the impact of solid objects carried by the surge. The velocity 'v' of a surge on a dry bed can be expressed as $v = k\sqrt{gd}$ where 'd' is the water depth in front of the surge and 'k' is a coefficient varying from about 0.7 to 2, corresponding to high bed resistance and corresponding to frictionless bed. For the Alaska tsunami having a surge depth of 6.4 m and assuming $k = 1$ one gets a surge velocity of $\sqrt{9.81 \times 6.4}$ = 7.92 m/s. Velocity of this magnitude can cause major structural damage and can carry off objects like cars, boats, aeroplanes, etc. Figures 14.6 and 14.7 show the destructive image of tsunami waves generated as a result of severe earthquake of magnitude 8.5 in Japan.

Figure 14.6 Destructive effect of tsunami in Japan, 2011.
(*Source*: http://www.tsunamijapan.com)

In tsunami prone country like Japan has in its place measures to counter the effect tsunami flooding. Japan has built many tsunami walls of upto 4.5 m height to protect populated coastal areas (Fig. 14.8). In other localities channels and flood gates have been built to redirect the water coming from tsunamis, however, their effectiveness is questioned in view of much higher wave heights that are generated by tsunami, i.e., as high as 30 m. However, the protective wall helps in slowing down and moderating the height of the tsunami. Design structures (geometry, orientation, strength) to withstand tsunami surge construction of offshore barriers can be ahead [55].

Figure 14.7 Damages due to tsunami in Japan earthquake 2011.
(*Source*: http://www.tsunamijapan.com)

Figure 14.8 Tsunami walls to protect coastal areas.
(*Source*: http://en.wikipedia.org/wiki/Tsunami Generation_mechanisms)

14.4 TSUNAMI FORECAST AND WARNINGS

As a direct result of the Indian Ocean tsunami, 2004 a reappraisal of the tsunami threat for all coastal areas is being undertaken by national governments and the United Nations Disaster Mitigation Committee. As a result, a tsunami warning system is being installed in the Indian Ocean. Computer models can predict tsunami arrival usually within minutes of the arrival time.

Bottom pressure sensors relay information in real time. Based on these pressure readings and other seismic information and the bathymetry of the sea floor and coastal topography, the models estimate the amplitude and surge height of the approaching tsunami. Figure 14.9 shows a deep water bouy used in tsunami warning system.

In order to make accurate forecasts of tsunami wave in coastal waters it is necessary to construct tsunami refraction diagram. A technique for constructing tsunami refraction diagram has been presented [56].

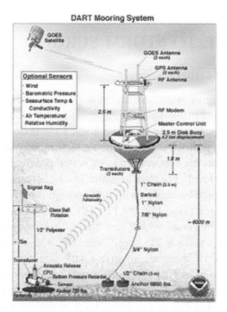

Figure 14.9 A deep water bouy for tsunami warning system.
(*Source*: http://www.tsunami.noaa.gov/tsunami_story.html)

For carrying out wave refraction studies it is necessary to measure bathymetry in the near shore coastal zone, by carrying out a hydrographic survey. For this purpose a low cost hydrographic surveying system can be adopted [57]. This surveying system can be deployed in a variety of locations where it is difficult to deploy a heavier, engine equipped vessel. The primary components of the Kayak surveying system consists of an echo-sounder to measure the sub-aqueous bottom topography. A GPS receiver provide horizontal positions and vertical elevations, as they are required in a wave environment. Further a data acquisition system is used to synchronously store the GPS & eco-sounder data strings.

Figure 14.10 shows a Kayak hydrographic surveying system. For details of its equipments on board and operations reference be made to the authors paper.

Figure 14.10 Kayak hydrographic surveying system.
(*Curtesy*: Jourl. of coastal Research vol 27, No3 2011)

14.5 FLOODING DUE TO STORM SURGES

14.5.1 Introductory Remarks

A storm surge is an offshore rise of water associated with a low pressure weather system typically tropical cyclones. They are caused primarily by high winds pushing on the ocean's surface, the wind causes the water to pile up higher than ordinary sea level. Low pressure at the center of a weather system also has a small effect, besides the bathymetry of the body of water. So truly speaking a storm surge can be considered as an abnormal rise of water generated by a storm over and above the predicted astronomical tide level. The rise in water level can cause extreme flooding in coastal areas particularly when storm surge coincides with normal high tide resulting in water level reaching up to 6 m or more in some cases (vide Fig. 14.11). For prediction of the storm surge magnitude requires weather forecasts to be accurate within few hours. Factors that determine the surge heights for land falling tropical cyclones include the speed, intensity, size of the radius of maximum wind, radius of the wind fields, angle of the track relative to the coast line, the physical characteristics of the coast line and the bathymetry of the offshore water. As mentioned already a severe cyclonic disturbance in near shore/shallow water region usually results in a large water level fluctuation as a result there can occur a rise or a fall in the water level at different location and times.

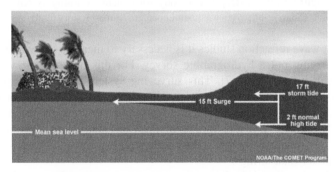

Figure 14.11 Flooding due to storm surge.
(*Source*: http://www.nhc.noaa.gov/ssurge)

The forces responsible for water level change during a storm are as follows:

1. Atmospheric pressure differentials
2. Wind wave set up
3. Long wave generation by the moving disturbance due to pressure.
4. Coriolis acceleration
5. Surface wind shear as well as corresponding bottom stress due to current generated by surface wind.

The pressure effects of tropical cyclone will cause the water level to rise in the open sea in regions of low atmospheric pressure and fall in the regions of high atmospheric pressure. The rising water level will counteract the low atmospheric pressure such that the total pressure at some plane beneath the water surface remains constant. For every millibar drop in atmospheric pressure this effect is estimated to cause 10 mm approximately rise in sea level. Figures 14.12 to 14.13 show the processes that generate storm surge and wind–pressure components of Hurricane storm surge.

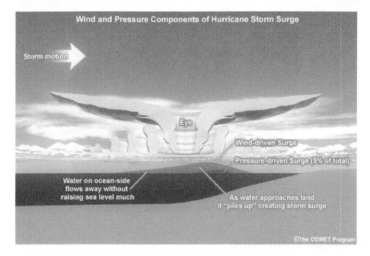

Figure 14.12 Figure to show the processes the generate storm surge.
(*Source*: http://www.nhc.noaa.gov/ssurge)

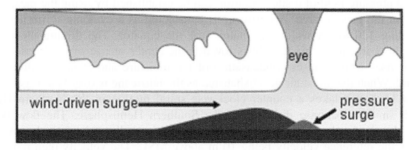

Figure 14.13 Showing wind-pressure components of hurricane storm surge.
(*Source*: http://en.wikipedia.org/wiki/storm_surge)

14.5.2 Storm Surge Vulnerable Areas

The destructive power of surge, battering waves may increase damage to buildings directly along the coast. The weight of water approximately 1050 kg per cubic meter and extended pounding by frequent waves can demolish any structure not specifically designed to withstand such forces. Figure 14.15 show the damages in house, road and boats damaged in a harbour.

Figure 14.15 Damage due to storm surge.
(*Source*: http://www.nhc.noaa.gov/ssurge)

14.5.3 Storm Surge Estimation

For calculation of storm surge detailed information of the spatial and temporal distribution of current speed, its duration and surface air pressure for the design storm conditions are required. The wind and pressure field of a site condition can be established by using the measured record of the storm in the general area. A hurricane is considered as a cyclonic storm having current speed exceeding 120 kmph which originates near the equator. The driving mechanism in a hurricane is the warm moist air that flows towards the center or eye of the hurricane. It gives off heat as it rises which causes condensation of the moisture and at higher altitudes, the air flows outward. When the moist air gets exhausted as the hurricane is over land it gets dissipated. The air flow in the air takes a counter clockwise spiral motion in the Northern Hemisphere whereas it is in the clockwise direction in the Southern Hemisphere. The flow field in the cyclone viz. the wind velocity increase to a maximum at a radius 'r' relatively small distance from the eye, the reference velocity being 10 m above MSL. The velocity then reduces rapidly to a low value at the eye. The air pressure on the contrary continually drops from the ambient pressure at the outer edge to the lowest pressure at the eye. The pressure is designated as the

CPI, i.e., Central Pressure Index expressed in terms of mercury. The other parameter under reference are the forward speed, i.e., the speed of its movement and the direction of its travel.

The Standard Project Hurricane (SPH) represents a series of hurricanes so designed as to represent the most severe combination hurricane parameters, viz. radius, wind speed, and central pressure index, i.e., R, V_S, CPI which reasonably represents the characteristics of a region.

Meyers proposed an empirical relationship for estimation of surface pressure distribution [58]

$$p_a - p_r = (P_a - \text{CPI})(1 - e^{-R/r}) \qquad (14.2)$$

where

p_r = pressure at any radius r from the eye

p_a = ambient pressure = 76.2 m of Hg for SPH.

Simplified Storm Surge calculation can be made by estimating separately the rise in water level due to (a) initial setup, (b) pressure setup, (c) wave setup, (d) long wave setup, (e) wind and bottom stress setup, (f) Cariolis setup.

14.5.3.1 Initial setup

It has been observed that the near shore water level often rises around 50 cm or higher above, astronomical tide level prior to the arrival of a storm. The pressure set up due to a surface pressure variation S_p between two points over a central body of water can be expressed as

$$S_p = \frac{\Delta p}{\gamma}$$

where

γ = Specific weight of water.

14.5.3.2 Pressure setup

If we consider Δp is the pressure drop from the periphery of a hurricane to a point within the hurricane, then S_p can be calculated from the following expression

$$S_p = \frac{p_a - p_r}{\gamma} = \frac{p_a - \text{CPI}}{\gamma}(1 - e^{-R/r}) \qquad (14.3)$$

14.5.3.3 Wave setup

Wind waves generated due to storm causes a nearshore setup S_{near} because of mass transport due to a small forward movement of mass as water particles advance slightly during each orbit. An expression of S_{near} has been given by U.S. Army Coastal Engg. Research Center [59].

$$S_{\text{near}} = 0.19 \left[1 - 2.82 \sqrt{\frac{H_b}{gT^2}} \right] H_b \qquad (14.4)$$

where

H_b = breaker height in the surf zone.

14.5.3.4 Long wave setup

Due to moving surface disturbance waves are generated. These waves achieve their greatest amplitude when the disturbance speed equals the speed of a shallow water wave ($\sqrt{gd}$), provided the disturbance persist at the required speed for sufficient time for complete development of the wave. A long wave and resulting water level rise and fall can be generated by a hurricane moving at the right speed for a sufficient duration. Hardly any information are available for predicting the peak amplitudes of these long wave surges.

In order to make accurate forecasts of tsunami wave in coastal waters it is necessary to construct tsunami refraction diagram. A technique for constructing tsunami refraction diagram is presented [56].

14.5.3.5   Wind and bottom stress setup

The water surface stress τ_s exerted by the wind is given by an equation of the form

$$\tau_s = C_d \rho_a U^2$$

where

ρ_a is the air density,

U is the wind velocity at reference elevation (usually at 10 m) and

C_d is a drag coefficient that depends upon the surface roughness and boundary layer characteristics. Wilson evaluated and summarized the results of 47 references that presented values for C_d. He concluded that the most reasonable value for C_d varied from 1.5×10^{-3} for light winds asymptotically to 2.4×10^{-3} for strong winds [60].

More convenient form of equation is in terms of the water density

$$\rho, \text{ so } \tau_s = k\rho U^2 \tag{14.5}$$

where the wind stress coefficient $k = \{1.19 \times 10^{-3}\}C_d$.

Van Dorm relation for estimating k is $1.21 \times 10^{-6} + 2.25 \times 10^{-6}\left(1 - \dfrac{5.6}{U}\right)^2$ where U is the wind speed in m/s [61].

The surface wind stress generates a current that in turn, develops a bottom stress. Saville [62] has provided the following relationship

$$\tau_b + \tau_s = (3.3 \times 10^{-6})\, \rho u^2$$

and suggested that $\tau_b / \tau_s = 1$.

Considering a section of the nearshore water column of length Δ_x, normal to shore unit width and depths d and $d + \Delta S_w$, where ΔS_w is the set up due to wind and bottom stress acting over the length Δ_x. The water surface slope $\Delta S_w / \Delta_X$ is very flat. So a static balance between the forces yields the following expression

$$\Delta S_w = d\left[\sqrt{\frac{2kU^2\Delta x}{gd^2} + 1} - 1\right] \tag{14.6}$$

when the wind blows at an angle θ to the x-direction the effective stress is $k\rho U^2 \cos\theta = k\rho U U_x$, where U_x is the component of the wind velocity in the x direction. Accordingly Eq. (14.6) becomes

$$\Delta S_w = d\left[\sqrt{\frac{2kU U_x \Delta x}{gd^2} + 1} - 1\right] \tag{14.7}$$

14.5.3.6 Coriolis setup

Coriolis acceleration causes a moving water mass to deflect to the right in the northern hemisphere, perpendicular to the direction of motion. The current velocity V coming out normal to page due to Coriolis acceleration causes a set up along the shore line. Ignoring water surface and bottom stress a static balance between the hydrostatic forces and the Coriolis acceleration or force per unit mass yields.

$$\frac{1}{2}\rho g d^2 + [2\omega V \sin\phi]\rho g d \Delta x - \frac{1}{2}\rho g (d + \Delta S_c)^2 = 0 \qquad (14.8)$$

The term in brackets is the Coriolis force per unit mass, ϕ is the latitude, ω is the angular speed of rotation of the earth (7.28×10^{-3} rad/sec), and ΔS_c, is the Coriolis setup force or

$\Delta S_c = \dfrac{2\omega}{g} V \sin\phi \, \Delta x$.

Away from the shore wind generated currents are free to respond to the Coriolis acceleration, it is only nearshore that this response is restricted and set up (or setdown) develops. It is very difficult to predict the magnitude of nearshore wind generated current as well as the degree of restriction and resulting setup that occurs. Bretschneider solved the equation of motion for the alongshore direction including only the wind and bottom stresses and the resulting local acceleration, to obtain an equation for longshore current velocity [63]. For the resulting steady state condition (i.e., sufficient wind direction)

$$V = U d^{1/6} \sqrt{\frac{k}{14.6 n^2}} \sin\theta \qquad (14.9)$$

where n is Manning's roughness co-efficient (typical value 0.035), θ is the angle between the wind direction and a line perpendicular to the coast and/or bottom contours, and V is the resulting longshore velocity. Equation (14.9) can be used to calculate the current velocity field in the nearshore zone for a given wind field. From this, ΔS_c can be estimated by Eq. (14.8).

With the help of procedure outlined above and with the estimated initial set up and the calculated pressure, wind and bottom stresses and Coriolis set up for each of a sequence of offshore storm portions, it is possible to develop the total storm hydrograph for a given coastal location.

14.5.4 Storm Surge Estimation by Numerical Modelling

Nowadays however one uses a computer code, based on numerical finite difference approach using bathymetric storm tide theory. One approach is to use vertically integrated horizontal equations of continuity and motion neglecting vertical component of motion, with x-axis normal to shore and y-axis parallel to shore (Fig. 14.16). Assuming precipitation rate as negligible and considering flow towards shore is small, instantaneous sea surface elevation as constant and parallel to the shore, finally neglecting convective acceleration.

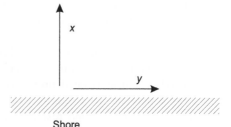

Figure 14.16 Co-ordinate system.

The simplified equations can be written as

$$gd\frac{\partial s}{\partial x} = -fq_y + \frac{\tau_{sx}}{\rho} \tag{14.10}$$

$$\frac{\partial q_y}{\partial t} = \frac{\tau_{sy} - \tau_{by}}{\rho} \tag{14.11}$$

where

q_y = discharge/unit width along y-axis

s = Surge elevation above SWL (Sea Water Line)

f = Coriolis parameter $2\omega \sin \phi$; ϕ = latitude

τ_{sy}, τ_{by} = wind shear at surface and bed along y-axis.

The equators are sequentially solved for q_y and s for small increment of time and space. The time and space increments may be non-uniform to allow for changes in rate of water level rise as storm approaching shore and irregularity in hydrography.

To take care of real situations such as irregular bays, estuary and shoreline where flooding of low lying terrain occurs it is necessary to use two-dimensional velocity integrated finite difference modes, accordingly the differential equations have to be expressed in a different manner.

$$\frac{\partial q_x}{\partial t} + \frac{q_x}{d}\frac{\partial q_x}{\partial x} + \frac{q_y}{d}\frac{\partial q_y}{\partial y} = -fq_y - gd\frac{\partial s}{\partial x} + KUU_x - cqq_x \tag{14.12}$$

$$\frac{\partial q_y}{\partial t} + \frac{q_x}{d}\frac{\partial q_y}{\partial x} + \frac{q_y}{d}\frac{\partial q_y}{\partial y} = -fq_y - gd\frac{\partial s}{\partial y} + KUU_y - cqq_y \tag{14.13}$$

where c is bottom stress coefficient and $q = \sqrt{q_x^2 + q_y^2}$.

The affected zone is divided into square segments in horizontal plane and an average depth and bottom stress co-efficient are assigned to each square. The continuity equation and equations of motion are then written in finite difference form for applications to each square segment in the mesh.

They are then solved with appropriate boundary conditions, which are water level boundary, where no flow will be there, adjacent low lying square which will flood when it will reach certain height, inflow from rivers surface run off, seawalls and offshore boundary having a prescribed water level, time boundary due to the astronomical tide. The unknown at each square segments, i.e., q_x, q_y and s are then solved over each row segments in the mesh for a given boundary condition. Time is advanced and the spatial calculations are repeated. The calculations of storm surge level is required to plan coastal defence works specially to determine the height of sea dykes and associated protection device. It has already been mentioned the damage in the coastal defence work due to a super cyclone hitting the part of West Bengal.

For reconstruction and restoration, new embankment is proposed on the country side as well as suitable drainage structures to ensure proper drainage of the upstream areas.

There could be various alternating in the shape, of the dykes depending our availability of material and use of construction machines and other local factors such as availability of infrastructural facilities.

In the present situation considering all aspects suitable design is proposed by the Irrigation Department of the Government of West Bengal for reconstruction the washed out embankments.

14.5.4.1   Wave modelling in coastal water

There are many locations along the coast which are vulnerable to tropical storms and frequent intense cyclones. One such area is the coastal waters along the U.S. north-east coast.

Here the water storm system originating at mid latitude, entrain maritime polar air from the North Atlantic Ocean resulting in heavy precipitation in the form of rain and surge. The wind system associated with the storm system can attain magnitudes comparable to tropical storms.

Sometime the combined affect of long wind duration and large fetch can generate wave that those produced by hurricanes. The wave generated current erode beaches, dams, endanger structures and reduce protection against flooding and wind damages. Breaking waves result in local setup along the coast which can increase the SWL (Sea Water Level) upward of 20 per cent of the deep water wave height [64].

Estimating near shore wave transformations is a critical input for coastal engineering investigation. Wave influence in the coastal zone affects bathymetric evolution, channel shoaling, long- and short-term shore line changes and the design and in maintenance of coastal structures.

A reasonably accurate wave characteristics could be calculated or predicted from wave models. Wave models can provide the spatial and temporal resolution of the incident wave fields necessary to investigate setup along the coast.

The two wave models are the (STWAVE) steady state wave, developed by the Waterway Experimental station. U.S. Army crops of Engineers at Vicksburg and the other (SWAN). Simulating waves nearshore is developed by Delft Technological University. STWAVE version 2 is a steady state finite difference model based on wave action balance [65]. The model simulates depth and current induced refraction and shoaling, depth and steepness induced wave breaking, diffraction, wave generation by wind and wave-wave interactions. The model is forced by spatially homogeneous offshore wave and wind conditions. SWAN is a third generation wave model and also a finite difference model based on the wave action balance estimate which can be run in steady state. It simulates the following physical phenomenon, wave propagation in time and space, shoaling, refraction due to current and depth, diffraction, reflection wave generation by wind, nonlinear wave, wave interactions, depths and steepness induced breaking, blocking of waves by current. A representation of the coastal morphology is critical for any numerical investigation wave transformation as breaking processes are controlled by water depth and bathymetric variations. Wave energy will focus or converge around headlands and diverge or spread through bays and canyons. The bathymetric data can be obtained by carrying out survey of the area using survey system such as Kayak as mentioned earlier. Also it can be obtained from scanning hydrographic operational air borne like radar survey which can be obtained from suitable agencies. The bathymetric and topographic data are then interpolated into the STWAVE.

SWAN modelling grids are at a resolution of 50 meters. The spacing is sufficient to provide adequate spatial distribution of wave properties. The sea ward boundary usually extended into water depth of about 50 meters encompassing the entire area under investigation. Wave observation can be obtained by various agencies which collect the data such as NOAA (National Oceanographic and Atmospheric Administration), NDBC (National Data Buoy Centre) USACE (US Army Crops and Engineers) in USA. In India data can be obtained from records maintained by different Ports, CWPRS (Central Water and Power Research Station, Pune), NIO (National Institute of Oceanography, Goa), etc.

The latest numerical model to calculate storm surge at a designated area is designated as SLOSH model. SLOSH stands for the Sea, Lake and Overland Surges from hurricanes is a computerized numerical model developed by NWS (National Weather Service) to estimate surge heights resulting from historical hypothetical or predicted hurricane by taking into account the atmospheric pressure, size, forward speed and track data. The parameters are used to create a model of the wind field which drives the storm surge. They consist of a series of physical equations which are applied to a specific local shore line incorporating the unique bay and river configuration water depths, bridges, roads, levees and other typical features.

Deterministic approach forecasts is based on solving physical equations and uses a single simulation based on perfect forecasts which results in a strong dependence on accurate meteorological input. The location and timing of a hurricane's landfall is crucial in determining which areas will be inundated by the storm surge. Storm changes in track, intensity, size, forward speed and landfall location can have huge impact on storm surge. At the time of emergency, managers must make an evacuation decision, the forecast track and intensity of a tropical cyclone are subject to large errors thus a single simulation of the SLOSH model does not always provide an accurate depiction of the true storm surge vulnerability. The probabilistic surge product is the newest edition to a size of available storm surge products which incorporates of the past forecast performances to generate an ensemble of SLOSH runs based on distribution of cross track, along track, intensity and size errors. Composite approach products surge by running SLOSH several thousand times with hypothetical hurricanes under different storm conditions. The product generated are the maximum envelope of water and maximum of MEOs which are generated by the National Hurricane Center, USA are the best approach for determining the storm surge vulnerability of an area.

15

Flooding due to Collapse of Dams
Breaching of Flood Levees and Urban Coastal Areas

15.1   INTRODUCTION

Phenomena like dam break occurs due to massive landslides, soil mass falling in the reservoir generating surface waves, which may be responsible for failure of dams. The failure will in turn cause devastation to the people residing in the downstream side. The failure of dams can also be attributed due to torrential rain and consequent flood generated as a result of it, especially the old ones, which are under threat. The consequences of dam failure whatever be the cause results in loss of life and damage to property. It may be mentioned here that the failure of Teton dam in USA located at Idaho in the year 1976 caused about four hundred million-dollar loss, eleven people died, about 25,000 people were rendered homeless. The dam was a 93 m high earthen dam with 914.4 m long crest. The breach developed rapidly and subsequent erosion washed away 3 million cubic meters of earth while releasing a peak flow of 70,000 cumecs. The failure of the dam is the result of complex concourse of causes and mechanisms. Generally one can say the failure of dam depends mainly on the type of dam, nature of breach formation and its development, foundation and embankments, type of external disturbances to which the dam is subjected like the forces acting on the dam, incoming flood and the existing operation conditions. Dam failure is classified as sudden or gradual depending on the duration of failure of the dam. If the duration is of the order of 10 to 15 minutes the failure can be termed as sudden otherwise it can be termed as gradual. Arch and gravity dams fail by sudden collapse, overturning or sliding away of the structure due to overstress caused by inadequate design or excessive forces that may result in overtopping of flood flows, earthquakes, deterioration of abutment or foundation materials.

15.2 CASES OF FAILURE OF EARTH DAMS

The major causes of earth dam failure are sliding, overtopping, seepage, human intervention and earthquake. Sliding can occur to the reservoir banks, embankment or foundation when the shear stress due to external loads along a plane in the soil mass exceeds the shear strength that can be sustained in that plane. In such a situation the failure along the plane is imminent. Periodic cycle of the saturation of the porous material and the increased pore water pressure reduce the shear strength which leads to loss of stability and produce major landslides. Due to this sudden arrival of soil mass in the reservoir, generally produces a high solitary wave propagating against the earth dam. An initial breach can form immediately in the dam and progressive erosion caused by flowing water will lead to a partial or total failure of the dam. Improvements in the analysis of probable maximum storm have caused significant increase in the predicted probable maximum flood (PMF). Many earlier dams have not been designed on PMF values. As a result many dams once considered safe are now considered unsafe due to inadequate spillway capacity. In the year 1982, United States Army Corps of Engineers inspected 8,639 high hazard dams under Federal Dam Safety Programme. Of these 2,884 dams were found to be potentially unsafe due to inadequate spillway capacity. Seepage of water can be through the embankment, foundation or abutment of the dam. Uncontrolled or controlled seepage through the body of the dam or foundation may lead to piping or sloughing and the subsequent failure of the dam. Due to continuous seepage breach will be formed in the embankment resulting in the outflow through the breach. The breach size will continuously grow as material is removed by outflow from the storage and storm water runoff. The size, shape and time required for the development of breach are dependent on the embankment material and the characteristics of the flow forming the breach. Breaches of this type can occur fairly rapidly or can take several hours to develop. Earthquake at the dam site can produce waves or landslides and consequently leads to a partial or total failure of dam. Depending on the severity of the earthquake the entire dam may be washed off or only a part of it may be removed due to breach failure.

Such failure is sudden in nature. Other causes are due to differential settlement of the foundation of the dam. Londe has reported the findings of Middlebrooks who provided a statistical analysis of earth dam failure using 200 case histories. The record covers a period of hundred years. Figure 15.1 provides the result in terms of percentages of each category of failures. From the diagram it is obvious that only 15 per cent of the total failure numbers is by sliding which could be evaluated by usual concept of factor of safety. The remaining 85 per cent do not come under conventional stability analysis and are relevant to design, construction and operation procedures, which can hardly be computed in the conventional manner.

To properly describe the gradual breaching of an earth dam, the geometry of the breach must be related to the hydraulics of the flow and bed material properties, i.e., the energy of the flow through the breach at any instant of time must be compatible with work required to be expended in the erosion of the bank material and its transport. It is to be noted that rate of erosion and mode of failure of dam determine the shape and duration of the flood wave to a large extent. Fread assumes the rate of growth of the breach to be time-dependent, with either rectangular, triangular or trapezoidal shape [66].

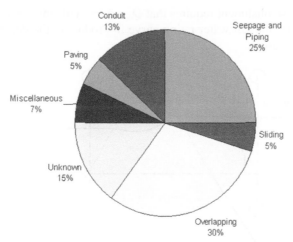

Figure 15.1 Statistics of earth data failures.

The outflow discharge through the breach will have to be compatible with the stage-discharge hydrograph (tail water rating curve) at the downstream side of the dam during the establishment of the corresponding flow depths. The outflow discharges are governed by St. Venant's equations while passing through the river channel. Fread observes that the outflow hydrograph is sensitive to the rate of vertical erosion of the breach, but assumes any error in prediction to be dampened as the sharp waves move downstream.

15.3 METHOD OF ANALYSIS [67]

15.3.1 Upstream Boundary Condition and Flow through Breach

During a given computational time step Δt, the rise/drop in reservoir level can be given by [68]

$$\Delta H = (Q_{out} - Q_{in})\Delta t/A_n \tag{15.1}$$

where ΔH = rise/drop in reservoir level from time level n to time level $n + 1$; Q_{out} = flow rate at upstream section of breach at time level n; Q_{in} = inflow to the reservoir at time level n and A_n = reservoir surface area at time level n. Equation (15.1) amounts to a local linearisation of the upstream boundary condition and is accurate enough for small durations of time steps.

The flow conditions through the breach at any instant of time are schematically shown in Fig. 15.2. Considering the flow through breach to be approximated as flow over a broad-crested weir, the outflow discharge can be given by [69]

$$Q_{out} = Ck_1k_2k_3B\sqrt{g}H_1^{3/2} \tag{15.2}$$

where C is a flow constant as applicable for a suppressed weir, k_1, k_2, k_3 are constants. k_1 accounting for the effects of viscosity and surface tension, k_2 accounting for curvature and frictional effects and k_3 accounting for the influence of submergence); B is the width of the breach and H_1 is the head over the weir.

The channel flow establishment requires that Q_{out} along with any discharge over the spillway must be compatible with the tail water rating curve in producing the depth of flow downstream of the dam in the river gorge.

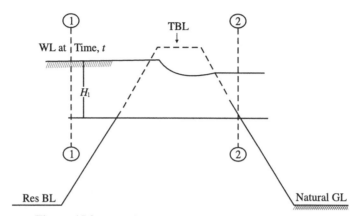

Figure 15.2 Overflow through Beach-definition section.

15.3.2 Routing of Flow through Channel

The weighted time average implicit method using Preissman scheme with adaptive time grid has proved to be a very efficient method of routing flood flows in a river channel [70, 71]. However, the success of this method is dependent on the evaluation of the friction slope, S_f in Eq. St. Venent Eq. (3.16) of unsteady flow. It is very sensitive to the value of Manning's roughness coefficient. The calibration for the combined Manning's roughness value from the available raw/untested data will be a prerequisite for successful implementation of the available numerical scheme as mentioned above.

A suitable method for such calibration, which, infact, amounts to an approximate method for routing of the flow is contained in the method of finite increments due to Thomas [72]. In the method of finite increments, the channel is divided into reaches of finite length Δx and routing period is denoted by Δt. The notation in Fig. 15.3 is used to express the flow variables in the unsteady flow equations. of continuity and momentum with due consideration for the time weightage through the factor Ψ.

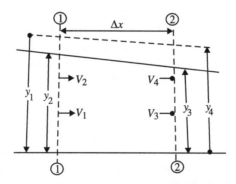

Figure 15.3 Definition section for Thomas method.

$$y = \frac{\Psi(y_2 + y_4) + (1 - \Psi)(y_1 + y_3)}{2} \tag{15.3}$$

$$A = \frac{\Psi(A_2 + A_4) + (1 - \Psi)(A_1 + A_3)}{2} \tag{15.4}$$

$$T = \frac{\Psi(T_2 + T_4) + (1 - \Psi)(T_1 + T_3)}{2} \tag{15.5}$$

$$R = \frac{\Psi(R_2 + R_4) + (1 - \Psi)(R_1 + R_3)}{2} \tag{15.6}$$

$$V = \frac{\Psi(V_2 + V_4) + (1 - \Psi)(V_1 + V_3)}{2} \tag{15.7}$$

The partial differential equations of continuity momentum equations can be replaced by

$$\frac{\partial y}{\partial x} = \frac{\Psi(y_4 - y_2)}{\Delta x} + \frac{(1 - \Psi)(y_3 - y_1)}{\Delta x} \tag{15.8}$$

$$\frac{\partial V}{\partial x} = \frac{\Psi(V_4 - V_2)}{\Delta x} + \frac{(1 - \Psi)(V_3 - V_1)}{\Delta x} \tag{15.9}$$

$$\frac{\partial Q}{\partial x} = \frac{\Psi(Q_4 - Q_2)}{\Delta x} + \frac{(1 - \Psi)(Q_3 - Q_1)}{\Delta x} \tag{15.10}$$

$$\frac{\partial y}{\partial x} = \frac{(y_2 - y_1)/\Delta t + (y_4 - y_3)/\Delta t}{2} \tag{15.11}$$

$$\frac{\partial V}{\partial x} = \frac{(V_2 - V_1)/\Delta t + (V_4 - V_3)/\Delta t}{2} \tag{15.12}$$

Substitution of Eqs. (15.4) to (15.12) into continuity equation yields:

$$V_4 = [\{\Psi(T_2 + T_4) + (1 - \Psi)(T_1 + T_3) (y_1 - y_2 + y_3 - y_4)/2\Delta t\}\Delta x$$
$$+ \{(1 - \Psi)(A_1 V_1 - A_3 V_3)\Psi A_2 V_2\}]/\Psi A_4 \tag{15.13}$$

Similarly, substitution of Eqs. (15.4) to (15.13) into momentum equations yields

$$S_0 = \sum_{i=1}^{4} S_{f_i} + \{\Psi(y_4 - y_2) + (1 - \Psi)(y_3 - y_1)\}/(2\Delta x) + [\{\Psi(V_2 + V_4) +$$
$$(1 - \Psi)(V_1 + V_3)\}/(2g)\}] \times [\{\Psi(V_4 - V_2) +$$
$$(1 - \Psi)(V_3 - V_1)\}/(\Delta x)] + \{(V_2 - V_1 + V_4 - V_3)/(2g\Delta t) \tag{15.14}$$

If Ψ is chosen as 0.5, the above equations will be the same as those in Thomas method of finite increments procedure [72]. Numerically, this is the same as Crank-Nicholson central difference scheme. If Ψ is chosen as 1.0, the scheme reduces to fully forward finite difference scheme.

In the following problem, the quantities y_1, A_1, V_1, y_2, A_2, V_2, y_3, A_3, V_3 are known from the initial conditions or from the computations on previous reaches. The unknowns y_4, and V_4 can be obtained by solving the Eqs. (15.13) and (15.14) through a process of successive approximation. Repeating the computations for the subsequent reaches, the complete water surface profile and discharges along the river at various times can be obtained.

15.4 BREACHING OF FLOOD LEVEES

15.4.1 Introduction

Engineering techniques such as dykes, levees, marginal embankments have been adopted to control flood inundation of the flood plain area to a certain extent. As the embankment height increases the greater is the risk of breaking. So it is necessary to study the risk involved once the embankment breaches. This type of investigation requires accurate flood plain topography data, which is not easily available. However, the flood inundation phenomenon can be suitably modelled using the available flood plain topography and using the numerical river models.

15.4.2 Numerical Model

The numerical model for the simulation of hydro-dynamical behaviour of river and its flood plains should have the capacity to tackle the complex topography, ability to simulate steady/unsteady, both sub- and super-critical flow, continuous and discontinuous flow, abilities to simulate flow through tributaries and existing hydraulic systems.

A numerical model code named CCHEZ2D (Center for Computation and Hydro Science and Engineering) has been developed by National Center for computational Hydro-Science and Engineering. Units of Mississippi, USA has been widely accepted to deal with the type of situation that we are discussing.

The model has been used by Dutta to study the flood hazard due to embankment breaching in the river Brahmaputra for flooded area in Lakhimpur district of Assam [73].

15.4.3 Case Study Area

The study area (93°55′E, 27°15′N) is the part of Lakhimpur district situated on the North East part of Assam (Fig. 15.4). The flood plain is the part of river Brahmaputra and situated just upstream of the confluence with Subansiri river and Majuli island. It covers an area of over 600 km² which mainly comprises of rural and agricultural area. The heavy monsoonal rainfall ranges from 2840 mm in some places of its valley to 6350 mm or more in the hilly terrain. Nearly 60-70 per cent of this rainfall occurs during monsoon period from June to September, and is responsible for frequent and damaging floods (Singh and Sarma, 2004).

The Matmara embankment breaching area has been facing acute erosion problem after the great earthquake of 1950 due to subsequent morphological changes in the river Brahmaputra (Sharma, 2005). Because of continuous changing of morphology in the main river and its tributaries, the embankments have been now very prone to breaching during flood seasons. Breaches on the Brahmaputra dyke from Sissikalghar to Tekeliphuta area have occurred many times.

15.4.4 Data Used and Methods

The model was initially calibrated with flood inundation at Matmara embankment breaching, located in Lakhimpur district, Assam, for the year 2008 high flood period. The computational mesh has been constructed from SRTM (Shuttle Rader Topographic Mission) topographic data which provides digital elevation model of 90 m raster size. The study reach of the river bed was mapped onto a grid with 271 cells in horizontal direction and 264 cells in the vertical direction. In field scale it resulted in an average horizontal cell size of 175 × 185 m. Field survey of the flood plain was carried out using Differential Global Positioning System (DGPS) in the pre-monsoon period and an accuracy of SRTM DEM was found to be less than 1 m. The embankments on the either side of the Brahmaputra river were identified from satellite image and were modified with the locally available data from DGPS survey.

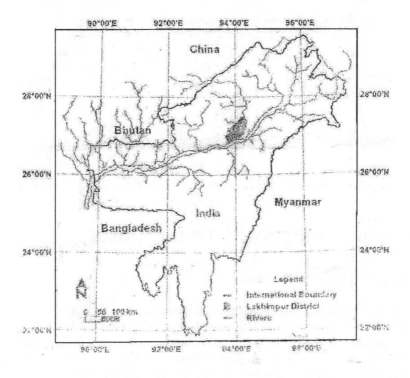

Figure 15.4 Location map of the study area at the Lakhimpur district, Assam (vide ref. 73).

15.5 ANALYSIS AND RESULTS

Simulations were carried out for different discharges representing flood waves of different return periods. Assumption were made to consider the river bed and the soil as impermeable or completely saturated, so that no significant loss in terms of recharge and evaporation occurs.

Area of flood inundation for different discharges at Matmara Embankment, Assam is shown in Fig. 15.5.

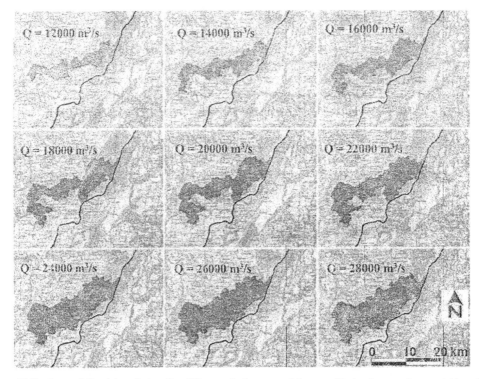

Figure 15.5 Area of flood inundation for different discharges at Matmara Embankment, Assam (vide ref. 73).

15.6 FLOODING IN URBAN COASTAL AREAS

15.6.1 Introductory Remarks

Flooding in Coastal Urban areas are due to heavy precipitation accompanied by high tidal level. This has become a big concern for major cities located in coastal areas due to loss of life and property, disruptions of transport services, power supply and health related diseases. In India, the cities that are affected most are Kolkata, Mumbai, Surat, etc. The annual disasters from urban flooding will affect more than 60 crore people in India by 2020.

There are various reasons for urban flooding and it could be due to inadequate drainage system, increase in the paved areas resulting in of increased runoff, from intense precipitation due to sudden cloud blasts. In coastal areas tidal variation severely causes drainage congestion specially the areas where draining of storm water from the areas to the discharge location depends on tide.

To study the problem there is a need to have a urban flood simulation tool which will provide an indication in advance the flooding areas or the flooding nodes in the drainage channel of the catchment for a given precipitation scenario. Numerical methods forms the main model building task for such study. Mark simulated urban flooding by 1 D (one dimensional) hydrological modelling with GIS for the city of Dhaka, capital of Bangladesh [74].

15.6.2 Concept in Model Formation

For the urban flow modelling the urban area is considered as a watershed and the result is considered with respect to overflow and channel flow component. So, the basic component of the model consists of one overland flow and two channel flow model.

15.6.3 Formulation for Overland Flow

The overland flow for any sub-region can be evaluated using the continuity equation [75]:

$$I - Q = \frac{\partial(VOL)}{\partial T} \tag{15.15}$$

where

I = Inflow,
Q = outflow,
VOL = the detention storage over the sub-region.

Equation (15.15) can be expressed as

$$r \cdot A_c - q \cdot L = \frac{\Delta VOL}{\Delta t} \tag{15.16}$$

where r is the excess rainfall which is the inflow for the catchment and q is the overland flow from the catchment into the stream element, expressed as flow per unit length of the stream. This flow is assumed to be perpendicular to the length of the element. Considering Δt is the time step in seconds, L is the length of the stream element in meters, A_c is the area of sub-region in km^2, r is expressed in mm/hour and d is flow depth in mm obtained by using Manning's empirical formula, Eq. (15.16) can be written as,

$$K_1 d_{t+\Delta t}{}^{5/3} + 100 d_{t+\Delta t} = K_2 \tag{15.17}$$

where

$$K_1 = \frac{LS^{1/2} \Delta t}{2nA_c} \quad \text{and} \quad K_2 = 100 d_t + \Delta t \left[\frac{r_t + r_{t+\Delta t}}{72} \right] - K_1 d_t^{5/3} \tag{15.18}$$

so the overland flow q during the time t is

$$q = \frac{1}{2}(q_1 + q_{t+\Delta t}) \tag{15.19}$$

where n is the Manning's roughness coefficient and S is the slope of the overland flow.

15.6.4 Formulation for Channel Flow

The channel flow can be represented by the following one-dimensional gradually varied unsteady flow equations based on Saint Venant's equations [76]

$$\frac{\partial A}{\partial t} + \frac{\partial Q}{\partial x} - q = 0 \tag{15.20}$$

$$\frac{\partial Q}{\partial t} + \frac{\partial \left(\dfrac{Q^2}{A} \right)}{\partial x} + gA \left(\frac{\partial H}{\partial x} + S_f \right) = 0 \tag{15.21}$$

Let

$$m = \frac{\dfrac{\partial Q}{\partial t} + \dfrac{\partial}{\partial x} \left(\dfrac{Q^2}{A} \right)}{gA} \tag{15.22}$$

From Eqs. (15.20) and (15.21), we have,

$$S_f = -\left(\frac{\partial H}{\partial x} + m \right) \tag{15.23}$$

Manning's formula can be written as

$$Q = \frac{1}{n} AR^{2/3} S_f^{1/2} = -k \left(\frac{\partial H}{\partial x} + m \right) \tag{15.24}$$

where

$$k = \frac{(A)R^{2/3}}{n \left(\dfrac{\partial H}{\partial x} + m \right)^{1/2}} \tag{15.25}$$

In diffusion wave model, the local and convective acceleration terms in the momentum equation are neglected. Finally, the diffusion wave from, where m is 0 can be expressed as [77]:

$$W \frac{\partial H}{\partial t} - \frac{\partial}{\partial x} \left[K \left(\frac{\partial H}{\partial x} \right) \right] - q = 0 \tag{15.26}$$

where W is the width of the channel.

The overland flow model and channel flow model, discussed above, are coupled to have an integrated model for event based rainfall runoff simulation of urban watershed. The integrated model simulates water levels (stage) and runoff at any location on the main channel for a watershed for the given rainfall and tidal flow condition. The inputs for the model are given from the database, which is generated from GIS thematic maps.

15.6.5 Tidal Effects

In coastal urban areas, the drainage channel joins the creek and the flow depth in the channel depends upon the tidal variation at end node. Tides around the catchment are semidiurnal (two high waters and two low waters each day). the tidal boundary condition is considered based on a semidiurnal sinusoidal equation in time t, such as given:

$$d = d_m + h \times \sin\left(\frac{2\pi t}{t_p}\right)$$
(15.27)

Here d is the tidal stage; d_m is the mean tidal stage; h is half the tidal oscillation range; t is time; t_p is the period for one complete tidal cycle. The values at the boundary condition are based on the above mentioned equation.

Case Study 1

Results of study carried out by Shahapure for the Panvel area located in Navi Mumbai is briefly mentioned below [77]:

It is surrounded by Sion-Panvel expressway on North–West side, Parsik foothills on the eastern side. The catchment flows into the Panvel Creek the western side. The G.L. variation is from 0.0 to 251.0 m above MSL. The dominant flow direction is to the west.

Use of GIS and Remote Sensing

The topographic map of the catchment area was geo-registered in Arc GIS 9.3. The contour lines were digitized and Arc GIS as a polyline file. The data on spot levels, for the catchment obtained from CIDCO was used to prepare point files. A Digital Elevation Model (DEM) was prepared using the above mentioned polyline and point shape files, in ArcGIS software.

The DEM of the study are the slope map. In the context of urban watershed the flow in the mainland is based on the man-made drainage system. Hence, the actual delineation of the urban watershed is done subjectively using natural delineation (ArcGIS Software) and storm water network layout.

For the purpose of modelling the catchment area is divided into 38 overland flow elements. Maximum and minimum area of the overland flow elements are 103.48 ha and 1.26 ha respectively. The slope values are obtained from slope map and grid coverage map using zonal statistics option of Arc GIS. Land use classification for the catchment is carried out using remotely sensed data.

The main channel of the catchment has been longitudinally divided into 57 elements with 58 nodes. The tidal length of the channel is 4750 m with the maximum and minimum elements lengths of 100 m and 120 m, respectively. The details of bed level and details of channel sections were obtained for CIDCO. The average longitudinal slope of the channel is 1 in 1011. The channel Manning's roughness coefficient is taken as 0.03. Two rainfall events were simulated and the simulated water levels compared with the measured ones. A summary of the rainfall and runoff for the simulated events is shown in Table 15.1.

Table 15.1 Summary of Rainfall and Runoff for the Simulated Events.

Event no.		1	2	3
Event date		July 23, 2009	September 6, 2009	July 26, 2005
Rainfall start time		11.00 a.m.	11.58 a.m.	3.00 a.m.
Total volume	mm	33	38	745
Total rainfall duration	mins,	810	180	1260
Average intensity	mm/hr	2	13	35
Max. rainfall intensity (RI)	mm/hr	34	40	76
Duration of max. RI	mins	15	15	60
Initial Tidal Conditions				
Tide level at start simulation (above MSL)	m	1.19	1.285	1.639
Phase of tide		Rising	Rising	Rising
Results				
Total simulation period from start of rainfall	mins.	1440	1440	1440
Total run-off volume	mm	34.54	36.04	739.57
Max. stage (at Ch. 0.00 m)	m	6.14	6.53	8.12
Time of max. stage w.r.t. rainfall start time	mins	735.5	45.5	600
Peak discharge (at Ch 4750.00 m)	m³/sec	24.09	22.95	164.78
Time of peak w.r.t. rainfall time	mins	269.5	116	608
Flooding at channel node		NIL	NIL	0 to 700 m

Selected References

1. *Atlas of (frequency), Flood runoff coefficients*, Applicable to the River systems in the various climatic zones of India, CW and PC Flood Estimation Directorate, Design Office, Report No. 20, 1968.

2. Fisher, R.A. and Tippett, L.H. C, Limiting forms of a frequency distribution of the largest or smallest number of a sample, *Proc. Cambridge Phil. Soc.*, Vol. 24, 1928, pp. 180-190.

3. Gumbell, E.J., Floods estimated by the probability method, *Engg. New Record*, Vol. 134, 1945, pp. 97-101.

4. Horton, R.E., An approach towards a physical interpretation of infiltration capacity, *Proc. Soil. Sci. Soc. Am.*, Vol., 5, 1940, pp. 399-417.

5. Sherman, L.K., Stream flow from rainfall by the unitgraph method, *Engg. News Record*, Vol. 108, 1932, April 7, 1932, pp. 501-505.

6. Bernard, M.M., An approach to determine stream flow, *Trans. ASCE*, Vol. 100, 1935, pp. 347-395.

7. Collins, W.T., Runoff distribution graphs from precipitation occurring in more than one time unit, *Civil Engg.* (NY), Vol. 9, No. 9, 1939, pp. 559-561.

8. Snyder, F.F., Synthetic unit-graphs, *Trans. Am. Geophysical Union*, Vol. 19, 1938 pp. 447–454.

9. *Manual of estimation of design flood*, CWP and C, India, 1961.

10. Nash, J.E., The form of the instantaneous unit hydrograph, *Intern. Assoc. Sci. Hydrology* Pub. 451, Vol. 3. 1957, pp. 114-121.

11. Diskin, M.H., evaluation of segmented IUH from derivatives, *Jourl. of Hy. Div.*, Jan. 1969, Proc. ASCE, Vol. 95, HY 1, pp. 329-346.

12. Puls, L.G., *Flood Regulation of the Tennesse River*, 70th Cong. 1st session, H.D., 185 pt. 2, Appendix B, 1928.

13. Murthy, K.V.R., *Electric Analogue Methods of Flood Forecasting*, Symp. on flood forecasting control and flood damage protection, 43rd Annual Session CBI and P. New Delhi, 1970. Pub. 107, Nov. 1970, pp. 113-124.

14. Chow, V.T., *Handbook of Applied Hydrology*. McGraw-Hill, N.Y. 1964 pp. 25.53-25.54.

15. McCarthy, Methods of Flood-routing, *Report on Survey for Flood Control, Connecticut River Valley*, Vol. Set. 1, Appendix US Ar. Corps of Engrs., 1936.

16. Lighthill, M.J. and Whitlam, G.B., On kinematic waves: Flood movement in long rivers, *Proc. Roy. Soc. Lond.*, Vol. 229A, 1955.

17. Henderson, F.M., Flood waves in prismatic channels, *Procl. ASCE*, Vol. 89, No., Hy, 4 July 1963, pp. 39-67.

18. Abott, N.B., *An introduction to the method of characteristics*, Thames and Hudson, London, 1966.

19. Ghosh, S.N., Flood wave modification along a prismatic river, *Water Resources Research Am. Geophysical Union*, Vol. 17, No. 3, June 1971, pp. 697-703.

20. Rippl, W., The capacity of storage reservoirs for water-supply, *Proc. Institution of Civil Engineers*, Vol. 71, 1883, pp. 270-278.

21. Brune, G.M., Trap efficiency of reservoirs, *Trans. A.G.U.*, Vol. 34, No. 3, June 1953, pp. 407-418.

22. Maddock, T., Reservoir problems with respect to sedimentation, *Proc. FIASC, USDA (Washington)*, Jan. 1948.

23. Shen, H.W., *River Mechanics*, Vol. II, P.O. Box 606, Fort Collins, Colorado, USA 80521, 1971.

24. Borland, W.M. and Miller, C.R., Distribution of sediment in large reservoirs, JHD *Proc. ASCE*, Vol. 84, No. HY-2, pt.-l, April. 1958.

25. Graf. W.H., *Hydraulics of Sediment Transport*, McGraw-Hill, New York, 1971.

26. Raudkivi, A.J., *Loose Boundary Hydraulics, Pergamon Press*, 1976.

27. Chow, V.T., *Open channel hydraulics*, McGraw-Hill-Kogakkush Ltd. 1959. International Student Edition.

28. Parikh, V.N. and Iyer, G.R.R., *Economics of Flood Control: Analysis of Benefit Cost*, Symp. on flood forecasting control and flood damage protection 43rd session, CBI and P. 1970. Pub. 107, Nov, 1970, pp. 191-121.

29. Luthin, J.N., *Drainage Engineering*, Wiley Eastern Pvt. Ltd.,

30. Lacey, Gerald, Stable channels in alluvium, *Proc. Inst. of Civil Engrs, Lond.*, Vol. 229, 1930, pp. 259, 384.

31. Linsley, R.K. and Franzini, J.B., *Water Resources Engg.*, McGraw-Hill Book Co., N.Y. 2nd Ed. 1972. International Student Edition.

32. Dronkers, J.J., *Tidal Computations in Rivers and Coastal Waters*, North Holland Pub. Ca. Amsterdam, 1964.

33. Khosla, A.N., Silting of Reservoirs, Central Board of Irrigation and Power, Publication number, 51, 1953.

34. Phyususnin, I.I., Reclamative Soil Science, Foreign Language Publication House, Moscow.

35. Pilgrim, D.H and Cordary, I., Flood run-off, Chapter in Handbook of Hydrology. Tata-McGraw Hill Inc., New York. 1993.

36. Chow, V.T., Handbook of Applied Hydrology, section 8-1, Mc-Graw Hill, New York 1964.

37. US Soil Con servation Service, Hydrology, National Engineering Handbook Section 4, Washington, D.C. 1972.

38. Mishra, S.K. and Seth, S.M., Use of hysterisis for determining the nature of flood wave propagation in natural channels, Hydrological Sciences, Vol. 41 (2), 1996 pp. 153-170.

39. National Weather Service (NWS): The NWS dam break flood forecasting model, Users manual, Davis, California.

40. Goodland, R., Watson, C. and Ledec, G., Environmental management of tropical Agriculture, Westview Press, Boulder, Colorado, 1984.

41. Srivastava, D.K., Real time reservoir operation, Scientific Contribution, INCOH/SAR-9/95, NIH, Roorkee, 1995.

42. Paudyal, G.N., Forecasting and Warning of water of related disaster in complex hydraulic setting, the case of Bangladesh, ICIWRM-2000, Proc. of International Conference on Integrated Water Resources Management for sustainable development, 19-21 Dec. 2000, New Delhi.

43. Danish Hydraulic Institute, MIKE 11, Flood Forecasting Model, DHI, Denmark, 1998.

44. Kirkham, D., Seepage of Steady rainfall through soil into drains, Transactions, Am. Geo. Union, 1958, pp. 892-908.

45. Freeze, R.A. and Cherry, J.A., Groundwater, Prentice Hall, Inc., New Jersey, 1979.

46. Raghunath, H.M., Groundwater 2nd Edition, Wiley Eastern Limited, New Delhi.

47. Todd, D.K., Groundwater hydrology, John Wiley and Sons, New York.

48. Voss, C.I., A finite element simulation model for saturated-unsaturated fluid density dependent groundwater flow with energy transport or chemically reactive single species solute transport, USGS Water Resources Investigators, Report, 84.

49. Norman, N.S., Nelson James, E. and Zumdel, Alam, K. Review of automated flood plain delineation, from digital terrain models, Jourl. of Water Resources Planning and Management, Nov-Dec. 2001, ASCE, Vol. 127, No. 6.

50. Islam, M.M. and Sado, K., Development of priority map for flood countermeasures by remote sensing data with GIS, Proc. ASCE, Jourl. of Hydrologic Engg, Vol. 7, No. 5, Sept. 1, 2002.

51. Rumer and Harleman, Intruded salt water wedge in porous media, Jourl. ASCE, HYD. Divn, Hy-6, Vol. 89, 193-220, 1963.

52. Kirpich, P.Z., Time of concentration of small Agricultural water shed, Civil Engg, 10, 362, 1940.

53. Wilson, B.W. and Torum A. (1968) The Tsunami of the Alaskan Earthquake 1964. Engineering Evaluation "Tech Memo 25, US Army Coastal Engineering Research Center.

54. Magoon, O.T. (1965) "Structural damage by Tsunamis "Proceedings Santa Barbara Coastal Engineering speciality conference ASCE, May. 1965.

55. Kamel, A. (1970) Laboratory Study for design of Tsunami Barrier, Journal Waterways and Harbour and Coastal Engineering ASCE Nov 1970.

56. Keulegan, G.H. and Harrison, J., Tsunami Refraction Diagram by Digital Computer. Trans *Journal of Waterways and Harbour Division*, ASCE, May, 1970.

57. Hampson, R., MacMahan, J. and Kirby, J.T. 2011, A Low Cost Hydrographic Kayak Surveying System, *Journal of Coastal Research* 27(3), West Palm Beach, Florida.

58. Myers, V.A. (1954) Characteristics of U.S. Hurricanes Pertinent to Levee Design for Lake Okuchobee Florida, M. R. Report 32, U.S Weather Bureau.

59. U.S. Army Coastal Engineering Research Center (1973) Shore Protection Manual, (3 vols), USA Govt. Printing Press, Washington DC.

60. Wilson, B.W. (1960) Note on Surface Wind Stress over water at low and high wind speeds, *Journal of Geographical Research*, vol. 64.

61. Van Dorn, W.G. (1953) Wind Stress on an Artificial Pond. *Journal of Marine Research*, Vol. 12, pp. 249-276.

62. Saville, T. (1952) Wind set up and waves in shallow water, Tech Memo 27 U.S. Army Beach Erosion Board, Washington, D.C.

63. Bretschneider, C.L. (1967) Storm Surges vol. 4, Advances in Hydro-science, Academic Press, New York.

64. Komar, P.D. (1998) Beach process and sedimentation 2nd ed. Upper Saddle Press, Prentice Hall Inc., New Jersey.

65. Smith, J.M. and Eber Sole, B.A. 2000. Modeling and Analysis of short waves chapter 5 Studies of Navigation channel feasibility Willapa Bay Washington, U.S. Army Corps of Engineers, Vicksburg, Mississippi.

66. Fread, D.L. (1977) The Development and testing of a Dam-break Flood Forecasting Model. Proc. of the Dambreak Flood Routing workshop. Water Resources Council, USA.

67. Murthy, J.S.R. (1998) Gradual Dam Breach Flow Routing, ISH Journal of Hydraulics, Vol 4, no. 2 Sept. 1998.

68. Ponce, V.M. and Tsivoglou, A.J. (1981) Modeling Gradual Dam Break JHD, Proc. ASCE, Vol. 107, No, Hy 7.

69. Ranga Raju, K. (1993) Flow through open channels Tata McGraw-Hill Publishing Company Ltd., New Delhi.

70. Amein, and Chu F.L. (1975) Implicit Numerical Modeling of Unsteady Flow JHD Proc. ASCE, Vol. 101, No Hy 6.

71. Fread D.L. (1971) Discussion on Paper. Implicit Flood Routing in Natural Channels JHD Proc ASCE, Vol. 97 No. Hy 7.

72. Chow, Van Te (1961) Open Channel Hydraulics McGraw Hill Book Company, New York.

73. Subhashis Dutta, Medhi, H., Karmaker, T., Singh, Y., Prabu, I. and Dutta, U. (2010) Probabilistic Flood Hazard Mapping for embankment breaching, Journal of Hydraulic Engg., Vol. 16 Nov 2010. ISH.

74. Mark, O., Wasakul, S., Apurmaueleal, C., Aroonnet, S.B. and Dgordjavic S. (2004) Potential and limitations of 1-D modeling of urban flooding, J. Hydrology, Vol. 299.

75. Rao, B.V. and Rao, E.P. (1980) Surface runoff modeling of small water sheds Proc, Intl seminar on Hydrology of Extremes, Roorkee.

76. Ross, B.B., Contractor, D.N. and Shauholts, V.O. (1979) A finite element model of overland flow and channel flow for assessing the hydrologic impact of land use change, *J of Hydrology* Vol. 4.

77. Hromadka II, T.V., DeVris, J. and Nestlenger, A.J. (1984) Comparison of Hydraulic Routing methods for one dimensional channel routing problems Proc. 2nd International Conference, Southampton, UK.

78. Shahpure, S.S., Kulkarni, A.T., Bherat, K.S.R., Eldho T.I. and Rao, E.P. (2010). Coastal Urban Modeling using FEM –GIS. Based Model 1SH, *J of Hydraulic Engg.* Vol. 16, Nov. 2010

Suggested Readings

Application of systems Techniques for water Resources Management in India - Current Trends and Future directions organised by NIH, Roorkee (Feb 13, 2003).http://www.tsunami.noaa.gov/tsunami_story

    Creager W.P., Justin J.D. and Hinds. J., *Engineering for Dams* (Vol. I), Wiley Eastern Pvt. Ltd., New Delhi, 1961.

Cross, R.H. (1967), Tsunami surge forces. *Journal of the Waterways and Harbors Division*, A.C.E., 93, 201–231.

Fundamentals of Irrigation Engineering: Bharat Singh, Nemchand and Bros; Roorkee, U.P. (India).

Garde, R.J. and K.G. Ranga Raju, Mechanics, of *Sediment transportation and Alluvial StreamProblems*, Wiley Eastern Ltd., New Delhi.

Handbook of Applied Hydraulics, edited by C.V. Davis and T.E. Sorensen, McGraw-Hill Company, 3rd Ed. 1969.

Handbook of Fluid Dynamics, edited by V.L. Streeter, McGraw-Hill Books Company, 1961.

————, Transactions, Institution of Engineers (India), Bengal Centre, Kolkata Dec, 1969, pp. 25-40.

Hiranandini, M.G. and S.V. Chitale, *Stream Gauging*, Ministry of Irrigation, CWPRS (Pune). 1960.

http://en.wikipedia.org/wiki/storm_surge.

http://eu.wikipedia.org/wiki/Tsunami.html # Generation mechanism.

http://www.nhc.noaa.gov/ssurge/

Hydrology, Pt. IV, Flood Routing, Irrigation Department, Govt. of Karnataka, Bangalore, 1978.

Indian Standard Code of Practice: IS 5477 (Pt. I to III - 1969, Pt. IV, 1971). Methods for *fixing the capacities of reservoirs*. IS: 4890 (1968). Methods for measurement of suspended sediment in open channel, Manak Bhavan, New Delhi.

International Seminar on Hydraulics of Alluvial Streams, New Delhi, CBI and P., 1973.

Lecture notes on 'Reservoir Operation': International Course in Hydraulic Engineering, 1967, Delft, Netherlands.

Principles of River Engg: Edited by P. Th. Jensen, Pitman, 1979.

Scientific Reports published by National Institute of Hydrology, Roorkee, India.

Speeth, M. G. and S.C. Berkman (1965). The Tsunami of March 28 1964 as recorded at Tide Stations U. S. Coast and Geodetic Survey Report.

Tsunami – Wikipedia the free encyclopedia.

Varshney, R.S., *Engineering, Hydrology*, Nemchand Bros. New Delhi, 1974.

Index

Printed and bound by CPI Group (UK) Ltd, Croydon, CR0 4YY

18/10/2024

01776252-0001